LABEL-FREE BIOSENSORS

Label-free biosensors use biological or chemical receptors to detect analytes (molecules) in a sample. They give detailed information on the binding selectivity, affinity, and, in many cases, the stoichiometry, kinetics, and thermodynamics of an interaction. Although they can be powerful tools in the hands of a skilled user, there is often a lack of knowledge regarding the best way to utilize label-free assays to screen for biologically active molecules and to accurately and precisely characterize molecular recognition events.

This book reviews both established and newer label-free techniques. It is intended to give both the expert user and the general reader insight into the field from expert opinion leaders and practitioners of these techniques. Chapters also contain worked examples that are written to guide the reader through the basics of experimental design, setup, assay development, and data analysis.

Matthew A. Cooper is Founder and Managing Director of Cambridge Medical Innovations and Distinguished Australia Fellow at the University of Queensland. Dr. Cooper has consulted widely for biosensor, biotechnology, and pharmaceutical companies in the United Kingdom, Europe, and the United States. He is a review panel member for Biotechnology and Biological Sciences Research Council, Engineering and Physical Sciences Research Council, National Institutes of Health, National Institute of Allergy and Infectious Diseases, and Science Foundation Ireland and a Fellow of the Royal Society of Medicine.

Label-Free Biosensors

TECHNIQUES AND APPLICATIONS

Edited by

Matthew A. Cooper
The University of Queensland

CAMBRIDGE
UNIVERSITY PRESS

University Printing House, Cambridge CB2 8BS, United Kingdom

One Liberty Plaza, 20th Floor, New York, NY 10006, USA

477 Williamstown Road, Port Melbourne, VIC 3207, Australia

314-321, 3rd Floor, Plot 3, Splendor Forum, Jasola District Centre, New Delhi - 110025, India

103 Penang Road, #05-06/07, Visioncrest Commercial, Singapore 238467

Cambridge University Press is part of the University of Cambridge.

It furthers the University's mission by disseminating knowledge in the pursuit of education, learning and research at the highest international levels of excellence.

www.cambridge.org
Information on this title: www.cambridge.org/9780521711517

© Cambridge University Press 2009

First published 2009

A catalogue record for this publication is available from the British Library

Library of Congress Cataloging in Publication data
Label-free biosensors : techniques and applications / edited by Matthew A. Cooper.
    p. ; cm.
Includes bibliographical references and index.
ISBN 978-0-521-88453-2 (hardback) – ISBN 978-0-521-71151-7 (pbk.)   1. Biosensors.
I. Cooper, M. A. (Matthew A.)
[DNLM: 1. Biosensing Techniques – methods. QT 36 L116 2009]
R857.B54L33   2009
610.28´4 – dc22   2008026569

ISBN 978-0-521-88453-2 Hardback
ISBN 978-0-521-71151-7 Paperback

Contents

Contributors   *page* vii

Preface   ix

1 **Label-free optical biosensors: An introduction**   1
   Brian T. Cunningham

2 **Experimental design**   29
   Robert Karlsson

3 **Extracting affinity constants from biosensor binding responses**   48
   Rebecca L. Rich and David G. Myszka

4 **Extracting kinetic rate constants from binding responses**   85
   Rebecca L. Rich and David G. Myszka

5 **Sensor surfaces and receptor deposition**   110
   Matthew A. Cooper

6 **Macromolecular interactions**   143
   Francis Markey

7 **Interactions with membranes and membrane receptors**   159
   Matthew A. Cooper

8 **Application of SPR technology to pharmaceutical relevant
   drug-receptor interactions**   179
   Walter Huber

9 **High-throughput analysis of biomolecular interactions and cellular
   responses with resonant waveguide grating biosensors**   206
   Ye Fang, Jack Fang, Elizabeth Tran, Xinying Xie, Michael Hallstrom,
   and Anthony G. Frutos

10 **ITC-derived binding constants: Using microgram quantities of protein**   223
    Richard K. Brown, J. Michael Brandts, Ronan O'Brien, and William B. Peters

11 **Electrical impedance technology applied to cell-based assays**   251
    Ryan P. McGuinness and Edward Verdonk

Index   279

Color plates follow page 36

Contributors

J. Michael Brandts
Vice President
MicroCal
Northampton, MA

Richard K. Brown, PhD
President and CEO
MicroCal
Northampton, MA

Matthew A. Cooper, PhD
Managing Director
Cambridge Medical Innovations
Cambridge, UK

Distinguished Australia Fellow
Institute for Molecular
  Bioscience
University of Queensland
St. Lucia, Australia

Brian T. Cunningham
Associate Professor
University of Illinois
Department of Electrical and
  Computer Engineering
Urbana, IL

Jack Fang, PhD
Strategic Analysis Manager
Corning® Epic® System
Corning, Inc.
Corning, NY

Ye Fang, PhD
Research Manager
Cellular Biophysics
Corning, Inc.
Corning, NY

Anthony G. Frutos, PhD
BioAssay Development Manager
Science & Technology
Corning, Inc.
Corning, NY

Michael Hallstrom
Assistant Product Line Manager
Corning® Epic® System
Corning, Inc.
Corning, NY

Walter Huber, PhD
Pharmaceutical Research Discovery
  Technology
F. Hoffmann-La Roche
Basel, Switzerland

Robert Karlsson
Research and Development Director,
  System and Applications
  Department
GE Healthcare
Uppsala, Sweden

Francis Markey
GE Healthcare
Uppsala, Sweden

Ryan P. McGuinness
Senior Scientist
Drug Discovery
MDS Analytical Technologies
Sunnyvale, CA

David G. Myszka, PhD
Director
Center for Biomolecular Interaction
   Analysis
School of Medicine
University of Utah
Salt Lake City, UT

Ronan O'Brien, PhD
Head of Applications Research
MicroCal
Northampton, MA

William B. Peters, PhD
Applications Scientist and Head of
   Training
MicroCal
Northampton, MA

Rebecca L. Rich, PhD
Senior Research Scientist
Center for Biomolecular Interaction
   Analysis
School of Medicine
University of Utah
Salt Lake City, UT

Elizabeth Tran, PhD
Senior Research Scientist
Science & Technology
Corning, Inc.
Corning, NY

Edward Verdonk, PhD
Senior Scientist
Drug Discovery
MDS Analytical Technologies
Sunnyvale, CA

Xinying Xie
Field Application Scientist
Corning® Epic® System
Corning, Inc.
Houston, TX

Preface

Over the past two decades the benefits of biosensor analysis have begun to be recognized in many areas of analytical science, research, and development, with analytical systems now used routinely as mainstream research tools in many laboratories in many fields. Simplistically, biosensors can be defined as devices that use biological or chemical receptors to detect analytes (molecules) in a sample. They give detailed information on the binding affinity and in many cases also the binding stoichiometry, thermodynamics, and kinetics of an interaction. Label-free biosensors, by definition, do not require the use of reporter elements (fluorescent, luminescent, radiometric, or colorimetric) to facilitate measurements. Instead, a receptor molecule is normally connected in some way to a transducer that produces an electrical signal in real time. Other techniques such as isothermal titration calorimetry (ITC), nuclear magnetic resonance (NMR), and mass spectrometry require neither reporter labels nor surface-bound receptors. In all cases detailed information on an interaction can be obtained during analysis while minimizing sample processing requirements. Unlike label- and reporter-based technologies that simply confirm the presence or absence of a detector molecule, label-free techniques can provide direct information on analyte binding to target molecules typically in the form of mass addition or depletion from the surface of a sensor substrate or via changes in a physical bulk property (such as the heat capacity) of a sample. Until recently, label-free technologies have failed to gain widespread acceptance due to technical constraints, low throughput, high user expertise requirements, and cost. Whereas they have proved to be powerful tools in the hands of a skilled user, they have not always been readily adapted to everyday lab use in which simple-to-understand results are a prerequisite. Despite this limitation, the potential today for label-free approaches to complement or even displace other detection technologies has never been higher.

This book covers established label-free technologies and emerging developments in label-free detection systems, their underlying technology principles, and end-user case studies that reveal the power and limitations of such biosensors. The chapters are intended to give both the expert user and the general reader interested in the technologies and applications behind label-free biosensors an insight into the field from expert opinion leaders and practitioners. As such, most chapters contain one or more worked examples that guide the reader through the basics of experimental design, setup, assay development, and data analysis. The book is heavily weighted toward applications using optical biosensors and surface plasmon resonance (SPR) instrumentation. This is primarily because of the overwhelming bias in the installed base of optical biosensors due to their early commercialization and uptake. Other label-free technologies conspicuous by their absence from this volume include analytical ultracentrifugation, nuclear magnetic resonance spectroscopy, and mass spectroscopy. These will be covered as part of a new cluster of titles on bioanalytical techniques and applications.

Matt Cooper
Cambridge, U.K.
May 2008

LABEL-FREE BIOSENSORS

1 Label-free optical biosensors: An introduction

Brian T. Cunningham

INTRODUCTION	1
RATIONALE FOR LABEL-FREE DETECTION	3
OPTICAL BIOSENSORS	4
PERFORMANCE METRICS	7
Sensitivity	7
Assay sensitivity	8
Resolution (or limit of determination)	8
Ease of use	9
Sensor cost	10
Detection instrumentation	10
Throughput	11
REVIEW OF OPTICAL BIOSENSOR METHODS	12
Surface plasmon resonance	12
HOLOGRAM BIOSENSORS	14
Reflectometric interference spectroscopy	16
Dual polarization interferometry	18
PHOTONIC CRYSTAL BIOSENSOR	19
WHISPERING GALLERY MODE RESONATORS	22
ACTIVE RESONATORS: DFB LASER BIOSENSOR	25
CONCLUSIONS	25
REFERENCES	27

INTRODUCTION

The ability to study interactions among biomolecules, to observe the activity of cells, and to detect analytes specifically from bodily fluids, manufacturing processes, or environmental samples are cornerstones of life science research, pharmaceutical discovery, medical diagnosis, and food/water safety assurance. These capabilities and many others are enabled by the ability to perform biochemical and cell-based assays that allow scientists to ask basic questions about whether one analyte interacts with another, how strong the binding affinity is between two proteins, whether a chemical compound will affect the proliferation rate of cancer cells, and the concentration of a biomarker for cancer within a patient's blood sample. The development of technology to meet these requirements is challenging because biochemical analytes, which can include drug compounds with molecular weights below 500 Da, DNA oligomers, peptides, enzymes, antibodies, and viral particles, are exceedingly small and sometimes present within a test sample at concentrations in the fg/ml to pg/ml concentration range that simultaneously contains thousands of other molecules at concentrations orders of magnitude greater. Larger biochemical analytes – such as bacteria, spores, cells, and cell clusters – are less difficult to observe if they can be stained with a colored or fluorescent dye. However, such treatments generally result in the death of the specimen, thus preventing the ability to study a single population repeatedly over a long time period. An additional challenge arises from the need to perform many thousands of individual measurements, as researchers working in pharmaceutical discovery and life science research seek to perform large numbers of assays in parallel as a means to understand how many molecular permutations affect the efficacy of a drug candidate and as researchers in the field of proteomics seek to characterize extremely complex protein interaction pathways at the heart of common human diseases. When assays must be performed in quantities in the tens of thousands to several millions, the cost of performing assays in terms of disposable labware, reagents, and scientist/technician time must be considered carefully – and the ability to translate instrument measurements into biologically relevant information efficiently is a key consideration for selection of a particular assay technology.

Due to the difficulty in detecting biological analytes directly through their intrinsic physical properties (such as mass, size, electrical impedance, or dielectric permittivity), biological research has historically relied upon attachment of some sort of "label" to one or more of the molecules/viruses/cells being studied. The label is designed to be easily measured by its color or its ability to generate photons at a particular wavelength and acts as a surrogate to indirectly indicate the presence of the analyte to which it has been attached. For example, many commercially available fluorescent dyes can be conjugated with DNA, proteins, or cells so that, when illuminated with a laser at the excitation wavelength of the fluorophore, light is emitted at a characteristic wavelength. Likewise, semiconductor quantum dots, fluorophore-embedded plastic beads, and metal nanoparticles can be functionalized with a surface that enables selective attachment to molecules or particular regions within a cell. In a similar fashion, a wide class of assays has been developed in which a molecule is labeled with a specific antibody that in turn can initiate an enzymatic reaction. The reaction generates a colored stain that can be visually observed.

Though molecular, nanoparticle, enzyme, and radioactive labels have been central to implementing nearly all biochemical and cell-based assays, the labels themselves pose several potential problems. For example, though the detection of radioactive labels can be performed with tremendous sensitivity, their usage requires specially outfitted "hot labs" and the generation of large quantities of contaminated reagents and labware that must be properly disposed. The excitation/emission efficiency of fluorophores is degraded by time and exposure to light (including room light and the laser excitation light) by photobleaching, reducing the ability of the technique to supply highly quantitative measurements and requiring that assays be read once only in an "end point" fashion so that kinetic information from an assay is lost. Quenching and self-quenching may also reduce the efficacy of fluorescent tags in an unpredictable manner. Nanoparticles, although as small as a few nanometers in diameter, are quite large when compared to biological molecules and cannot penetrate easily through cell membranes.

In practice, label-based assays require a high degree of development to assure that the label does not block an important active site on the tagged molecule or modify the molecule's conformation. Particularly for sandwich-type assays, intensive optimization of washing conditions, blocking conditions, and exposure protocols for a series of reagents must be performed. Labware for fluorescent assays must be constructed from materials that do not autofluoresce, as unintentional background fluorescence can overwhelm the signals being measured. The ultimate cost of using a label incorrectly may be the development of a drug whose side effects are not well characterized or an inaccurate assessment of the potency of a drug's interaction with its targeted protein receptor.

RATIONALE FOR LABEL-FREE DETECTION

Due to the above considerations, there has been a drive to reduce assay cost and complexity while providing more quantitative information with high throughput. For example, direct measurement of the affinity binding constant between two proteins provides fundamental information regarding kinetics of the interaction, and the ability to quickly quantify binding kinetic parameters enables direct comparisons to be made between a wide variety of protein–protein pairs. Likewise, a pharmaceutical company's typical screening campaign to investigate how large libraries of chemical compounds interact with a target protein representing a disease's biochemical pathway can involve several million assays performed over several months.[1] Limiting assay complexity and associated advantages in assay development time, reagent usage, and accuracy of results is commercially important, as these factors directly affect the time and cost required to bring a new drug through initial discovery, characterization, and validation phases. In later phases of drug discovery, assays determine how chemical compounds and proteins interact with cells to learn how cell attachment, ion channel activation, proliferation, and apoptosis are modulated. Screening the interaction between large libraries of biochemical compounds and cell panels representing different tissues in the body or different groups of patients is used to predict the efficacy of a treatment and its likely side effects before trials are conducted with animals or people. Therefore, the ability to inexpensively perform large numbers of assays is central for pharmaceutical companies as they seek to provide innovative therapeutic compounds while limiting the time and cost required to bring them to market.

There are now many methods that allow direct detection of biological analytes without labels. Label-free detection generally involves a transducer capable of directly measuring some physical property of the chemical compound, DNA molecule, peptide, protein, virus, or cell. For example, all biochemical molecules and cells have finite mass, volume, viscoelasticity, dielectric permittivity, and conductivity that can be used to indicate their presence or absence using an appropriate sensor. The sensor functions as a transducer that can convert one of these physical properties (such as the mass of a substance deposited on the sensor's active surface) into a quantifiable signal that can be gathered by an appropriate instrument (such as a current or voltage proportional to the deposited mass). Label-free detection removes experimental uncertainty induced by the effect of the label on molecular conformation, blocking of active binding epitopes, steric hindrance, inaccessibility of the labeling site, or the inability to find an appropriate label that functions equivalently for all molecules in an experiment. Label-free detection methods greatly simplify the time and effort required for assay development while removing experimental artifacts from quenching, shelf life, and background fluorescence.

OPTICAL BIOSENSORS

Though a biological analyte's mass is a familiar intrinsic property for label-free detection, all molecules also have the capability to interact with electromagnetic fields that pass through them because they contain atomic nuclei and a variety of electrons in various orbital states. Most fundamentally, electrons within molecules experience a force when they are exposed to the oscillating electromagnetic fields associated with the propagation of light. Molecules with an abundance of free electrons will become polarized by exposure to the light's electromagnetic field (one side of the molecule will be temporarily more negatively charged than the opposite side – resulting in the formation of an electric dipole), where the extent of the polarization may be different for any particular molecule depending on its size, shape, and orientation with respect to the electric field. A constant known as the *electric susceptibility*, χ_e, quantifies the extent of a molecule's "polarizability," and molecules with greater χ_e are more easily polarized. When a polarizable molecule is placed in an electric field, the induced electrical dipole produces a secondary electric field such that the resulting electric field (i.e., the sum of the originally applied field and the secondary field) is of lower magnitude than the applied field. Because an electromagnetic field associated with light is time-varying, the electrons within the molecule will experience a time-varying force so that electrons will oscillate within the molecule. Moving electrons, by definition, produce an electrical current, so the molecule actually experiences a "polarization current" as a result of this electron motion. The result of the polarization current is that light travels more slowly through the molecule than it would through free space. Though free space has a permittivity defined by the constant ε_0, (where $\varepsilon_0 = 8.85 \times 10^{-12}$ F/m), the permittivity of a dielectric material containing molecules is given by $\varepsilon = \varepsilon_r \, \varepsilon_0$, where ε_r is known as the *relative permittivity* of the *dielectric constant* of the molecule. ε_r is directly related to the polarizability of the molecule because it is mathematically defined as $\varepsilon_r = 1 + \chi_e$. Many people are more familiar with the term *refractive index* (*n*) to describe a dielectric material. In an ordinary dielectric material at optical wavelengths, *n* is defined as $n = \sqrt{\varepsilon_r}$, so the refractive index is directly related to the polarizability of the molecules within a dielectric material. Generally, the refractive index is a quantity defined for a bulk material, while electric susceptibility and dielectric permittivity can apply to individual molecules such as those adsorbed to optical biosensor surfaces.

The key behind optical biosensors' ability to detect biological analytes is that biological molecules, including proteins, cells, and DNA, all have dielectric permittivity greater than that of air and water. Therefore, these materials all possess the intrinsic ability to reduce the propagation velocity of electromagnetic fields that pass through them. Optical biosensors are designed to translate changes in the propagation speed of light through a medium that contains biological material into a quantifiable signal proportional to the amount of biological material present on the sensor surface. In the design of optical biosensors, the detected biological material is often modeled as a thin film with a finite refractive index, although this is a simplification. Several studies have been performed to characterize the dielectric properties of representative molecular monolayer films.[2–4] Therefore, if a biosensor transducer surface is covered with water, and if biological molecules can adsorb to the transducer surface, a small quantity of water molecules is displaced and replaced with a molecule that is more easily polarized by electromagnetic fields associated with light. Therefore, the design goal for all optical biosensors is to provide a transducer with some externally measurable characteristic that is modified by changes in dielectric permittivity on its surface. In this way, optical biosensors do not measure the mass of adsorbed material (as sometimes stated), although often the mass of deposited material is often related to the change in dielectric permittivity.

As shown in several examples within this chapter, the majority of optical biosensors are measured by illuminating them with light (from a laser, light-emitting diode, or incandescent light bulb) and detecting changes in some characteristic of the light reflected from or transmitted through the sensor. This type of excitation/collection method is convenient because the sensor can be measured without the detection instrument making any direct physical contact with the sensor itself. Without requirements for electrical contacts, many optical biosensors can be measured simultaneously or sequentially by illuminating sensors at different locations on a single multisensor surface, and integration with liquid samples (which must be generally kept separate from electrical contacts to prevent short circuits) is greatly simplified. Many types of optical biosensors can be fabricated inexpensively on glass or plastic surfaces, while others require more complex definition of patterned regions using photolithography and etching processes commonly used for integrated circuit manufacturing. Most optical biosensors are considered "passive" optical components from the standpoint that one measures changes in characteristics of light reflected/transmitted from them, and all the light is supplied externally. This is advantageous because the sensor itself consumes no power, and generally low illumination levels are required to produce measurable responses. More recently, "active" optical biosensors have been demonstrated that are capable of producing their own light output. We provide an example at the end of the chapter.

The transducer alone, however, is not sufficiently intelligent to specifically identify material placed on its surface. For example, an optical biosensor cannot tell if an adsorbed layer of material consists of protein, plastic, or silicon oxide. To use a transducer as a biosensor, the surface of the transducer must have the ability to selectively attach specific material from a test sample while not allowing undesired material to attach. Selective detection capability is provided through the attachment of a layer of receptor molecules to the surface of the transducer. For example, when a single strand of DNA is attached to the sensor surface, the surface has the capability to preferentially bind a complementary strand of DNA from a test sample. Likewise, when a protein attaches to the sensor surface, the sensor has the capability to preferentially bind antibodies to that protein from a test sample. The material attached to the sensor surface is referred to as the *receptor ligand*. The detected material is called the *analyte*. Thus, a biosensor is the combination of a transducer that can generate a measurable signal from material that attaches to the transducer and a specific recognition surface coating containing a receptor ligand that can bind a targeted analyte from a test sample. For specific detection of biomolecules within complex mixtures that contain many potentially interfering molecules, means for "blocking" a surface with immobilized receptors to prevent nonspecific attachment of interferents is an important component of performing successful assays. The surface chemistry procedures used to attach receptor ligands to biosensor surfaces with covalent bond linkages (so they are not easily washed away) while maintaining the biological activity of the ligands is an extremely important element of a useful biosensor (optical or otherwise). This is outside the scope of this chapter.

For most types of optical biosensors, a solid material medium confines an electromagnetic wave in such a way that the wave has the opportunity to interact with a test sample. The electromagnetic wave may be in the form of a traveling wave or a standing wave, depending on the sensor configuration. For light to be guided by the sensor structure but concurrently interact with the external environment, the structure must be designed so that the light wave can extend from the sensor surface into the test sample. Electromagnetic fields bound to an optical device that couple some energy to an external medium are called *evanescent fields*. The evanescent field intensity decays exponentially with distance from the transducer surface, with a decay length of approximately $\lambda/2\pi$, where λ is the wavelength of the light. When we consider light in the typical wavelength range for optical biosensors of $\lambda = 600$–900 nm, therefore, the evanescent field extends only ∼100–150 nm into the test sample. Because the evanescent field distance is so short, the biosensor only interacts with material in direct proximity to its surface, such as proteins that are chemically attached to the surface, while being unaffected by unbound material suspended in solution. Different designs for optical biosensors can manipulate the characteristics of the evanescent field so that it either extends deeply into the test sample or is confined tightly to the sensor surface. One key to high sensitivity sensor design is to match the regions of greatest biochemical binding to those regions with the highest evanescent field intensity.

PERFORMANCE METRICS

Optical biosensors can be used to address a wide variety of applications ranging from high-precision detection of changes in biomolecular conformation, rapid measurement of the presence of a contaminant in a liquid, low-throughput measurement of kinetic binding constants of biomolecules, and high-throughput cell-based assays. Though some methods are only capable of measuring adsorbed mass density on a small number of discrete sensors, others are capable of generating high-resolution images of biomolecule and cell binding density to a sensor surface. Likewise, there are large differences in the way optical biosensors can be packaged. Though sensors that are expensive to produce are packaged with microfluidic flow channels that enable a single sensor to be used several times with washing procedures between each experiment, other sensors can be inexpensively produced over large surface areas from plastic materials and incorporated into standard format microplates for single use disposability. There are also large differences in sensitivity and resolution between different detection approaches that are determined by the detailed characteristics of the biosensor evanescent field and optical losses in the sensor materials, among other considerations. The suitability of an optical biosensor for a particular application depends upon its performance across a variety of metrics. In this section, we will define some of the methods used to compare optical biosensors. Some metrics, such as sensitivity or cost, can be defined numerically. Others, such as ease-of-use or instrument robustness, are subjective but can have a significant impact on the commercial success of an approach.

Sensitivity

Sensitivity is defined as the amount of change in sensor output response resulting from a unit change in mass density (which is usually assumed to be directly proportional to changes in dielectric permittivity) on the sensor surface. Often, the sensitivity is reported in terms of response per mass/area. The mass sensitivity represents characterization of the sensor structure at a basic level that is not dependent upon the affinity between the analyte being detected and the immobilized ligand. For optical biosensors, sensitivity is fundamentally determined by how efficiently the electromagnetic field associated with the optical transducer couples to biomolecules in contact with the sensor surface. The electric fields associated with propagating light or optical standing waves will tend to concentrate themselves into materials with higher refractive index. Most optical biosensors contain a material or structure allowing electric fields to concentrate themselves at the surface of a transducer in contact with a liquid sample that has a refractive index higher than the surrounding material. Although the electric field may reside mostly within the transducer itself, the evanescent field extends out of the transducer and into the surrounding medium (typically water with $n = 1.33$). The part of the electric field *inside* the transducer structure has no opportunity to interact with the adsorbed biomaterial – therefore the design goal for achieving high sensitivity is to produce a structure that allows as much of the electric field as possible to reside *outside* the transducer and within the surrounding media.

Assay sensitivity

Assay sensitivity refers to the minimum detectable concentration of an analyte, such as a protein, in a test sample. Assay sensitivity is typically defined in units of moles/volume or mass/volume of the analyte. Unlike mass sensitivity, assay sensitivity depends on other factors besides the biosensor, such as surface chemistry, the affinity of the analyte for the immobilized ligand, buffer conditions, and the molecular weight of the analyte. The surface chemistry method used to covalently link the immobilized ligand to the sensor surface can positively influence assay sensitivity by achieving a high density of active binding sites while eliminating the likelihood that undesired molecules can nonspecifically attach to the surface. Likewise, large molecular weight molecules and molecules with high affinity for the immobilized ligand can be detected at lower concentrations than small molecules with low binding affinity.

Resolution (or limit of determination)

Sensitivity refers to the magnitude of sensor response to a given change in surface-adsorbed mass density, and *resolution* refers to the smallest change in mass density that can be measured. Resolution is an especially critical performance criteria for detection of analytes present at low concentration or detection of adsorbed molecules with low molecular weight. To determine the resolution of a sensor, one must characterize the noise of the sensor when operated with its detection instrument. Noise can be easily characterized at a basic level by allowing the sensor to reach a steady-state condition and recording the measured output many times in sequence without any intentional change to the sensor. The noise is thus defined as the standard deviation, σ, of all the repeated measurements. If one defines a signal to be detectable if the signal has a magnitude of 3σ, then the *limit of determination* (LOD), defined as the smallest measurable mass density change of the sensor, is LOD $= 3\sigma/$(sensitivity). The sensitivity and the resolution are related but are entirely different figures of merit. Sensitivity is used to define the lowest value determined above zero concentration while the resolution describes the minimum resolvable difference between two measurements at any concentration.

Many optical biosensors are constructed from various types of optical resonators in which the transducer surface is illuminated with a broad range of wavelengths (or incident angles) and only a narrow range of incident wavelengths (or angles) is reflected back or transmitted through. For these types of resonant optical biosensors, the detection instrument measures changes in the range of wavelengths (or angles) that occur as a result of biomolecular binding on the resonator surface. For example, when a photonic crystal biosensor is illuminated with a broad band of wavelengths (from a light bulb or light-emitting diode [LED]), only a narrow band of wavelengths with a bandwidth of 2–3 nm reflects back. Similarly, when a surface plasmon resonance biosensor is illuminated with a broad range of angles (by focusing light from a monochromatic light source to a small point at the sensor surface), a small band of angles couples to a metal film and is *not* reflected. In either case, as the resonance becomes more narrow, the resolution improves because it is easier to measure small changes in the reflected/transmitted wavelength peak. The narrowness of the resonant peak is mathematically defined as the "Quality Factor" or "Q Factor," where $Q = \lambda_0/\Delta\lambda$ for a wavelength-based resonator. Here, λ_0 is the center wavelength of the resonance, and $\Delta\lambda$ is the spectral width of the resonance determined at one half of the peak maximum. The value of Q represents a measurement of how "pure" of an optical wavelength will couple to the resonator, and greater Q values represent structures with more narrow bandwidth response – and thus better resolution. Q values for optical biosensors can range from Q~15 for surface plasmon resonance to $Q\sim10^8$ for microtoroid biosensors. Of course, improving resolution will not improve the LOD if the sensitivity is simultaneously reduced, so both sensitivity and resolution must be considered.

Ease of use

A feature critical to the commercial success of an optical biosensor is ease of use for its intended application. Often, this subjective measure of sensor performance is not a factor of the biosensor technology but of how the technology is packaged and how the readout instrument is configured to interact with test samples. Several biosensors have been configured to interface with the test sample through a flow chamber that has the ability to provide a low-volume steady stream of liquid past the sensor surface while it is being measured. Though the flow chamber frees the operator from direct handling of the sensor, the sensor must be connected to a pump/valve system that, in turn, has the ability to draw material from sample reservoirs. A flow system's complexity can be eliminated by direct application of the test sample to a sensor embedded within a cuvette, a small volume reservoir without continuous flow. Though the flow cell can continuously provide fresh reagent to the sensor surface and can therefore provide fast binding with low volumes, the cuvette method relies on diffusion of analyte molecules to the sensor surface that can be enhanced through aspiration with a pipette or other mixing techniques. Cuvettes can provide much greater parallelism if the cuvette is a well within a standard 96-, 384-, or 1536-well microplate. Ward and Winzor have described the relative merits of the two approaches.[5] Biosensors can also be packaged onto the tips of optical fibers so the sensor can be dipped into the test sample. This method can provide multiplexing, because several probe tips can be operated in parallel and can also provide stirring when the detection instrument incorporates a motion stage that can perform shaking/swirling motions of the probe tips within microplate wells. Stirring of probe tips within the test sample has provided kinetic measurements similar to those provided by a biosensor packaged within a flow cell.

Sensor cost

Adoption of biosensor technology for most applications in diagnostics or pharmaceutical screening will be driven to some extent by the cost of performing an individual assay. For a primary screen used in the pharmaceutical research industry, for example, a screening campaign to determine a set of candidate chemicals that has a desired affinity level for a protein can involve over 1 million assays. Researchers working on high-volume industrialized assays describe the need to minimize the "cost per data point" in such a campaign. Though optical biosensors offer tremendous advantage over labeled assay technologies by not requiring the use of tag reagents, the cost of the transducer in each assay must be low enough to compensate the cost of additional reagents. This cost goal challenges the wide acceptance of optical biosensors, which are often high-precision optical components fabricated from expensive materials (such as glass, silicon, or optical fiber) using highly exacting processes such as photolithography, dielectric or metal deposition, and plasma etching. Even if a sensor is inexpensive to fabricate, the cost of packaging and testing it must be efficient.

Two main methods have been used to bring the cost/assay to acceptable levels. First, if a sensor is expensive to fabricate and package, it can be regenerated and used for several successive assays. Second, sensors can be designed to be compatible with mass production by using inexpensive materials and methods, so they can be used one time before disposal, much like typical labeled assays. Examples of both types will be shown in the following sections.

Detection instrumentation

To the same extent that the cost of the "consumable" part of an optical biosensor system can affect its likelihood for adoption, the cost of the instrument will also dictate its usage for different applications. The cost of a readout instrument is related to its complexity and the cost of the components required to reach a specified level of performance. Most optical biosensor readout instruments incorporate some sort of illumination source that can range from a highly stabilized laser to an inexpensive LED or light bulb. Most optical biosensor instruments also incorporate some method for gathering information about light reflected from or transmitted through the sensor. Often instruments incorporate a charge-coupled device (CCD) chip for measuring reflected (or transmitted) light's spatial position or wavelength content. The resolution (or LOD) of a biosensor will be driven by the accuracy of these components. Instrument cost can also be increased through the requirement for fine temperature control of the sensor and reagents and through integrated fluid handling devices that automate the introduction of samples. Detection instruments for biosensors used in drug discovery research must be compatible with standard-format microplates, and a variety of approaches have been demonstrated. In some systems, test samples are fed into the instrument within an ordinary microplate, which simply acts as a reservoir from which a probe will withdraw required volumes for flow within a microfluidic system that eventually flows past the biosensor. In other systems, the biosensor is incorporated within the microplate itself, and the detection instrument can be "fed" with biosensor microplates either by hand or by conventional robotic microplate handlers that interact with cell incubators, automated pipetting stations, and plate washers.

Throughput

The rate at which a biosensor system performs separate assays determines its throughput in units of assays/hour or assays/day. The throughput of a sensor system will determine its usefulness in large pharmaceutical screening campaigns or for diagnostic tests in which a test sample must be measured for the contents of many different proteins. A biosensor embedded within a flow chamber will have a throughput limited by the number of parallel flow channels and the time required to flush reagents away from a previous assay, to regenerate the sensor surface and introduce a new test sample. Cuvette-based systems will have throughput limited by the number of cuvettes that can be operated in parallel. Some biosensor approaches can measure high-resolution images of binding upon their surface. If the distance that evanescent waves are allowed to travel laterally across the sensor surface is intentionally limited, it is possible to take independent measurements of binding on the sensor surface from locations separated by distances greater than only a few microns. Using tools with the ability to deposit small spots of immobilized protein reagents (typical spot sizes are 100–500 μm in diameter) in an x–y grid across the sensor surface, each spot can be measured separately with an imaging biosensor readout instrument. With this method, a biosensor area of as little as 1 cm^2 can be used to perform several hundred assays in parallel, thereby dramatically improving the throughput of the system compared to single-channel flow cell-based systems. As we shall see, several optical biosensor approaches can be adapted to perform image-based assay analysis, a feature not shared by other label-free detection approaches. Image-based readout

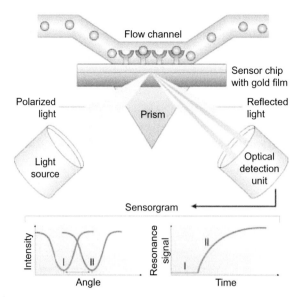

Figure 1–1: Typical set-up for an SPR biosensor. SPR detects changes in the refractive index in the immediate vicinity of the surface layer of a sensor chip. SPR is observed as a sharp shadow in the reflected light from the surface at an angle that is dependent on the mass of material at the surface. The SPR angle shifts (from I to II in the lower left-hand diagram) when biomolecules bind to the surface and change the mass of the surface layer. This change in resonant angle can be monitored non-invasively in real time as a plot of resonance signal (proportional to mass change) versus time. Reprinted with permission from Cooper, M. A., *Optical Biosensors in Drug Discovery*. Nature Reviews, 2002. **1**: pp. 515–528.

can be especially advantageous for obtaining high throughput when the imaged area can be increased.

REVIEW OF OPTICAL BIOSENSOR METHODS

In this section, we review a variety of optical phenomena that have been exploited to design biosensors. The methods described here are not an exhaustive list of all the approaches but are intended to be representative of approaches currently being developed for use as commercial products. However, this section is not intended to review all the commercial features of each technology but to focus mainly on the sensing phenomena. In addition, a handful of relatively new approaches have been highlighted that may find important applications in the next 5 yr. No effort is made to rank detection approaches, although direct comparative studies have been performed in some cases in the literature.

Surface plasmon resonance

Surface plasmon resonance (SPR) is an optical phenomenon that is sensitive to changes in the optical properties of the medium close to a metal surface.[6] The detection system, shown in Figure 1–1, typically consists of a monochromatic linearly polarized (polarization parallel with the plane of incidence) light source, a glass prism, a thin metal film in contact with the base of the prism, and a photodetector.[7] Obliquely incident light on the base of the prism will exhibit total internal reflection for angles larger than the critical angle, causing an evanescent field to extend from the prism to the metal film. The evanescent field can couple to an electromagnetic surface wave, a surface plasmon, at the metal–liquid interface. Coupling is achieved at a specific angle of incidence, the SPR angle. At this angle, the reflected light intensity goes through a minimum due to the SPR. The SPR angle position is sensitive to changes in the refractive index of a thin layer adjacent to the metal surface that is sensed by the evanescent wave so that when a protein layer is adsorbed onto the metal surface, an increase in the optical density occurs, resulting in a shift of the SPR angle to larger values. The evanescent electromagnetic field decays exponentially from the metal film surface into the interfacing medium so the impact of adsorbate in the probed volume depends on its distance from the surface.

The SPR sensor itself is simple to fabricate, as it consists of a thin gold film deposited upon a glass substrate. The gold film contains an abundance of free electrons able to counteract electric fields applied by the incident light. A general property of all highly conducting materials is that they efficiently exclude electric fields, except for a very thin "skin depth" of electric field present within a few nanometers of the conductor's outer surface. Because of this phenomenon, electric fields associated with surface plasmons reside primarily within the dielectric media (typically water) that is in contact with the gold surface, resulting in excellent sensitivity for SPR sensors. However, metals such as gold also provide a high degree of loss for photons, and therefore the Q-factor for SPR is generally as low as Q~15 as discussed earlier, which can limit detection resolution. Nonetheless, SPR has been demonstrated as an effective method for measuring small molecule analytes binding to large protein immobilized ligands through the use of three-dimensional surface chemistry to increase binding density, temperature control of reagents, and accurate reference sensors to compensate for environmental effects, nonspecific binding, and variability of the refractive index of sample buffer.

An SPR instrument using the Kretschmann configuration consists of an illumination system, a glass prism, and a detector array with imaging optics. The sensor chip, a gold-coated glass slide, is placed on the prism base with the coating facing upward. Optical contact between the prism and the glass side of the sensor chip is achieved by a refractive index matching fluid, and a flow cell is placed onto the sensor chip. A collimated light beam from an LED is focused onto the glass–gold interface through the prism, providing a wedge-shaped beam that simultaneously provides light at a range of incident angles, typically from 66–72 degrees. The reflected light wedge projects onto a photodiode array where SPR is observed as a minimum of reflected intensity of light (a dark band) as a function of the incident angle.

Since the first reports of protein detection via SPR,[8] the sensitivity of the method has been thoroughly characterized,[9] and commercial products based

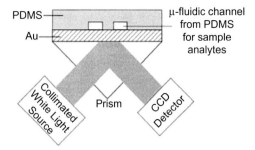

Figure 1–2: Optical arrangement for a detection instrument that performs imaging SPR measurements. Reprinted with permission from H. J. Lee, et al., *Anal. Chem.*, 73, p. 5525, 2001, Figure 4b.

upon this approach have been used in several industries, as evidenced by thousands of research articles published in recent years.[10–12]

One limitation of the Kretschmann SPR configuration described in the previous section is that the system can only take measurements from the point at which the laser light is focused. If greater throughput is desired from the ability to simultaneously measure multiple locations on the Au surface, it is possible to generate images of biomolecular binding using SPR. An alternative method for performing SPR is to measure the reflectivity of a wide-area collimated monochromatic source from the prism/Au film assembly at a fixed angle of incidence, as shown in Figure 1–2.[13] If the illuminating light is incident at the correct angle for surface plasmon coupling, the reflected image will be dark when imaged onto an inexpensive CCD chip. When a particular region of the imaged surface adsorbs material, the resonant coupling condition is no longer completely satisfied at the adsorbing location, so the location becomes brighter in the CCD image, as shown in Figure 1–3. Either visible or near-infrared illuminating wavelengths have been demonstrated,[14–17] and a wide variety of applications have been demonstrated, including DNA arrays and RNA arrays. Extremely high sensitivity has been obtained using enzymatic amplification of adsorbed analytes for some applications.[18]

The lateral resolution of method is determined by the propagation distance of surface plasmons, which can approach 25 µm at a wavelength of 830 nm.[14] Though the use of laser illumination produces interference fringes in the measured images due to interfering reflections from the sensor/sample, substrate/air, and sample/air interfaces, the substitution of an incoherent white light source and a narrow bandpass filter was found to provide cleaner images. Several important applications for imaging SPR have been demonstrated recently, including detection of protein biomarkers[19] and microRNA profiling.[20]

HOLOGRAM BIOSENSORS

Holographic biosensors have been demonstrated as test strips that provide a changing optical image (color, text messages) as the test result so that a biosensor

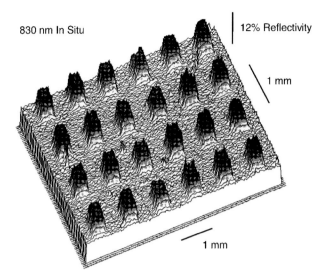

830 nm In Situ

12% Reflectivity

1 mm

1 mm

Figure 1–3: NIR SPR difference image of single-stranded DNA binding protein adsorbed onto a photopatterned DNA array. A single-stranded oligonucleotide was bound to the surface array elements and then exposed to a nanomolar solution of SSB. The image shown is the difference of two images taken immediately before and after exposure of the array to SSB. The image was taken at 830 nm with a 10-nm fwhm interference filter. The increased sharpness of NIR SPR enhances the contrast of the image. Reprinted with permission from Nelson, B. P., *et al.*, Analytical Chemistry, 1999. **71**: pp. 3928–3934.

experiment can be monitored visually without the requirement for readout instrument hardware (although an appropriate instrument may also be used for quantitation of data). Such devices can be inexpensive, durable, and disposable. A reflection hologram provides an image when it is illuminated with white light, where the image may be stored in a thin polymer film using photosensitizing methods. When the polymer film is chemically sensitized with an immobilized ligand that can selectively bind a target substance in a test sample, the presence of the target substance alters the image displayed by the hologram.

To produce such a hologram, a two-laser interference pattern exposes a photo-sensitive gelatin-thin film. Development of the film produces a modulated refractive index in the form of parallel planar fringes. On illumination with white light, constructive interference between partial reflections from each fringe plane gives a characteristic spectral peak with the wavelength peak described by the Bragg equation:

$$\lambda_{peak} = 2nD \cos \theta,$$

where the peak reflectivity is dependent on the number of fringe planes and the modulation depth of refractive index. The fringe separation is D, the average refractive index is n, and the angle of illumination is θ. Mechanisms for altering the reflected spectrum include changing the spacing of the fringes by weakening the swollen gelatin matrix and by progressive removal of gelatin into solution through interaction with an enzyme capable of cleaving gelatin.[21] Such sensors have been demonstrated to provide highly sensitive tests for the presence of proteases through the use of a spectrometer-based readout instrument that

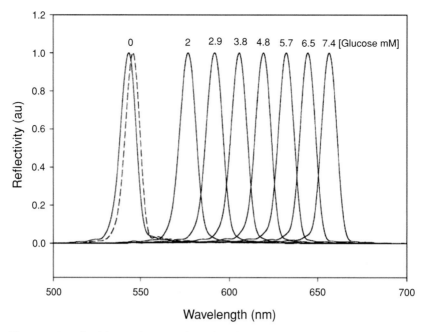

Figure 1–4: Reflectivity as a function of wavelength for a hologram biosensor exposed to a range of glucose concentration, where the reflected color spectrum is modulated over a wide band of wavelengths. A linear relationship between peak reflected wavelength and glucose concentration is obtained. Reprinted with permission from Kabilan *et. al.*, Biosensors and Bioelectronics, 20, p. 1602, 2005, Figure 2a.

records the reflected intensity as a function of wavelength as the protease interacts with the gelatin hologram. The Lowe Group at the University of Cambridge recently developed holographic base matrices that can be modified rationally to construct specific response mechanisms to target analyte. Holograms have been fabricated in a variety of defined polymeric matrices, including polyvinylalcohol, polyhydroxyethylmethacrylate and polyacrylamide. Such a capability opens up the prospect of constructing holographic sensors based on rationally designed synthetic polymers for almost any analyte. Recent demonstrations of holograms responsive to pH,[22] metabolites,[23] oligonucleotides, antigens, and whole cells have been performed in addition to multianalyte devices that respond to a range of measurands in a single biological sample. Figure 1–4 shows an example of measured output from this system for detection of glucose.[24]

Reflectometric interference spectroscopy

Several high-sensitivity optical biosensors have been demonstrated by using biomolecules to modulate optical interference effects, which can be measured with a high degree of precision. The basic effect is due to interference occurring within thin transparent films. A light beam passing a weakly reflecting thin film will be reflected in part at each of the interfaces, as shown in Figure 1–5. A phase difference is introduced as the two reflected beams travel different optical paths. If the film thickness is small (a few micrometers), the difference in optical paths

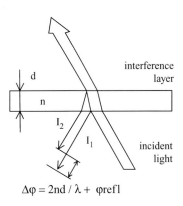
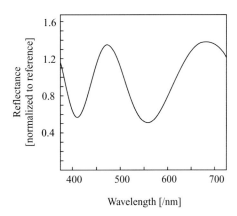

$$\Delta\varphi = 2nd\,/\,\lambda + \varphi\mathrm{refl}$$

Figure 1–5: *(Left)* Light reflection at weakly reflecting thin film. At thin films (1 μm), the reflected partial beams I_1 and I_2 show interference. Reflectance varies according to the phase difference $\Delta\varphi$ between I_1 and I_2. *(Right)* Resulting reflectance spectrum. Reprinted with permission from Piehler, J., A. Brecht, and G. Gauglitz, *Affinity Detection of Low Molecular Weight Analytes.* Analytical Chemistry, 1996. **68**: pp. 139–143.

is minute, and interference effects can be observed for noncoherent white light sources.[25]

In reflectometric interference spectroscopy (RIFS), a white light source illuminates a sensor surface containing a transparent thin film on a reflecting substrate. Constructive and destructive interference of the light beam reflected from the upper and lower surface of the thin film generates a modulation of the total reflected intensity that can be accurately measured with a spectrometer, as shown in the optical instrument setup in Figure 1–6. In the case of perpendicular incidence of light, the reflectance R of a thin nonabsorbing layer as a function of the wavelength is given by:

$$R = R_1 + R_2 + 2(R_1 R_2)^{1/2} \cos(4\pi nd/\lambda + \varphi_{ref\,l}),$$

where R_1 and R_2 are the Fresnel reflectances of the film interfaces, n is the refractive index of the film, d is the physical thickness of the film, λ is the wavelength of

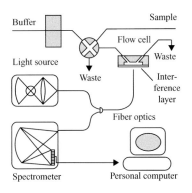
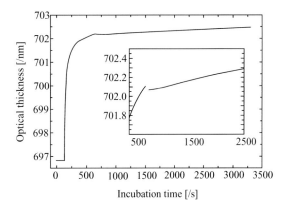

Figure 1–6: Setup for monitoring binding events with RIFS, and example data showing detection of a protein layer. Reprinted with permission from Piehler, J., A. Brecht, and G. Gauglitz, *Affinity Detection of Low Molecular Weight Analytes.* Analytical Chemistry, 1996. **68**: pp. 139–143.

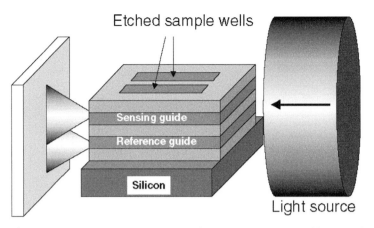

Etched sample wells

Sensing guide

Reference guide

Silicon

Light source

Figure 1–7: Dual slab waveguide interferometer. The sensor chip comprises five layers of deposited silicon oxynitride. A window is opened in the nal layer to expose the sensing waveguide. Reprinted with permission from Cross, G. H., et al.,. *Journal of Physics D: Applied Physics*, 2004. **37**: pp. 74–80.

incident light, and φ_{refl} is the phase change occurring upon reflection. Reflection at a medium of higher refractive index gives a phase shift of π.[26]

If dispersion effects are neglected, the reflectance pattern can be described by a cosine function in the wavenumber space. Any change in thickness of the thin film can be determined from the reflectance spectrum with high resolution. Using a spectrometer to measure the reflected spectrum, multiple measurements can be made rapidly (<100 msec), and multiple independent measurements can be made from different locations on the surface of a single sensor chip.[27] The RIFS transducer measures changes in the thickness of any thin transparent film. Binding analyte molecules to the surface of the interference film leads to the formation of an adlayer of organic matter. The formation of the adlayer can be monitored if the refractive index of the adlayer is similar to that of the interference film. The RIFS method has been demonstrated for detection of a wide variety of small molecule, protein, and DNA analytes.[25,27,28] Initially, RIFS sensors were implemented on the surface of a glass substrate that is illuminated by an external optical fiber probe, as shown in Figure 1–5. More recently, miniature RIFS sensors have been applied to the tip of the fiber probe itself (with the thin film oriented on the exposed surface end of the tip), thereby removing the need to have a separate sensor surface. Packaged in this way, RIFS sensors on the ends of optical fibers can be dipped into liquid test samples, and the dipped fiber probe can be agitated mechanically to provide stirring. Systems using this configuration have been developed commercially with multiple RIFS fiber probes that can obtain measurements within several microplate wells in parallel.

Dual polarization interferometry

Dual Polarization Interferometry (DPI) uses two thin film waveguides fabricated on top of each other, as shown in Figure 1–7.[29] In a DPI sensor, an upper waveguide thin film made of high refractive index dielectric is separated from a lower wave-guide film by a low refractive index dielectric material. The upper waveguide is in contact with the test sample and can couple light into an evanescent mode while the reference waveguide is buried in the interior of the chip and thus does not interact with the test sample. By coupling laser light through the edge facet of the chip to both the upper and lower waveguides simultaneously, the sensor and reference have the opportunity to interfere with each other over the length of the chip. When the two guided waves exit the chip at the opposite facet, their interference generates a Young's interference fringe pattern that can be imaged with a sensor array positioned in the far field region beyond the exit facet. The interference pattern represents the relative phase position of the upper and lower modes of light at the output face. The illuminating laser may either excite Transverse Electric (TE) or Transverse Magnetic (TM) modes in the wave-guides, depending on the orientation of an alternating half-wave plate. In either case, adsorption of biomolecules on the sensing waveguide surface results in a shift in the interference fringe pattern that can be quantified and recorded.[30]

One important aspect of the DPI sensor is the observation that TE and TM modes have different evanescent field profiles and therefore different effective refractive index profiles. Using information from measurements of the TE and TM modes, we can obtain accurate values for both thickness and refractive index of the adsorbed layer. Information about the density of an adsorbed protein monolayer can be used, for example, to detect changes in the conformation (i.e., folding) of the protein.[31]

PHOTONIC CRYSTAL BIOSENSOR

A class of optical biosensors based on the unique properties of optical device structures known as *photonic crystals* has been recently developed.[32–40] A sur-face photonic crystal consists of a periodic arrangement of dielectric material, in which a low refractive index periodic surface structure made of plastic is over-coated with a high refractive index film of TiO_2, as shown in Figure 1–8. With proper choice of period, symmetry, and material dielectric constants, the pho-tonic crystal will selectively couple energy at only one wavelength, known as the resonant wavelength, which will be strongly reflected. At the resonant wave-length, electromagnetic fields form an optical standing wave that is confined from propagating in lateral directions.

To adapt a photonic crystal device to perform as a biosensor, some portion of the structure must be in contact with a liquid test sample. By attaching biomolecules or cells to the photonic crystal surface, the resonant coupling of light into the crystal is modified so the reflected wavelength output is tuned. The sensor is measured by illuminating the surface with white light and collecting the reflected light with a noncontact optical fiber probe where several parallel probes can be used to independently measure different locations on the sensor. Although the sensor structure must contain surface features smaller than the

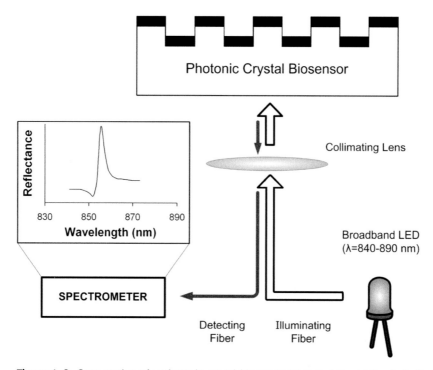

Figure 1–8: Cross section of a photonic crystal biosensor fabricated on sheets of plastic film. The surface structure is replicated from a silicon master wafer with a negative of the desired sensor shape, and a thin, noncontinuous layer of high refractive index dielectric material is deposited over the structure. The structure acts as a narrow bandwidth reflectance filter with a peak reflected wavelength that is modulated by the attachment of biomolecules of cells to the sensor surface. Printed with permission from B. T. Cunningham, University of Illinois at Urbana-Chyampaign.

resonant wavelength, and therefore in the 100–200 nm size scale, the photonic crystal sensor structure is inexpensively produced over large surface areas using a nanoreplica molding method that can be performed upon continuous sheets of plastic film. Both the replica-molded periodic grating structure and the deposition of TiO_2 are performed on continuous rolls of plastic. The sensor structures are cut from the sheet and attached to standard format microplates (96-, 384-, and 1536-well formats) so each microplate well has a sensor as the bottom surface. The detection instrument illuminates the underside of the plate, where the target material is loaded onto the bottom of the microplate well and can measure all the sensors within a microplate in several seconds. Photonic crystal biosensors have also been integrated with microfluidics, with microfluidic channels integrated into the microplate format.[41,42]

The sensor operates by measuring changes in the peak wavelength value (PWV) of reflected light as biochemical binding events take place on the surface. For example, when a protein is immobilized on the sensor surface, an increase in the reflected wavelength is measured when a complementary binding protein is subsequently presented to the sensor. Using low-cost components, the read-out instrument resolves protein mass changes on the surface with resolution

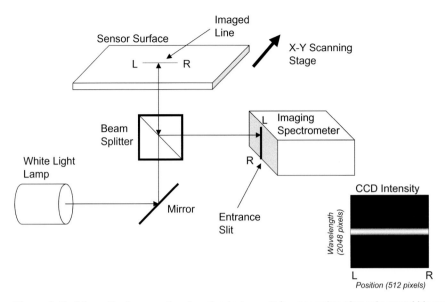

Figure 1–9: Schematic diagram of an imaging instrument for measuring photonic crystal biosensors. The fiber probe system from Figure 1–8 is replaced with free-space optics and an imaging spectrometer capable of making ∼512 spectral measurements simultaneously from one line of the sensor surface. The sensor is sequentially scanned across the image line to measure peak wavelength values with a resolution as low as 7 µm. Reprinted with permission from P. Li, B. T. Cunningham, *et. al.* Sensors and Actuators B, Vol. 99, pp. 6–13, 2004.

less than 1 pg/mm^2. Though this level of resolution is sufficient for measuring small-molecule interactions with immobilized proteins, the dynamic range of the sensor is large enough to also measure binding of larger biochemical entities including proteins, live cells, cell membranes, viruses, and bacteria. The sensor is especially sensitive to the capture of cells via specific protein interaction or to the response of the cell to certain chemical stimuli such as small organic test compounds. A sensor measurement requires ∼20 msec so large numbers of interactions can be measured in parallel and kinetic information can be gathered.

The reflected wavelength of the sensor can be measured either in "single point mode" (such as for measuring a single interaction within a microplate), or an imaging system can be used to generate an image of a sensor surface with ∼7-micron resolution, using an instrument based on the design shown in Figure 1–9. The imaging mode can be used for applications that increase the overall resolution and throughput of the system such as label-free microarrays, imaging plate reading, self-referencing microplates, and multiplexed spots/well.[43] The microplate format is especially advantageous for performing assays that involve cells, which would normally clog microfluidic channels. Figure 1–10 shows a label-free image of human breast cancer cells, where the attachment of each cell results in a locally increased PWV value compared to adjacent regions without cells.[40,44] Label-free optical biosensor imaging can quantify how many cell processes are modulated by challenges to their environment (such as the

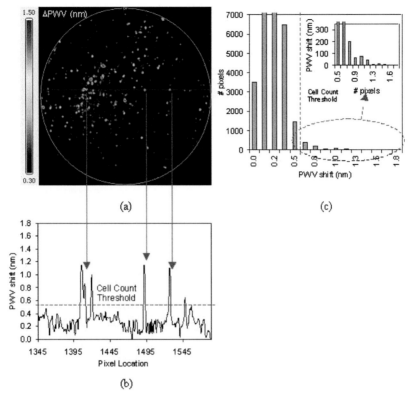

Figure 1–10: Label-free image of human breast cancer cells adsorbed to a photonic crystal biosensor surface captured with the imaging detection instrument shown in Figure 1–9. Because the photonic crystal does not allow lateral propagation of light at the resonant wavelength, each cell attached to the sensor increases the reflected wavelength only in a precise location. The method can be used to quantify the number of attached cells and capture kintetic images of a cell population as they undergo proliferation, apoptosis, and chemotaxis. Reprinted with permission from L. Chan, S. Gosangari, K. Watkin, and B. T. Cunningham, Apoptosis, Vol. 12, No. 6, pp. 1061–1068, 2007.

introduction of drug compounds) including proliferation, apoptosis, ion channel activation, chemotaxis, and wound healing. Because no stain is used, the cells can function in their culture environment while being observed nondestructively over long periods of time.

WHISPERING GALLERY MODE RESONATORS

Recently, many research groups have been working to improve the resolution of optical biosensors through the design of optical resonators with extremely high Q-factor. High Q is generally achieved with a resonator that has little optical loss (necessitating the use of dielectric materials), smooth surfaces (to avoid scattering light out of the optical cavity), and an efficient system for forming optical standing waves. These goals can be accomplished by constraining light waves to follow a circular path, where the only allowed optical modes are for wavelengths that

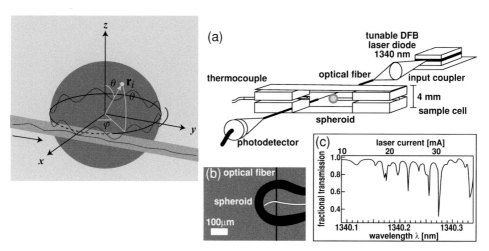

Figure 1–11: Schematic illustration and experimental setup for a whispering gallery mode micro-sphere resonator biosensor. A glass spheroid is created by melting the tip of an optical fiber. The cladding is stripped from a portion of a second optical fiber, which is positioned in close proximity to the perimeter of the spheroid with micropositioners. Light from a tunable laser diode enters one end of the second fiber, and a photodetector on the distal end monitors the intensity trans-mitted past the spheroid. When the resonant coupling condition is reached by the tunable laser, the transmitted intensity measured at the photodetector is decreased. The wavelength of most efficient coupling increased when biomolecules attach to the spheroid surface. Reprinted with permission from Arnold, *et. al., Applied Physics Letters,* 80, p. 4057, 2002 (Figure 1) and Arnold, *et. al., Applied Physics Letters,* 87, p. 223900, 2005 (Figure 1). *See color plates.*

can fit an integer number of half-wavelengths around the perimeter of the circle. For these specific modes, resonant optical resonances can be sustained and are commonly referred to as whispering gallery mode resonators for their analogous function to circular rooms that can transmit acoustic modes efficiently over large distances around the perimeter. Whispering gallery mode biosensors have been demonstrated in a variety of configurations, including glass microspheres[45] with $Q = 5 \times 10^6$ ring resonators on glass or silicon surfaces[46] with $Q = 20,000$, hol-low glass capillary tubes,[47] and silicon oxide toroids sitting upon silicon support posts[48,49] with Q up to 10^8. Figures 1–11 to 1–13 show illustrations of these approaches.

To excite the whispering gallery mode sensor to resonate, light at one of the allowed wavelengths must be supplied from the outside from an optical fiber or integrated waveguide that comes within close proximity (<300 nm) to the circular perimeter of the resonator without physically touching it. Light from a tunable-wavelength laser is introduced into one end of the fiber/waveguide, and the wavelength is adjusted until the laser wavelength matches the whis-pering gallery mode wavelength, and light is coupled to flow into the circular path. When the coupling condition is satisfied, a photodiode sensor on the distal end of the fiber/waveguide detects a decrease in transmitted intensity. The cou-pling wavelength will be modulated to longer wavelengths when biomolecules attach to the whispering gallery mode structure, as the dielectric permittivity of the biomolecules increases the effective refractive index of the optical path

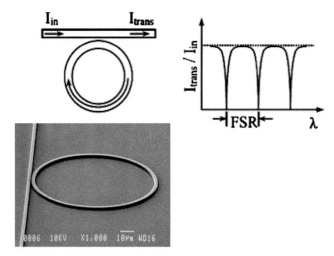

Figure 1–12: A replica molded polymer microring resonator fabricated in a glass substrate. Light from a tunable laser (I$_{in}$) passes through an integraded rib waveguide, and wavelengths that satisfy the whispering gallery mode resonant coupling condition will pass from the waveguide and couple to the ring structure, resulting in decreased transmitted light intensity (I$_{trans}$) at the resonant wavelengths. The wavelength of optimal coupling efficiency will modulate to longer wavelengths by the adsorption of biomolecules on the ring structure. Reprinted with permission from Guo, *et. al., IEEE J. Quantum Electronics,* 12, p. 134, 2006. (Figure 1).

experienced by circularly coupled light waves. As with other optical biosensors, a portion of the light resides within the circular waveguide structure, but a portion of the light also extends into the surrounding media in the form of an evanescent field. Though the high Q optical cavity provides a high degree of resolution, the necessity for the field to be strongly guided by the circular waveguide results in sensitivity that is generally much lower than SPR. It is anticipated that approaches in which multiple whispering gallery mode sensors can be integrated on a single chip will be more widely adopted for life science research and drug discovery.

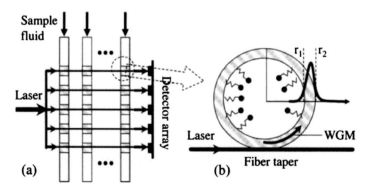

Figure 1–13: (a) Conceptual illustration of a Liquid Core Optical Ring Resonator (LCORR) sensor array. Glass capillaries with thin walls lie perpendicular to optical fibers on the surface of a substrate, where light from a tunable laser couples from the fiber to a whispering gallery mode of the capillary. **(b)** Cross section of a LCORR sensing element. The inner radius and the wall thickness are r_1 and r_2-r_1, respectively. Biosensing occurs on the inner surface of the capillary. Reprinted with permission from I. M. White, *et. al., Optics Letters,* 31, pp. 1319–1321, 2006 (Figure 1).

ACTIVE RESONATORS: DFB LASER BIOSENSOR

Each of the optical biosensor examples described thus far represents a different type of "passive" optical component, in which the sensed output is measured simply by measuring some characteristic of light reflected from or transmitted through the device. Within the family of whispering gallery mode biosensors, we have discussed how implementation of evanescent-coupled high Q-factor passive resonators provides higher levels of detection resolution at the expense of reduced sensitivity and the added requirement for highly accurate optical alignment for light coupling, providing potential limits to practical application in high-throughput biomolecule screening. Recently, "active" optical resonators that can produce their own narrow bandwidth light,[50] such as the distributed feedback (DFB) laser biosensor reported here, provide high Q-factor with simple coupling to the excitation source and collection optics for detecting small changes in wavelength while retaining excellent sensitivity as defined by the magnitude of the wavelength shift.

DFB laser biosensors have been fabricated and characterized as a fundamentally new type of label-free optical biosensor. Fabricated as a one-dimensional dielectric subwavelength grating incorporating a dye-doped polymer gain region, the DFB laser source features a second-order Bragg grating that supports a narrow linewidth,[51] single wavelength,[52] and vertically emitting mode by first-order diffraction.[53] In addition to simplifying the excitation of the laser and extraction of the output radiation, this monolithic structure is sensitive to surface refractive index change. Because the DFB laser structures can be fabricated uniformly and inexpensively over large surface areas by employing recently developed nanoreplica molding techniques[54,55] to produce the periodic grating structure, this biosensor is potentially inexpensive and amenable to mass production.

A DFB laser biosensor structure produced with a plastic-based replica-molding process is shown in Figure 1–14. The DFB laser surface is produced over large surface areas, within which laser emission occurs only at precise locations on the surface excited with a short (~10 nsec) excitation laser pulse that is absorbed by the dye within the polymer active region. As with other types of optical biosensors, the output laser wavelength is tuned to greater wavelengths by adsorption of biomaterial on the transducer surface, but because the Q factor is "artificially" boosted by the stimulated emission process of the laser (Q~10 000), a high degree of resolution is obtained without sacrificing sensitivity. The pulse from the excitation laser may illuminate the DFB surface from any angle, and the laser emission always exits the structure at a normal angle, so gathering the high-intensity emission into a spectrometer or interferometer instrument for measuring its wavelength is simple, using an optical fiber to gather the light.

CONCLUSIONS

As the foregoing examples demonstrate, optical biosensors have evolved from laboratory-based measurement approaches to practical commercial instruments

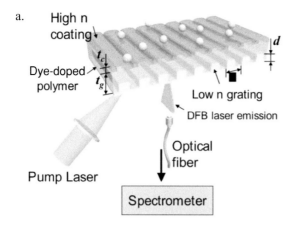

a.

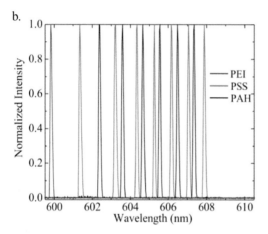

b.

Figure 1–14: (a) Schematic illustration of a DFB laser biosensor surface. The large area nanoreplica molded surface is comprised of a 1-dimensional linear grating structure formed from a polymer on a plastic substrate, coated with a dye-doped polymer layer and a thin (optional) high refractive index film of TiO_2. Laser output is activated by pumping with a ~10 nsec pulse from an excitation laser, and DFB emission occurs normal to the sensor surface, where it is captured by an optical fiber that feeds a wavelength measurement instrument such as a spectrometer. **(b)** Modulation of the laser wavelength due to the sequential adsorption of a series of positively and negatively charged polymer monolayers, where high intensity single mode laser output wavelength increases due to the addition of each layer. The laser wavelength peak width is <0.09 nm. Reprinted with permission from B. T. Cunningham, University of Illinois at Urbana-Champaign.

and devices routinely used for many important tasks in pharmaceutical research, diagnostic testing, environmental testing, and life science research. The fundamental advantages of optics-based methods, compared to other label-free detection approaches, include their sensitivity, ease of sensor fabrication, simplicity of excitation/readout instrumentation, and ease of integration with fluid handling systems. Within the family of optical biosensors, methods based on SPR, interferometry, waveguides, and photonic crystals have been developed into commercial products. Adoption of these methods tends to be driven by several factors, including sensitivity, cost/assay, assay throughput, compatibility with standard liquid handling systems, and ease of use. Recent trends have been the development of sensor structures that can be inexpensively fabricated using plastic materials, including fabrication on long, continuous sheets of plastic film, and incorporated into single-use disposable labware. Other important developments include the design of sensors and instrumentation that can perform high spatial resolution imaging of biomolecular or cellular binding to the surface, in the context of a closely packed x–y grid of immobilized receptor spots. Such systems are likely to offer the highest level of assay throughput while also minimizing the use of reagents. As label-free detection instrumentation becomes more simple and compact, its integration with other biochemical analysis methods, such as liquid chromatography, fluorescence assays, and mass spectrometry, will yield advanced multimodality systems capable of characterizing biomolecules with higher levels of information content. Most recently, sensor concepts using high Q-factor resonances and active resonators with high Q laser output promise to continue advancing this area further toward single molecule detection resolution. As detection sensitivity advances through improved sensor concepts, higher-resolution optical components, and higher-density surface chemistry approaches, label-free detection methods will continue to replace label-based assays based on fluorescence, radiolabels, and secondary reporters due to the inherent simplicity of assay development, validation, and implementation.

REFERENCES

1. M. A. Cooper, *Nature Reviews* **1**, 515 (2002).
2. T. Arakawa, Y. Kita, *Anal Biochem* **271**, 119 (1999).
3. T. Cole, A. Kathman, S. Koszelak, A. McPherson, *Anal Biochem* **231**, 92 (1995).
4. J. Wen, T. Arakawa, *Anal Biochem* **280**, 327 (2000).
5. L. D. Ward, D. J. Winzor, *Anal Biochem* **285**, 179 (2000).
6. R. M. Sutherland, C. Dahne, *Biosensors: Fundamentals and Applications*. A. P. F. Turner, I. Karube, G. S. Wilson, Eds., (Oxford University Press, New York, 1987).
7. E. Kretschmann, H. Raether, *Z. Naturforsch. A* **23**, 2135 (1968).
8. B. Liedberg, C. Nylander, I. Lundstrom, *Sens Actuators* **4**, 299 (1983).
9. E. Stenberg, B. Persson, H. Roos, C. Urbaniczky, *J Colloid Interface Sci* **143**, 513 (1991).
10. R. L. Rich, D. G. Myszka, *J Mol Recognit* **13**, 388 (2000).
11. D. G. Myszka, *J Mol Recognit* **12**, 390 (1999).
12. R. L. Rich, D. G. Myszka, *J Mol Recognit* **14**, 1 (2001).
13. H. J. Lee, T. T. Goodrich, R. M. Corn, *Anal Chem* **73**, 5525 (2001).
14. B. P. Nelson, A. G. Frutos, J. M. Brockman, R. M. Corn, *Anal Chem* **71**, 3928 (1999).
15. A. Frutos, S. C. Weibel, R. M. Corn, *Anal Chem* **71**, 3935 (1999).
16. A. J. C. Tubb, F. P. Payne, R. B. Millington, C. R. Lowe, *Sens Actuators B Chem* **41** (1997).
17. R. C. Jorgenson, S. S. Yee, *Sens Actuators B Chem* **12** (1993).
18. T. T. Goodrich, H. J. Lee, R. M. Corn, *Anal Chem* **76**, 6173 (2004).
19. Y. Li, H. J. Lee, R. M. Corn, *Anal Chem* **79**, 1082 (2007).
20. A. Wark, H. J. Lee, R. M. Corn, *Angew Chem Int Ed Engl* **47**, 644 (2008).
21. R. B. Millington, A. G. Mayes, J. Blyth, C. R. Lowe, *Sens Actuators B Chem* **33**, 55 (1996).
22. A. J. Marshall, J. Blyth, C. A. B. Davidson, C. R. Lowe, *Anal Chem* **75**, 4423 (2003).
23. A. J. Marshall, D. S. Young, J. Blyth, S. Kabilan, C. R. Lowe, *Anal Chem* **76**, 1518 (2004).
24. S. Kabilan *et al.*, *Biosens Bioelectron* **20**, 1602 (2004).
25. J. Piehler, A. Brecht, G. Gauglitz, *Anal Chem* **68**, 139 (1996).
26. G. Gauglitz, A. Brecht, G. Kraus, W. Nahm, *Sens Actuators B Chem* **B11**, 21 (1993).

27. O. Birkert, G. Gauglitz, *Anal Bioanal Chem* **372**, 141 (2002).
28. R. Tunnemann *et al.*, *Anal Chem* **73**, 4313 (2001).
29. G. H. Cross, Y. T. Ren, M. T. Freeman, *J Appl Phys* **86**, 6483 (1999).
30. G. H. Cross *et al.*, *J Phys* **37**, 74 (2004).
31. S. Freitag, I. L. Trong, L. Klumb, P. S. Stayton, R. E. Stenkamp, *Prot Sci* **1997**, 1157 (1997).
32. B. T. Cunningham, P. Li, B. Lin, J. Pepper, *Sens Actuators B Chem* **81**, 316 (2002).
33. A. J. Haes, R. P. V. Duyne, *J Am Chem Soc* **124**, 10596 (2002).
34. L. L. Chan, S. Gosangari, K. Watkin, B. T. Cunningham, *Sens Actuators B Chem*, in press (2008).
35. L. L. Chan, P. Y. Li, D. Puff, B. T. Cunningham, *Sens Actuators B Chem* **120**, 392 (2007).
36. B. T. Cunningham, L. Laing, *Expert Opinions in Proteomics* **3**, 271 (2006).
37. B. T. Cunningham *et al.*, *J Biomol Screen* **9**, 481 (2004).
38. B. T. Cunningham, J. Qiu, P. Li, J. Pepper, B. Hugh, *Sens Actuators B Chem* **85**, 219 (2002).
39. B. Lin, P. Y. Li, B. T. Cunningham, *Sens Actuators B Chem* **114**, 559 (2006).
40. L. L. Chan, S. Gosangari, K. Watkin, B. T. Cunningham, *Sens Actuators B Chem* **132**, 418 (2008).
41. C. J. Choi, B. T. Cunningham, *Lab Chip* **6**, 1373 (2006).
42. C. J. Choi, B. T. Cunningham, *Lab Chip* **7**, 550 (2007).
43. P. Li, B. Lin, J. Gerstenmaier, B. T. Cunningham, *Sens Actuators B Chem* (2003).
44. L. L. Chan, S. Gosangari, K. Watkin, B. T. Cunningham, *Apoptosis* **12**, 1061 (2007).
45. S. Arnold, M. Khoshsima, I. Taraoka, S. Holler, F. Vollmer, *Opt Lett* **28**, 272 (2003).
46. Z. Guo, H. Quan, S. Pau, *Appl Opt* **45**, 611 (2006).
47. I. M. White, H. Oveys, X. Fan, *Opt Lett* **31**, 1319 (2006).
48. A. M. Armani, K. J. Vahala, paper presented at the Optomechatronic Micro/Nano Devices and Components, 2006.
49. A. M. Armani, S. E. Fraser, K. J. Vahala, paper presented at the Optomechatronic Micro/Nano Devices and Components, 2007.
50. W. Fang *et al.*, *Appl Phys Lett* **85**, 3666 (2004).
51. H. Temkin *et al.*, *Appl Phys Lett* **46**, 105 (1985).
52. S. Balslev, T. Rasmussen, P. Shi, A. Kristensen, *J Micromechanics and Microengineering* **15**, 2456 (2005).
53. R. F. Kazarinov, C. H. Henry, *IEEE Journal of Quantum Electronics* **21**, 144 (1985).
54. M. Lu, S. S. Choi, C. J. Wagner, J. G. Eden, B. T. Cunningham, *Appl Phys Lett*, in press (2008).
55. M. Lu, J. G. Eden, B. T. Cunningham, *Opt Comm* **281**, 3159 (2008).

2 Experimental design

Robert Karlsson

INTRODUCTION	30
ASSAY FORMAT	30
IMMOBILIZATION	32
Ligand activity	33
Regeneration	34
Regeneration scouting experiments	35
Analysis of regeneration data	35
SIGNAL CORRECTIONS	36
Reference subtraction	36
Solvent correction	37
BUFFER SCOUTING	38
MISCELLANEOUS	39
Wash procedures	39
Temperature	40
Positive and negative controls	40
Startup and maintenance procedures	41
Software wizards	41
Sample preparation	41
SMALL-MOLECULE SCREENING	42
PROTEIN–PROTEIN KINETICS	44
General principles	44
Ligand saturation	45
Association data	45
Dissociation data	45
Ligand activity	45
Systematic errors and blank injections	45
Analysis order	46
Reference surface	46
Flow rate	46
Simulation as a tool for experimental design	46
ACKNOWLEDGMENTS	46
REFERENCES	47

INTRODUCTION

Label-free detection is a key factor contributing to the increasing popularity of biosensors, in particular those based on surface plasmon resonance (SPR). By measuring changes in refractive index close to a sensor surface, SPR biosensors allow the user to study the interaction between immobilized molecules (often referred to as ligands) and analytes in solution, in real time and without analyte labeling. Observed binding rates and levels can be interpreted in different ways to provide information on the specificity, kinetics, and affinity of the interaction or for determination of the concentration of the analyte. The ease by which this information is obtained has changed customer workflows in antibody and small-molecule interaction analysis, and in screening. There is now a clear shift from label-based and affinity/IC_{50}-based workflows to a label-free and kinetic-based workflow.

Over the last 20 years I have obtained my knowledge and experience in the biosensor field in development and use of Biacore systems (GE Healthcare Companies), but the information provided in this chapter is general and therefore relevant to many label-free platforms.

Biacore systems commonly integrate detection, sensor surface, and liquid handling technologies and have multispot capability, that is, they allow independent and simultaneous measurements on several discrete spots on a single surface. Ligands can be attached to the sensor surface through a variety of chemistries, and the liquid handling system makes it possible to inject reagents and samples into different flow cells. Samples can be injected in sequence to build multimolecular complexes, but in most cases formation of the complex between the immobilized ligand and one analyte is investigated.

In assay development, multispot detection can be used to look at several ligands in parallel or to look at one ligand immobilized through different chemistries or to different ligand densities. In addition, one or several spots can be used for referencing. By using the addressing capabilities of the flow system, buffer conditions and regeneration solutions can be tested on selected ligands without affecting other ligands. The final assay is about injecting different samples or different concentrations of samples and also includes system and surface conditioning steps, positive and negative controls, and signal correction tools.

In most cases, experimental design is straightforward and supported by software that guides the user through the workflow. There are also reagent kits for immobilization and capture purposes where the instructions for use contain the information required to get started with real experiments. Regardless of the application, many steps in the experimental design are common as illustrated by the Figure 2–1 workflow. The workflow elements and their application to small-molecule screening and kinetics are the focus of this chapter.

ASSAY FORMAT

Direct binding, surface competition, and solution competition assays are three useful assay formats. Figure 2–2 illustrates the basic features of these formats. In

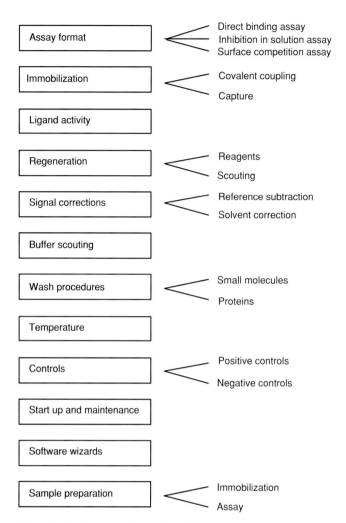

Assay format	Direct binding assay
	Inhibition in solution assay
	Surface competition assay

| Immobilization | Covalent coupling |
| | Capture |

| Ligand activity | |

| Regeneration | Reagents |
| | Scouting |

| Signal corrections | Reference subtraction |
| | Solvent correction |

| Buffer scouting | |

| Wash procedures | Small molecules |
| | Proteins |

| Temperature | |

| Controls | Positive controls |
| | Negative controls |

| Start up and maintenance | |

| Software wizards | |

| Sample preparation | Immobilization |
| | Assay |

Figure 2–1: Assay development workflow.

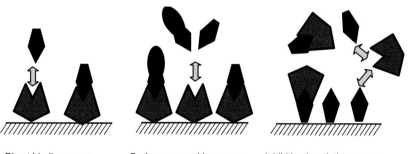

Direct binding assay   Surface competition assay   Inhibition in solution assay

Figure 2–2: Schematic illustration of assay formats. In direct binding and surface competition assays sample molecules interact with the immobilized biomolecule. No reactions take place in solution. In the inhibition assay format the reaction, which takes place in solution, is monitored indirectly by measuring the free concentration of the detecting molecule.

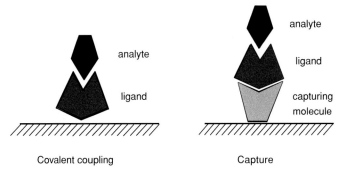

Figure 2–3: Schematic illustration of assays with covalently coupled and captured ligands.

the *direct binding assay* (DBA), the sample interacts directly with the immobilized ligand. This assay format is widely used in kinetic analysis and mapping studies.[1,2]

The *surface competition assay* (SCA) looks at binding site specificity but can also be used in kinetic analysis.[3-5] In both cases an analyte with known binding site specificity competes with varying concentrations of a second analyte. No reactions take place in solution, but the two analytes compete for the immobilized ligand.

The *solution competition assay*, also called the inhibition in solution assay (ISA), is typically used in concentration analysis[6] and in particular when the analyte is of low molecular weight. In this type of assay the analyte or an analogue of the analyte is immobilized to the sensor surface. A detecting molecule, usually an antibody, reacts with analyte in solution. When this reaction reaches equilibrium the sample is injected over the sensor surface where free detecting molecule binds to the sensor surface. This type of assay is also frequently used to determine the affinity of an interaction that takes place in solution[4] and to screen small-molecule inhibitors to protein activity.[7]

IMMOBILIZATION

Ligands are either covalently immobilized to the sensor surface or captured by using reagents immobilized to the sensor surface that recognize a specific part or tag on the ligand of interest (Figure 2–3).

Protein targets can be covalently immobilized to a sensor surface utilizing different functional groups on the protein including:

- amine groups
- carboxyl groups
- thiol groups
- cis-diols on carbohydrate moieties

Amine coupling is direct; coupling through other groups is a multistep process where the protein is first reacted with a bifunctional reagent or oxidized. Covalent immobilization on carboxymethylated dextran surfaces is normally performed with a ligand in a low (10 mM) ionic strength buffer with a pH of between

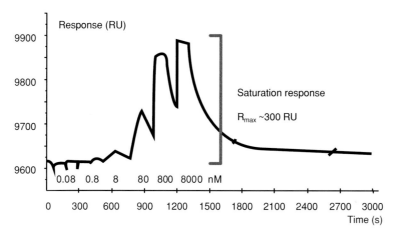

Figure 2–4: Successive injections of analyte at increasing concentration for rapid determination of the saturation response. The extended dissociation phase provides a first estimate of the stability of the interaction.

0.5 and 1.5 pH units below the isoelectric point of the ligand. Capture has the advantage of more oriented immobilization and can often be performed at neutral pH, which is useful for ligands sensitive to low pH. Immobilization levels can be controlled within wide borders using both covalent immobilization and capture. Very high capacity surfaces, however, can normally be obtained only through covalent coupling procedures.

Ligand activity

When the ligand has been immobilized or captured to the sensor surface the next step is to determine the activity of the immobilized ligand. This is done by injecting the analyte in increasing concentration order until reaching saturation level (Figure 2–4).

A good approximation of ligand activity is obtained by comparing analyte saturation R_{sat} and ligand immobilization R_{imm} levels:

$$Ligand \ \text{activity} = \frac{R_{sat}}{Mwt_{analyte}} \cdot \frac{Mwt_{ligand}}{R_{imm}}.$$

Ligand activities above 20% are often sufficient. For pure ligands, activities over 80% are not uncommon. In kinetic and affinity analysis, saturation levels should ideally fall in the range between 10 and 100 RU for an SPR biosensor. For concentration analysis high saturation levels are beneficial, in particular when a low detection limit is required and can often exceed several thousand RU.

If the ligand is histidine or GST-tagged, a capture procedure can be tried first. The number of tags and their availability on the analyte govern capture efficiency and stability of the capture interaction. Both capture levels and stability vary considerably for different ligands expressing the same tag.

Amine coupling is the next choice for a ligand that is captured to too low a level or that dissociates rapidly from the capturing molecule. If none of these methods can be used, coupling through a carboxyl group or biotinylation followed by

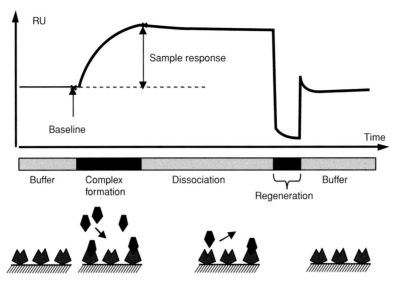

Figure 2–5: The complex formed during sample injection is stable and does not dissociate freely. By injecting a regeneration solution dissociation is achieved and the biosensor signal returns to the baseline.

capture on streptavidin may prove successful. When a protein is biotinylated, a long chain biotin derivative is preferable. A molar ratio of 0.5 to 1 between biotin and protein is recommended, as well as careful removal of free biotin. By passing the biotinylated protein two times through a buffer exchange column any interference from free biotin can effectively be minimized.

In some cases a ligand can be mildly cross-linked after covalent immobilization or capture to improve stability.[8]

Regeneration

When the complex on the sensor surface is stable and throughput is important, regeneration is necessary to dissociate the complex (Figure 2–5). Regeneration speeds up dissociation either by moderating the forces of interaction or by denaturing the analyte, the ligand, or both. The trick is to find regeneration conditions that give complete dissociation while the activity of the immobilized molecule is maintained. Regeneration conditions often depend on ligand density, and there is no guarantee that regeneration conditions optimized for one ligand density are valid if the ligand density is drastically changed. When stable ligands such as small molecules and peptides have been immobilized, rather harsh conditions can be used. In these cases we recommend regeneration solutions with low or high pH sometimes in combination with detergents. With oligonucleotide or DNA ligands, regeneration with 1 to 100 mM sodium hydroxide, sodium dodecyl sulfate (SDS), or sodium chloride solutions is frequently successful. Antibody ligands can often be regenerated with low or high pH. For other protein ligands it can be more difficult to find suitable regeneration conditions. A systematic scouting of regeneration conditions may then be helpful. Scouting

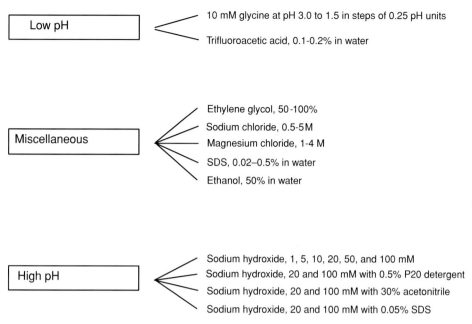

Low pH
- 10 mM glycine at pH 3.0 to 1.5 in steps of 0.25 pH units
- Trifluoroacetic acid, 0.1-0.2% in water

Miscellaneous
- Ethylene glycol, 50-100%
- Sodium chloride, 0.5-5M
- Magnesium chloride, 1-4 M
- SDS, 0.02–0.5% in water
- Ethanol, 50% in water

High pH
- Sodium hydroxide, 1, 5, 10, 20, 50, and 100 mM
- Sodium hydroxide, 20 and 100 mM with 0.5% P20 detergent
- Sodium hydroxide, 20 and 100 mM with 30% acetonitrile
- Sodium hydroxide, 20 and 100 mM with 0.05% SDS

Figure 2–6: Reagents for regeneration scouting. The reagents are grouped into three categories, low pH, miscellaneous, and high pH solutions. In each category an attempt has been made to list the solutions in order from mild to harsh regeneration conditions.

requires a library of regeneration solutions (Figure 2–6), and the success rate of scouting experiments can be improved by using a standardized experimental procedure.

The standardized scouting procedure described here uses an automated system with addressable flow cells, but the principles can be applied to other instrumentation and the procedure can be performed manually.

Regeneration scouting experiments

Immobilize the ligand to the desired level in one flow cell on the sensor surface.

Inject analyte at a high enough concentration to saturate or near saturate the surface.

Inject the first low pH regeneration solution for 30 s.

Inject the analyte at the same concentration as previously.

Repeat the injections of regeneration solution and analyte four more times.

Continue with the next low pH regeneration solution until all the low pH solutions have been injected.

If no suitable low pH regeneration conditions have been found, immobilize the ligand to a new flow cell and repeat the procedure with the next group of regeneration solutions.

Analysis of regeneration data

Plot the analyte response obtained prior to injection of the regeneration solution versus injection cycle number (Figure 2–7). A suitable regeneration condition is identified when the analyte response is constant and is the same or nearly the

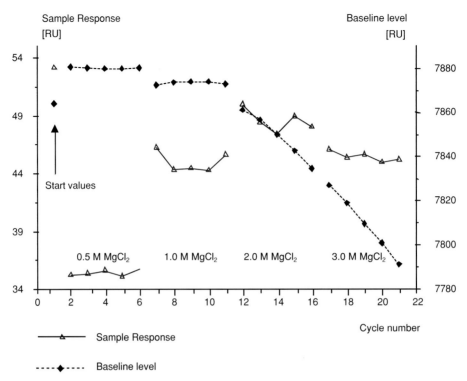

Figure 2–7: Plots of sample response (*left axis*) and baseline level (*right axis*). The sample response is high after regeneration with 1–3 M MgCl$_2$ whereas the baseline level drops significantly after injections of 2 and 3 M MgCl$_2$. These scouting experiments suggest that 1–2 M MgCl$_2$ solutions are potentially useful for regeneration.

same as in the first cycle. A baseline plot (Figure 2–7) is useful for diagnostic purposes. An increase in the baseline level is indicative of incomplete regeneration. A small overall drop in the baseline level is normally of no significance. If the drop is large it is a warning sign and may suggest that the ligand itself is not stable under the conditions tested.

When promising regeneration conditions have been identified the long-term effects on ligand stability and activity can be further tested by repeating analyte/ regeneration injections 20 to 30 times. This type of experiment can either confirm the usefulness of selected regeneration conditions or suggest further optimization. This often involves a variation of the injection time or a small adjustment in the concentration of the regeneration agent. A complete regeneration scouting can identify multiple regeneration conditions.[9] This can be useful when a common regeneration has to be used for several ligands in the same flow cell.

SIGNAL CORRECTIONS

Reference subtraction

The main use of a reference surface is in signal correction (Figure 2–8). An SPR signal is proportional to changes in refractive index close to the sensor surface.

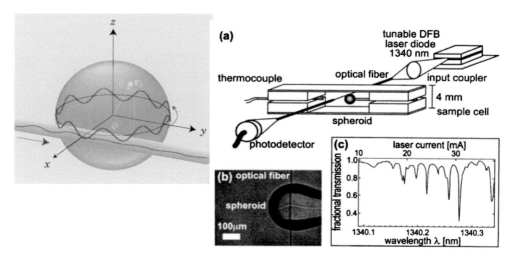

Figure 1–11: Schematic illustration and experimental setup for a whispering gallery mode micro-sphere resonator biosensor. A glass spheroid is created by melting the tip of an optical fiber. The cladding is stripped from a portion of a second optical fiber, which is positioned in close proximity to the perimeter of the spheroid with micropositioners. Light from a tunable laser diode enters one end of the second fiber, and a photodetector on the distal end monitors the intensity transmitted past the spheroid. When the resonant coupling condition is reached by the tunable laser, the transmitted intensity measured at the photodetector is decreased. The wavelength of most efficient coupling increased when biomolecules attach to the spheroid surface. Reprinted with permission from Arnold, *et. al., Applied Physics Letters*, 80, p. 4057, 2002 (Figure 1) and Arnold, *et. al., Applied Physics Letters*, 87, p. 223900, 2005 (Figure 1).

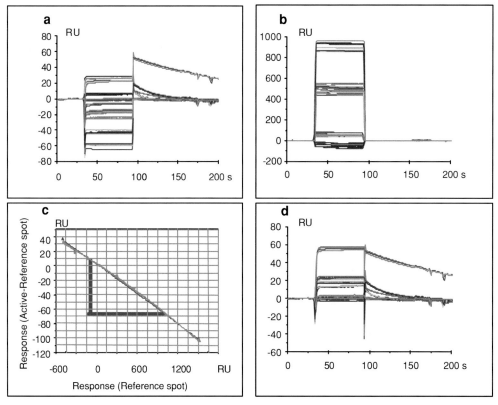

Figure 2–9: Solvent correction. **(a)** Reference subtracted binding data before solvent correction. **(b)** Refractive index signals obtained on the reference surface are in the range from −100 to 1000 RU. **(c)** By using the calibration curve, correction factors in the range from −70 to 10 RU are determined. **(d)** Corrected data with injection and dissociation phase aligned.

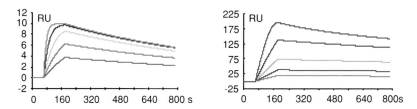

Figure 2–15: Simulation of interaction data on surfaces with low and high levels of immobilized ligand. In both cases analyte concentrations were 1–32 nM. The difference in the shape of the sensorgrams is due to transport limitations on the surface with high ligand density.

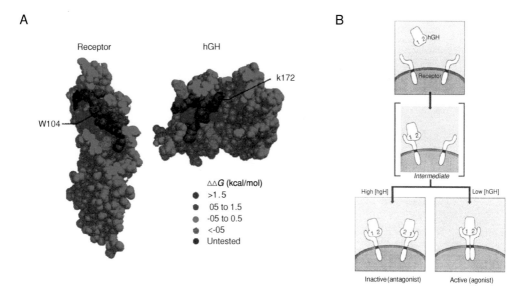

Figure 3-1: Human growth hormone (hGH) binding to its receptor (hGHr). **(A)** Hormone/receptor binding interface with contact residues highlighted. $\Delta\Delta G$ values (determined from affinity measurements) revealed which residues are critical for binding. To align the overlay of the binding faces, two residues that pack together (hGHr W104 and hGH K172) are highlighted. **(B)** Cartoon illustrating the concentration-dependent stoichiometry of the hGH/hGHr interaction. Panels A and B were reproduced from references 3 and 4 with permission from the American Association for the Advancement of Science (copyright 1995 and 1992, respectively).

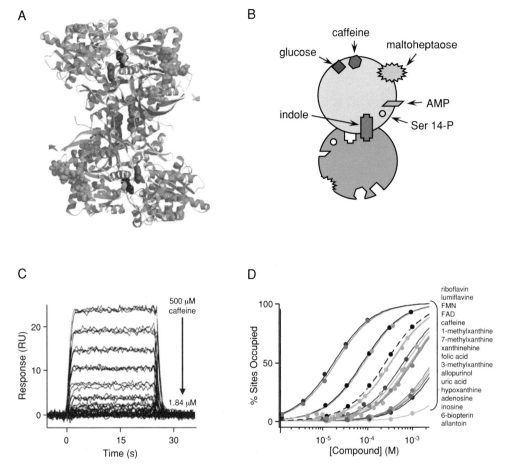

Figure 3–2: Small-molecule ligands binding to human liver glycogen phosphorylase a (HLGPa) homodimer. **(A)** Ribbon diagram of HLGPa showing the binding sites of different ligands. Helices and strands of one subunit are colored light and dark green; the helices and strands of the twofold symmetry related subunit are colored blue and purple. Ligands are shown as space-filling Corey-Pauling-Koltun (CPK) models; maltoheptaose in green, glucose in blue, adenosine 5′-monophosphate (AMP) in turquoise, caffeine in red, phosphorylated serine 14 in yellow and the indole inhibitor in pink at the dimer interface. **(B)** Cartoon showing the interactions targeted in this study. **(C)** Sensorgrams obtained for 0–500 μM caffeine binding to HLGPa. Responses at equilibrium were plotted against the tested caffeine concentrations and fit to a simple binding model to obtain an affinity of 108 ± 10 μM. **(D)** Isotherms generated from a simple model for compounds binding to HLGPa. The isotherm for caffeine is shown as a dashed line. Calculated enzyme affinities of these compounds ranged from 17 to 8000 μM. Panels A and B provided by VL Rath (unpublished results). Panels C and D were adapted from reference 37 with permission from Elsevier Science (copyright 2002).

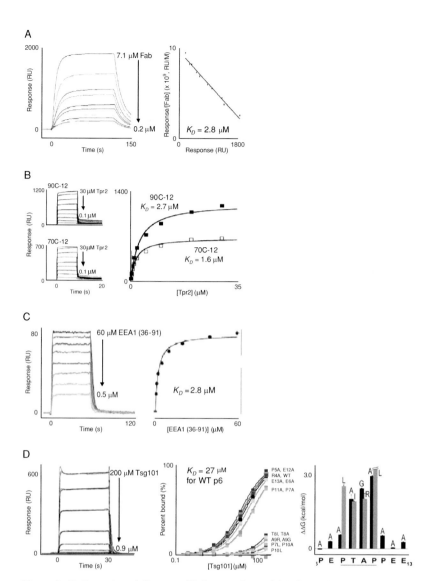

Figure 3–7: Examples of direct equilibrium analyses. **(A)** *Left panel*: concentration series of soluble Fab HyHEL-63 V$_L$N32A injected across immobilized hen egg white lysozyme. *Right panel*: Scatchard analysis of the equilibrium binding responses. Adapted from reference 55 with permission from the American Chemical Society (copyright 2003). **(B)** *Left panel*: concentration series of soluble cochaperone Tpr2 injected across 12-mer C-terminal Hsp70 and Hsp90 peptides immobilized on the surfaces of two flow cells. *Right panel*: binding responses plotted on a linear scale versus injected Tpr2 concentration and fit to simple isotherms. Adapted from reference 56 with permission from the European Molecular Biology Organization (copyright 2003). **(C)** *Left panel*: concentration series of soluble early endosome antigen 1 (EEA1) injected across GST-Rab5c captured on an anti-GST antibody surface. *Right panel*: binding responses plotted on a linear scale versus injected EEA1 concentration and fit to a simple isotherm. Adapted from reference 57 with permission from the American Society for Biochemistry and Molecular Biology (copyright 2003). **(D)** *Left panel*: concentration series of soluble Tsg101 injected across GST-p6 captured on an anti-GST antibody surface. *Middle panel*: Epitope mapping of the Tsg101 binding site on p6. For a panel of p6 proteins, binding responses were plotted on a log scale versus injected Tsg101 concentration and fit to simple isotherms. *Right panel*: Changes in free energy of Tsg101 binding caused by mutations in p6. Dark bars represent the changes due to single alanine substitutions in residues 5–13 and light bars represent alternative substitutions. Left and right panels adapted from reference 58 with permission from Cell Press (copyright 2001).

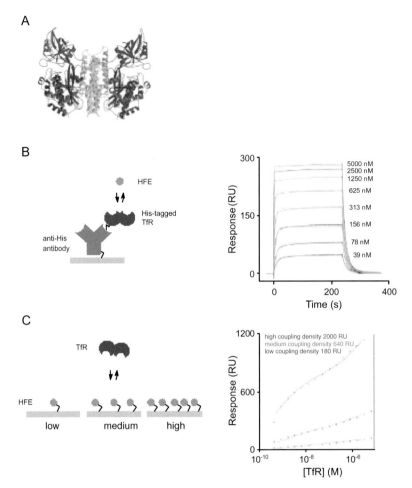

Figure 3–10: (A) Ribbon diagram of transferrin receptor. The HFE binding site on the helical domain closest to the viewer is highlighted in green. The HFE binding site on the other helical domain is omitted for clarity. **(B)** Concentration series of soluble HFE injected across His-tagged transferrin receptor captured on an anti-His antibody surface. Affinities determined from this orientation were $K_{D1} = 98$ nM and $K_{D2} = 520$ nM (gray lines represent the best fit to a bivalent ligand kinetic model). **(C)** Equilibrium binding responses plotted on a log scale versus TfR concentrations injected across HFE immobilized at three densities. HFE was immobilized at low, medium, and high densities to vary the avidity effect on apparent TfR binding. All three binding curves were globally fit to a bivalent analyte model, yielding affinities of $K_{D1} = 46$ nM and $K_{D2} = 210$ nM. Adapted from reference 62 with permission from Academic Press (copyright 2001).

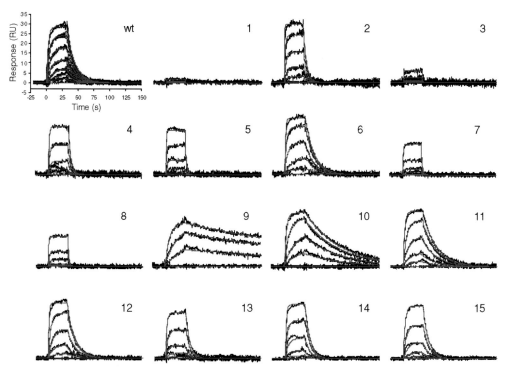

Figure 4–2: Kinetic analysis of p35/MC3 caspase interactions. Concentration series of p35 (wild-type or one of fifteen mutants) were flowed across immobilized MC3. All sensorgram data were fit to a simple 1:1 interaction model (shown as red lines) to obtain kinetic and affinity parameters. Table 4.1 summarized the parameters obtained for each binding pair. Adapted from reference 5 with permission from the American Society of Biochemistry and Molecular Biology (copyright 2003).

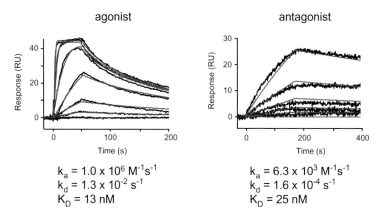

agonist

$k_a = 1.0 \times 10^6 \text{ M}^{-1}\text{s}^{-1}$
$k_d = 1.3 \times 10^{-2} \text{ s}^{-1}$
$K_D = 13 \text{ nM}$

antagonist

$k_a = 6.3 \times 10^3 \text{ M}^{-1}\text{s}^{-1}$
$k_d = 1.6 \times 10^{-4} \text{ s}^{-1}$
$K_D = 25 \text{ nM}$

Figure 4–3: Representative data sets (black lines) for the SPR kinetic analysis of ligand/estrogen receptor interactions. Concentration series of the agonist ligand (estriol, left set of sensorgrams) and the antagonist ligand (nafoxidine, right set of sensorgrams) were flowed across the immobilized receptor. Red lines represent the global fits of the data to a 1:1 bimolecular interaction model. Adapted from reference 8 with permission from the National Academy of Sciences USA (copyright 2002).

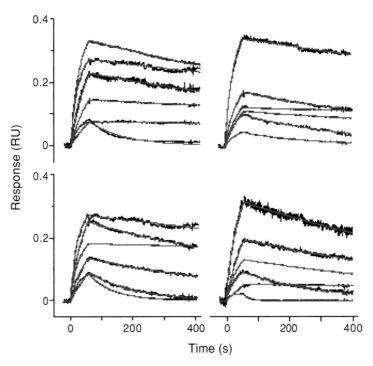

Figure 4–5: Kinetic screening of antibodies. Antigen responses (black lines) are shown in groups of six interactions per plot. The responses were all simultaneously fit with a 1:1 interaction model (red lines) using a single [B] (or R_{max}) for the entire data set. Reproduced from reference 9 with permission from Elsevier Science (copyright 2004).

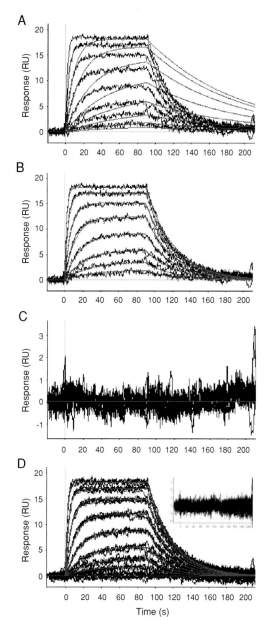

Figure 4–14: Fitting an experimental data set to a 1:1 interaction model. For illustrative purposes, sensorgrams of the buffer blank and replicate analyte injections are omitted from panels A–C. **(A)** Simulation of the data based on initial guess parameters of $k_a = 1 \times 10^4 \ M^{-1}s^{-1}$, $k_d = 1 \times 10^{-2} \ s^{-1}$, and [B] = 20 RU. **(B)** Data set overlaid with the final fit of the model. **(C)** Residuals of the fit shown in **(B)**. **(D)** The complete data set (blanks and replicates included) fit to a 1:1 bimolecular interaction model (red lines) to yield $k_a = (4.47 \pm 0.01) \times 10^4 \ M^{-1}s^{-1}$, $k_d = (3.416 \pm 0.006) \times 10^{-2} \ s^{-1}$, $K_D = 764 \pm 2$ nM, and [B] = 19.54 RU. The inset shows the fit residuals for the complete data set.

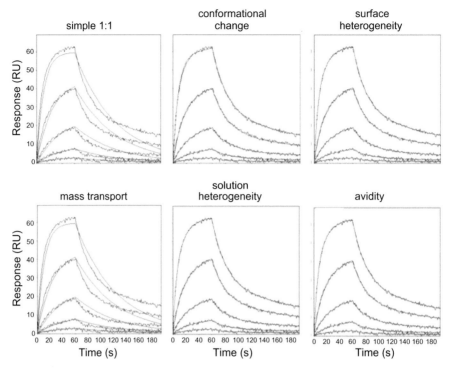

Figure 4–16: Binding responses for a complex interaction (black lines) overlaid with six different models (red lines).

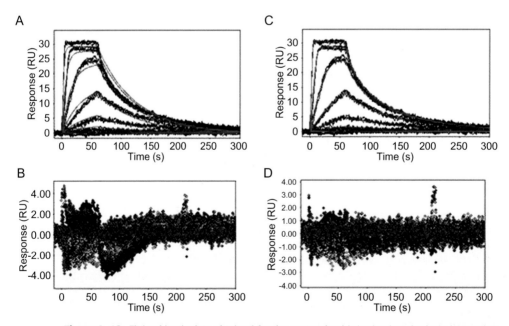

Figure 4–18: Fitting kinetic data obtained for the acetazolamide/carbonic anhydrase interaction to a 1:1 mass transport. **(A)** Final fit of the data set to a simple 1:1 interaction model. **(B)** Residuals of the fit shown in panel A. **(C)** Final fit of the data set to a 1:1 interaction model that includes a parameter to account for mass transport. **(D)** Residuals of the fit shown in panel C.

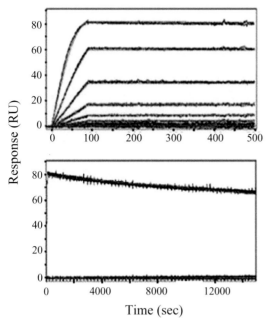

Figure 4–20: Two-phase kinetic analysis of the slowly dissociating antigen/antibody complex. The top panel shows the data collected in phase 1: the binding responses for triplicate injections of antigen (0, 0.18, 0.36, 0.72, 1.44, 2.88, 5.75, 11.5, and 23.0 nM) flowed across immobilized antibody. In each binding cycle, the association and dissociation phases were monitored for 90 and 450 s, respectively. The bottom panel shows the data collected in phase 2: triplicate binding responses of 23.0 nM antigen having a dissociation phase of 4 hr. The experimental data (black lines) from both panels were globally fit (red lines) to determine the kinetic rates, $k_a = 2.7 \times 10^6$ $M^{-1}s^{-1}$; $k_d = 1.6 \times 10^{-5}$ s^{-1}; $K_D = 6.1$ pM. Adapted from reference 5 with permission from Elsevier Science (copyright 2004).

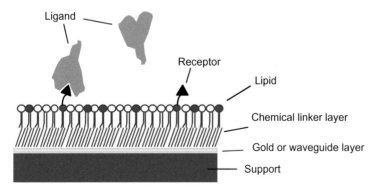

Figure 7–1: A supported lipid monolayer that has been formed on top of a hydrophobic self-assembled monolayer on a gold surface or a waveguide layer.

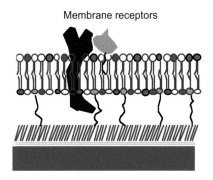

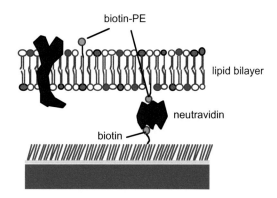

Figure 7–2: Two examples of tethered lipid bilayers that contain an integral (transmembrane) receptor. The bilayer is either captured on the surface using synthetic phospholipids that are tethered to the support by flexible, hydrophilic linkers (*left*) or via immobilized neutravidin in conjunction with biotinylated lipids or a biotinylated receptor.

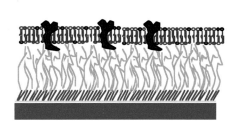

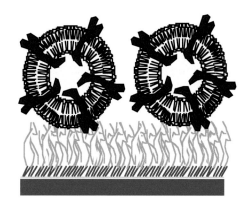

Figure 7–3: Flexible, amphipathic polymer cushions support membranes as either supported lipid bilayers (*left*) or captured proteoliposome layers (*right*).

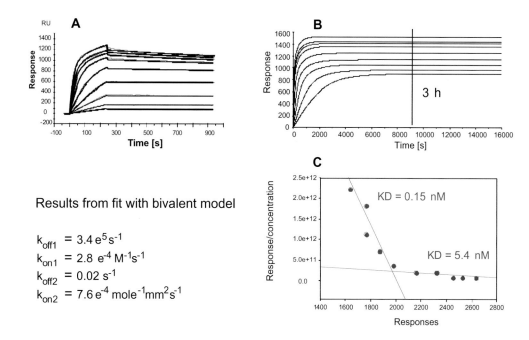

Results from fit with bivalent model

$k_{off1} = 3.4\,e^5\,s^{-1}$
$k_{on1} = 2.8\,e^{-4}\,M^{-1}s^{-1}$
$k_{off2} = 0.02\,s^{-1}$
$k_{on2} = 7.6\,e^{-4}\,mole^{-1}mm^2s^{-1}$

Figure 8–2: Determination of K_D values for antibodies with subnanomolar affinity binding in mono- and bivalent binding mode to antigen on the surface. **A**: Concentration-dependent binding of the antibody to immobilized fibers. The experimental curves are overlaid with the curves emerging from a fitting procedure based on a bivalent model. **B**: Extrapolation of the curves for long contact intervals using the kinetic rate constants determined by the fitting procedure. **C**: Evaluating high- and low-affinity component by Scatchard analysis.

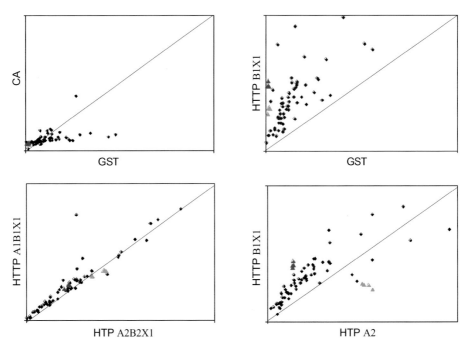

Figure 8–8: Correlation of sensor responses observed for compounds in contact with different constructs and dummy proteins. Black spots represent the responses observed for the test compounds, the red and green spots represent those monitored for the references R1 and R2, respectively.

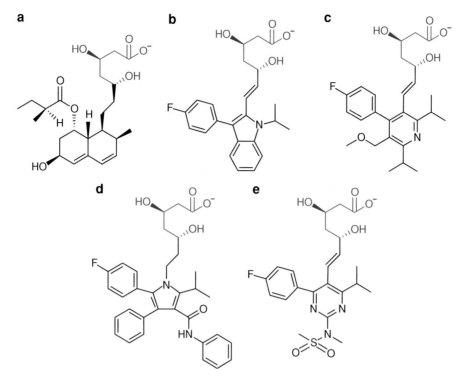

Figure 10–5: Chemical structures of the statins considered in these studies: **(a)** pravastatin, **(b)** fluvastatin, **(c)** cerivastatin, **(d)** atorvastatin, and **(e)** rosuvastatin. The HMG moiety common to all statins is colored red and the variable hydrophobic region black.

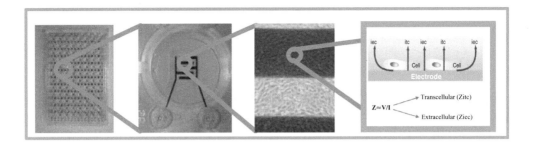

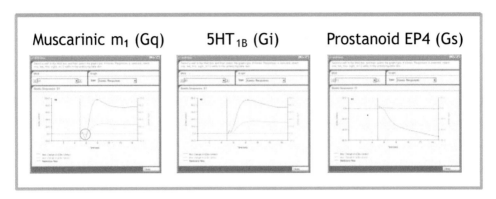

Figure 11–1: The CellKey^{TM} System measures changes in the impedance (Z) of a cell layer to an applied voltage. Cells are seeded onto a CellKey^{TM} Standard 96W microplate **(A)** that contains electrodes at the bottom of the wells. The instrument applies small voltages at 24 discrete frequencies, from 1 kHz to 10 MHz, once every 2 s. At low frequencies, these voltages induce extracellular currents (iec) that pass around individual cells in the layer. At high frequencies, they induce transcellular currents (itc) that penetrate the cellular membrane. Changes in impedance due to extracellular currents (dZiec) and to transcellular currents (dZitc) are reported kinetically for each well. When cells are exposed to a stimulus, such as a receptor ligand, signal transduction events occur that lead to cellular events which include changes in cell adherence, cell shape and volume, and cell-cell interaction. These cellular changes individually or collectively affect the flow of extracellular and transcellular current, and therefore, affect the magnitude and characteristics of the measured impedance. **(A)** CellKey cell plate (*left*), an individual assay well with close-up of interdigitated gold electrodes (*center left*), and live cells plated on and between the electrode fingers (*center right*). A diagram highlighting the flow of extracellular (iec) and transcellular (itc) current is shown (*right*). **(B)** In response to ligand mediated activation of GPCRs, the system generates response profiles that are characteristic of $G_{\alpha s}$, $G_{\alpha q}$ and $G_{\alpha i}$-coupled GPCRs. Examples are provided which illustrate the response profiles for the transfected muscarinic M_1 (typical $G_{\alpha q}$-GPCR), endogenous serotonin 5HT1B (typical $G_{\alpha i}$-GPCR), and prostanoid EP4 receptors (typical $G_{\alpha s}$-GPCR) in the context of CHO cells. $G_{\alpha i}$-GPCRs typically demonstrate an increase in impedance, while $G_{\alpha s}$-GPCRs exhibit a decrease in impedance following receptor activation. What distinguishes the $G_{\alpha q}$-GPCRs from the $G_{\alpha i}$-GPCRs is an initial transient dip in impedance (circled) that is followed by a later increase in impedance.

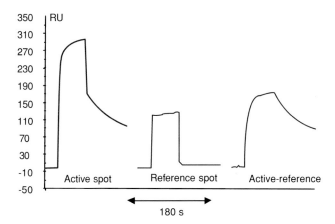

Figure 2–8: By subtracting data from a reference spot where no binding occurs, bulk refractive index effects are reduced and binding events are visualized more clearly.

The signal reflects binding events and differences in buffer composition between sample and running buffer. The latter are often referred to as *bulk refractive index* effects. A surface where analyte is not expected to bind will pick up the signal related to bulk refractive index effects. By subtracting this signal from the signal obtained on the active surface, binding is displayed more clearly.[10]

A *reference surface* is often a surface with an irrelevant protein immobilized, an activated and blocked surface, or an unmodified surface. The best reference surface is that to which the analyte shows negligible binding. Further improvements include subtraction of a blank cycle, where buffer is injected instead of sample. The combined procedure is referred to as double referencing.[11]

Solvent correction

Reference subtraction works fine as long as bulk refractive index effects are small (typically <100 RU). When these differences are larger and when there are large differences in immobilization levels on active and reference surfaces, the refractive index signal obtained on active and reference surfaces starts to differ. This is normal and can be tested by injecting a buffer with a refractive index that differs significantly from that of the running buffer. If the difference in the sample's refractive index can be linked to a particular cause (e.g., varying salt concentration, varying glycerol concentration, or varying DMSO [dimethylsulfoxide] concentration) then it is possible to calibrate the system for these effects and to correct the signal.

A correction for varying DMSO concentration is almost always necessary in small-molecule screening where compounds are diluted in a single step from stock concentrations in DMSO to screening concentration in buffer. In this case a calibration curve can be constructed by injecting buffer solutions with varying DMSO concentration over active and reference surfaces. The relative response, that is the difference in response values obtained on active and reference surface, is plotted versus the response obtained on the reference surface. These data points are fitted to a mathematical expression, usually a second order polynomial, to

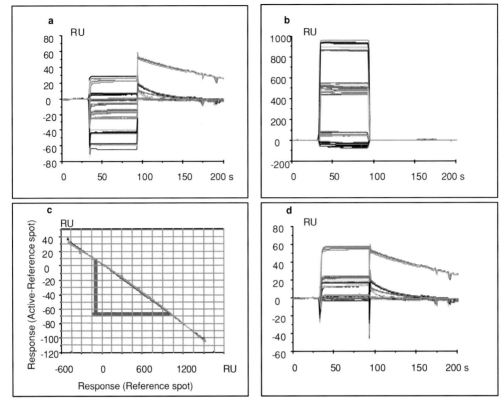

Figure 2–9: Solvent correction. **(a)** Reference subtracted binding data before solvent correction. **(b)** Refractive index signals obtained on the reference surface are in the range from −100 to 1000 RU. **(c)** By using the calibration curve, correction factors in the range from −70 to 10 RU are determined. **(d)** Corrected data with injection and dissociation phase aligned. *See color plates.*

obtain a mathematical expression for the calibration curve. When an unknown sample is injected the refractive index signal on the reference surface is measured and a correction factor is obtained from the calibration curve (Figure 2–9). The corrected response is found by subtracting the correction factor from the relative response. By applying this procedure it is possible to extract low (∼5 RU) signal binding events from very large (500–1000 RU) bulk refractive index effects.[12,13]

BUFFER SCOUTING

Typical running buffers in SPR experiments are (4-(2-hydroxyethyl)-1-piperazineethanesulfonic acid) (HEPES) 2-amino-2-hydroxymethyl-propane-1,3-diol (Tris), and phosphate buffers with ionic strength of 10–50 mM. They are normally supplemented with 150 mM sodium chloride to mimic physiological conditions. Other additives suppress nonspecific binding, stabilize interaction partners, or may be required for interaction to take place. Common additives include detergents, ethylenediaminetetraacetic acid (EDTA), reducing agents, and ions such as Ca^{2+}, Mg^{2+}, and Mn^{2+}. In some cases these additives reduce

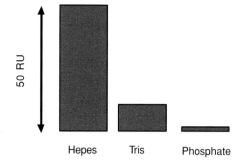

Figure 2–10: Average binding levels for selected small molecular weight compounds injected in different buffers over a carboxymethylated dextran surface.

nonspecific binding, in others they are an essential cofactor to activate binding activity of the ligand.

We cannot always predict interactions between buffer components and support reagents or surfaces. One recent observation relates to small-molecule screening (Figure 2–10). Compounds injected at high (≥ 10 μM) concentrations in HEPES buffer frequently bind to unmodified carboxymethyl-dextran. Binding to the dextran surface became negligible when HEPES was replaced with phosphate buffer.

Buffer scouting has been suggested as a means to predict the kinetics of peptide–antibody interactions.[14] For this purpose we advocate a multivariate approach involving modifications in peptide sequence and variations of buffers. Buffer scouting experiments have been absolutely essential to determine binding conditions for many immobilized enzymes such as kinases[15] and phosphatases[16] and for finding detergent mixtures that solubilize membrane proteins.[17] Buffer scouting experiments are strongly recommended in any assay development. In many cases there are unfounded assumptions on the use of certain buffers and buffer additions; hence scouting experiments are a more robust approach to optimized interaction conditions via a systematic and unbiased approach.

MISCELLANEOUS

Wash procedures

A sensor surface is contacted by samples, by support reagents, and by regeneration solutions. To avoid carryover from one injection to the other, the injection system is normally washed with buffer between injections. The need for efficient wash routines is easily understood when we consider a few examples.

1. A low pH regeneration solution can influence the binding of the following sample if the interaction is pH sensitive.
2. A mouse monoclonal antibody captured on a RAM (rabbit antimouse) antibody surface may give rise to a second signal on the RAM surface or may react with its antigen in solution if it is not removed from the injection system before the antigen injection.

3. When an analyte with a 10 nM affinity for the immobilized target is injected at 100 μM concentration a memory effect corresponding to 1 part of 100 000 may be detrimental because it can give rise to binding in the next injection.

4. If the sample is adsorbed to the surfaces of the flow system it may slowly leach from such surfaces and give rise to a memory effect at the sensor surface.

An insufficient wash can normally be detected during assay development by looking at repeated sample and blank injections. If washing is insufficient, a prolonged buffer wash or an extra buffer injection may solve the problem. In some cases tubing and other parts of the flow system require more efficient wash procedures. A wash with 50% DMSO in water followed by a buffer wash is useful to remove sticky low molecular weight compounds. Sticky proteins may be washed out or deactivated with 0.5% SDS followed by a wash with 10 mM glycine at pH 8.5 to remove residual SDS.

Temperature

Temperature can have a dramatic effect on reaction rates and on interaction affinity. In spite of this, temperature is seldom used as an experimental parameter, and most experiments are performed at 25°C. Even when a thermodynamic study is not the issue, there are many instances where temperature variation is practical. For instance, temperature increase during immobilization can facilitate a more rapid immobilization. A capture experiment may be easier to perform at low temperature because ligand dissociation is often reduced at low temperature. A higher temperature may also increase the dissociation rate and assist regeneration.

Positive and negative controls

By injecting positive and negative controls, the activity of the immobilized ligand and the level of background binding can be followed over time. In screening assays the positive control is normally injected at a concentration high enough to rapidly saturate the ligand during injection.

The negative control is often performed by injecting an irrelevant sample or the assay buffer. The response of the negative control defines a threshold value that must be exceeded for a sample to be scored positive. The activity of the surface is followed by making trend plots of the relative responses of positive and negative controls versus cycle number (Figure 2–11). The surface can be used as long as the remaining surface activity is high enough to allow discrimination between positive and negative samples. In kinetic analysis a more thorough control of ligand activity is performed by injecting the positive control at a concentration that allows inspection not only of binding levels but also of curve shape. In this case an overlay plot of the injections is useful for judging surface activity.

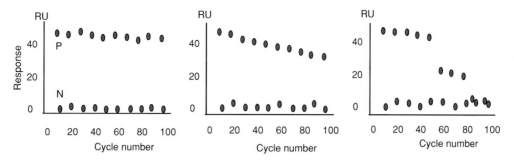

Figure 2–11: Trend plots of positive (P) and negative (N) controls injected at regular intervals during a screening experiment. The left and middle figures are examples of acceptable data where the assay is able to discriminate between positive and negative controls. The right trend plot indicates a jump in assay performance after 50 cycles and total loss of assay performance after 80 cycles.

Startup and maintenance procedures

Without startup and maintenance procedures biosensor system performance will not be optimal. The system should be thoroughly equilibrated with the selected buffer before start. The detector elements should be normalized. Buffer and regeneration solutions should be injected to condition the sensor surface. When the assay is ready the system should be kept under flow conditions. Regular cleaning of the system is necessary to maintain system and assay performance. Read system manuals to find out when and how.

Software wizards

In recent biosensor instrumentation, software wizards support common applications (immobilization, kinetic analysis, and screening). The user generates experimental protocols by interacting with the software. This makes the systems easy to operate in particular as startup and maintenance tools are integrated. In addition, data entered into software wizards are carried over from the system control software to the evaluation software. In this way user input during the data evaluation process can be kept to a minimum. However, the design of an assay, including the choice of sensor surface, immobilization chemistry, reference surface, buffer type, wash solutions, and assay temperature, and data analysis is still the responsibility of the user.

Sample preparation

Sample preparation is seldom viewed as a critical step, but there are a few things to consider.

1.   Protein ligands are normally immobilized at concentrations in the range from 10 nM to 1 mM. If amine coupling is used, the ligand solution can not hold high ano-levels of micro-buffers with reactive amines or additions such as azide. If this is the case and the protein sample already is at a low concentration, a rapid buffer exchange on a small gel filtration device is recommended.

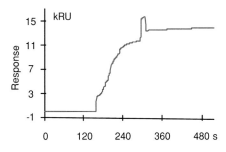

Figure 2–12: Spikes and irregular jumps in the SPR signal are indicative of sample precipitation. The increase in signal level can vary from 10 to thousands of resonance units.

2. If possible dissolve or dilute samples in running buffer to reduce bulk refractive index effects.
3. Proteins frequently stick to surfaces (and not only to sensor surfaces). This may lead to loss of sample, and the assumed concentration may no longer be valid. By adding detergent (for instance 0.05% surfactant P20) to all buffers and samples the effect of this problem can be reduced. For proteins present at very low concentrations (<5 nM), it may be necessary to increase the concentration of detergent or to add a support protein such as bovine serum albumin.

In small-molecule screening the compounds are usually provided as 5 to 10 mM stock solutions in DMSO. When these compounds are stored frozen or at non-ambient temperature ample time should be allowed to equilibrate the samples before assay. The stock solutions should then be mixed prior to dilution. Even when the screening buffer contains 1% to 5% of DMSO some compounds may precipitate when they are diluted. Precipitation may not occur immediately, so visual inspection after dilution can not be taken as a guarantee that the sample will be in solution at the time it is analyzed. This can be several hours or even a day later. Precipitation often shows up as spikes or jumps in a sensorgram (Figure 2–12) and can be confirmed by dynamic light scattering or high-speed centrifugation. Note that precipitation is not always linked to the sample but may be due to undue mixing of running buffer and regeneration solutions. Thus, regeneration with high pH can be problematic when the running buffer contains Zn^{2+} or Mg^{2+} ions.

Serum samples and other complex sample matrixes frequently contain components that bind to the dextran surface and nonspecifically to the immobilized ligand. Adding soluble carboxymethylated dextran to the sample can reduce binding to the dextran matrix. When the immobilized ligand is an antibody, addition of irrelevant antibody from the same species to the sample can be tested, that is, if a mouse antibody is immobilized add mouse immunoglobulins to the sample.

SMALL-MOLECULE SCREENING

The purpose of a screen is to identify compounds that bind to the target of choice with specificity over secondary targets or unrelated proteins. SPR systems where one sample interacts simultaneously with several immobilized targets are

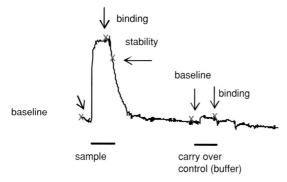

Figure 2–13: Sensorgram and report points from a screening cycle. Sample and buffer are typically injected for 30 to 60 s. The report points indicated by arrows and their associated signal levels are used in data analysis and for quality control.

therefore ideal for screening because selectivity data can be obtained from a single injection of drug candidate.[18] Throughput is also essential as directed libraries or fragments libraries can contain from a few hundred up to several thousand compounds. For this reason parallel injection systems are often used in screening. Biacore A100 is an example of a screening system. With this system four samples can be injected in parallel. Each sample can interact with up to four different ligands and a reference surface at the same time. The system is automated and can handle up to ten microtitre plates in one session.

In screening, compounds are injected at relatively high concentration (10–1000 μM) and typically for 30 to 60 s followed by a short dissociation phase, usually 10 to 60 s. High concentrations are necessary to detect low affinity binders. In a recent Alzheimer's drug candidate investigation it was suggested that binding assays such as SPR are superior to enzymatic assays for the identification of weakly binding fragments.[19]

Sample preparation, DMSO solvent correction, wash procedures, injection of positive controls for monitoring of ligand activity, and injection of negative controls for identification of threshold levels are all important in a small-molecule screen. Regeneration may not be necessary because most compounds can be expected to be nonbinders or of low affinity. Figure 2–13 illustrates a typical injection sequence in screening. The purpose of the buffer injection is to identify and mitigate any remaining memory effects.

A full screen in Biacore A100 generates close to 5000 sensorgrams. Data analysis is highly automated and takes both sensorgram quality (baseline stability, the shape of injection, and dissociation phases) and report points into consideration to provide filtered, solvent-corrected, and molecular weight-adjusted data as starting data for plot functions or for output to external databases.

It is often useful to start with two plots to obtain a first view of the results (Figure 2–14). A plot of the solvent corrected response level from the report point "binding" versus injection number makes it possible to follow trends and to identify compounds with response levels significantly higher than the negative controls. In the next step a plot of molecular weight-adjusted response values for these compounds is used to identify compounds with credible stoichiometry.

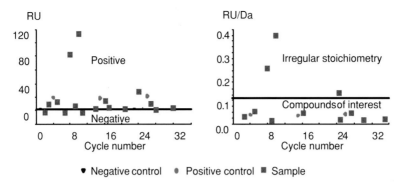

Figure 2–14: Plots of response versus injection cycle (*left*) and molecular weight adjusted response versus injection cycle (*right*).

Compounds with unrealistically high stoichiometry (>3 to 5) are often false positives[20] and may reflect micelle formation or aggregation. In Figure 2–14, right panel three compounds with positive scoring are deselected for this reason.

Selectivity data can be displayed by plotting approved molecular weight adjusted data from main and secondary targets.

PROTEIN–PROTEIN KINETICS

When two proteins A and B react, the rate of complex, AB, formation is given by:

$$\frac{d[AB]}{dt} = k_a \cdot [A][B] - k_d \cdot [AB].$$

For protein–protein interactions, association rate constant, k_a, values are normally in the range from 10^3 to 10^9 M^{-1}s^{-1} and dissociation rate constant, k_d, values are often larger than 10^{-4} s^{-1}. The range of association rate constants that can be determined by SPR sensors is determined by the sensitivity of the system and by the efficiency by which analyte molecules can be transported to the sensor surface. For protein analytes it is theoretically possible to determine association rate constants covering the range 10^3 to 10^8 M^{-1}s^{-1} provided that ligand saturation levels are below 50 RU and that the flow rate is 30 μl/min or higher.

The range of dissociation rate constants that can be determined is easy to estimate. If the reaction is stable a decrease in signal level may be masked by instrument drift. On the other hand, if the complex is unstable it may dissociate during the time it takes to switch from analyte to buffer conditions. Based on typical drift and buffer exchange rates k_d values from 10^{-4} s^{-1} to 0.1 s^{-1} can routinely be determined. In many cases these ranges can be extended five to ten times in each direction.

General principles

To determine the kinetics of the interaction the experiment should be designed so that:

1. Sufficient association and dissociation data including reliable information on the ligand saturation level, R_{max}, can be obtained.
2. The activity of the immobilized ligand over time can be monitored.
3. It is possible to estimate and reduce the influence of systematic errors.

Ligand saturation

The saturation response, R_{max}, indirectly determines the concentration of the immobilized ligand, B. In the kinetic experiment at least one but preferably two analyte concentrations should be high enough so the response approaches R_{max} during injection.

Association data

The shape of the pre-equilibrium phase partly depends on the association rate constant. It also depends on the concentration of analyte, the dissociation rate constant, and the concentration of the immobilized ligand. An appropriate data set should include concentrations that give response curves with clearly different initial binding rates and distinct curvature, and at least a few binding curves should start to approach equilibrium during sample injection.[21] It is often better to keep injection times short and to include higher concentrations than to inject lower concentrations for an extended time. For mechanistic studies[22,23] and for steady-state analysis [24] longer injection times should be considered. A convenient experimental design is to inject samples with increasing concentration in direct sequence.[25] This results in a sensorgram similar to that displayed in Figure 2–4. By using this approach it is possible to obtain kinetic data directly without using regeneration.

Dissociation data

For dissociation rate constants $>10^{-3}$ s^{-1} dissociation time can be short, but for interaction where the dissociation rate constant approaches 10^{-4} s^{-1} the dissociation time must be extended to 15–30 min. It is seldom meaningful to monitor dissociation events for longer than 60 min.

Ligand activity

The long-term activity of the immobilized ligand is followed by replicate injections and by the introduction of a standard sample injected at regular intervals between different analytes/compounds.

Systematic errors and blank injections

There may still be artifacts in the remaining data when the signal from the reference surface has been subtracted. Artifacts can arise from small (0.01°C) temperature changes or from pressure jumps. These artifacts will usually be present

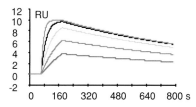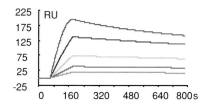

Figure 2–15: Simulation of interaction data on surfaces with low and high levels of immobilized ligand. In both cases analyte concentrations were 1–32 nM. The difference in the shape of the sensorgrams is due to transport limitations on the surface with high ligand density. *See color plates.*

also when buffer is used instead of a protein sample. If they occur systematically subtraction of a blank sensorgram can improve binding data by reducing or even eliminating these disturbances.[11]

Analysis order

Replicate runs should be spaced apart in the concentration series. When memory effects occur it is good to start with the blanks and to use analyte solutions in increasing concentration order. If the system is well behaved concentrations can be used in random order.

Reference surface

In kinetic analysis the most important demand on the reference surface is low degree of analyte binding. If the level of unwanted binding can't be neglected reference subtracted data will not only reflect analyte/ligand binding. This will make kinetic analysis more difficult.

Flow rate

Ideally the sample should reach active and reference surfaces at the same time. This situation is approached in Biacore systems by using flow rates of 30 μl/min or more. A high flow rate is beneficial for two other reasons; it reduces sample dispersion and improves mass transport.

Simulation as a tool for experimental design

Simulation of binding data is useful for understanding how parameters such as saturation levels, analyte concentration, and association and dissociation time influence the kinetics of an interaction (Figure 2–15). Simulations are readily performed with BIAevaluation software.

ACKNOWLEDGMENTS

I would like to thank my colleagues in GE Healthcare working with Biacore products. Their help and support have been invaluable.

REFERENCES

1. R. Gambari, *Am J Pharmacogenomics* **1**, 119 (2001).
2. Z. Wu *et al.*, *AIDS Res Hum Retroviruses* **19**, 201 (2003).
3. M. Alterman *et al.*, *Eur J Pharm Sci* **13**, 203 (2001).
4. R. Karlsson, *Anal Biochem* **221**, 142 (1994).
5. R. Karlsson *et al.*, *Anal Biochem* **278**, 1 (2000).
6. G. A. Baxter *et al.*, *Analyst* **124**, 1315 (1999).
7. S. Geschwindner *et al.*, *J Med Chem* **50**, 5903 (2007).
8. P. O. Markgren *et al.*, *Anal Biochem* **291**, 207 (2001).
9. K. Andersson *et al.*, *Anal Chem* **71**, 2475 (1999).
10. R. Karlsson, R. Stahlberg, *Anal Biochem* **228**, 274 (1995).
11. T. A. Morton, D. G. Myszka, *Methods Enzymol* **295**, 268 (1998).
12. J. Deinum *et al.*, *Anal Biochem* **300**, 152 (2002).
13. Å. Frostell-Karlsson *et al.*, *J Med Chem* **43**, 1986 (2000).
14. K. Andersson *et al.*, *J Mol Recognit* **12**, 310 (1999).
15. H. Nordin *et al.*, *Anal Biochem* **340**, 359 (2005).
16. P. Stenlund *et al.*, *Anal Biochem* **353**, 217 (2006).
17. I. Navratilova *et al.*, *Anal Biochem* **339**, 271 (2005).
18. M. Hämäläinen *et al.*, *J Biomol Screen*, E-pub ahead of print (2008).
19. A. Kuglstatter *et al.*, *Bioorg Med Chem Letters* **18**, 1304 (2008).
20. A. Giannetti *et al.*, *J Med Chem* **51**, 574 (2008).
21. K. Andersson, A. Önell, *J. Mol Recognit* **18**, 307 (2005).
22. C. A. Lipschultz, Y. Li, S. Smith-Gill, *Methods* **20**, 310 (2000).
23. R. Karlsson, A. Falt, *J Immunol Methods* **200**, 121 (1997).
24. M. Abrantes *et al.*, *Anal Chem* **73**, 2828 (2001).
25. R. Karlsson *et al.*, *Anal Biochem* **349**, 136 (2005).

3 Extracting affinity constants from biosensor binding responses

Rebecca L. Rich and David G. Myszka

INTRODUCTION	48
USING SPR TO OBTAIN AFFINITY CONSTANTS	49
DEFINING AFFINITY CONSTANTS	53
METHODS FOR DETERMINING AFFINITY CONSTANTS USING SPR	54
Direct equilibrium analysis	55
Generating equilibrium data	55
Data processing	58
Applying a simple binding model	59
Applying complex binding models	62
Indirect affinity determination	67
SUMMARY	73
WORKED EXAMPLES	73
Example 1: Direct equilibrium analysis of a small-molecule/ protein interaction	73
Example 2: Direct equilibrium analysis of a protein–protein interaction with one binding partner captured from crude sample	75
Example 3: Indirect competition analysis of a small-molecule/protein interaction	79
REFERENCES	82

INTRODUCTION

At the most fundamental level, biological processes are controlled by the interactions of macromolecules. Complex formation between molecules depends on their proximity, conformations, orientations, concentrations, and affinity. Affinity, a measure of the strength of the interaction, is dictated by electrostatic, van der Waals, and hydrophobic forces, as well as hydrogen bonding. The study of affinity constants provides information about specificity, structure/activity relationships, and binding mechanism.

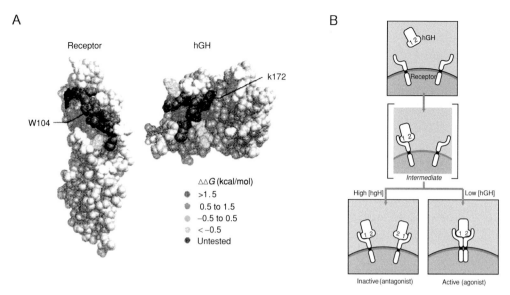

Figure 3–1: Human growth hormone (hGH) binding to its receptor (hGHr). **(A)** Hormone/receptor binding interface with contact residues highlighted. $\Delta\Delta G$ values (determined from affinity measurements) revealed which residues are critical for binding. To align the overlay of the binding faces, two residues that pack together (hGHr W104 and hGH K172) are highlighted. **(B)** Cartoon illustrating the concentration-dependent stoichiometry of the hGH/hGHr interaction. Panels A and B were reproduced from references 3 and 4 with permission from the American Association for the Advancement of Science (copyright 1995 and 1992, respectively). *See color plates.*

Experiments using human growth hormone (hGH) and its receptor elegantly illustrate the range of information that can be obtained from affinity measurements. The high-affinity interaction of growth hormone with its cognate receptor compared to other proteins helped confirm that this receptor was specific for hGH.[1] Examining how the affinity of this interaction was altered by single amino acid substitutions revealed which residues in the hormone/receptor binding interface were critical for complex formation (Figure 3–1A).[2,3] Also for this system, complex affinity profiles demonstrated the hormone binds sequentially to two receptor molecules and the stoichiometry of the receptor/hormone complex is hGH concentration-dependent (Figure 3–1B).[4,5] Together, these studies show how affinity measurements can be employed in progressive stages to characterize a biomolecular interaction – from the initial verification of binding partners to the detailed mapping of the binding interface.

USING SPR TO OBTAIN AFFINITY CONSTANTS

Surface plasmon resonance (SPR) biosensors are now routinely employed to determine affinities in a wide variety of biological systems.[6,7] Although SPR has traditionally been applied to examine antibodies,[8,9] receptors,[10,11] and other proteins,[12,13] as well as peptides[14,15] and oligonucleotides,[16,17] the technology's contribution in lipid and self-assembled monolayer,[18,19] extracellular matrix,[20,21] carbohydrate,[22,23] and small-molecule[24,25] studies has grown steadily over the past few years. In addition to structure/function analysis of these different systems, SPR is increasingly used to optimize one or both binding partners. For example, SPR impacts antibody selection and engineering (e.g., mapping binding epitopes[26,27] and designing higher-potency therapeutic antibodies [28,29]). SPR-based affinity measurements have also streamlined drug discovery efforts by improving target characterization,[30,31] identification and optimization of lead compounds,[32,33] and drug bioavailability studies (both the drug's affinity for serum transport proteins[34,35] and its ability to permeate membrane barriers [18,36]).

The data in Figures 3–2 and 3–3 illustrate how the biosensor can characterize both different biological systems and analytes of different masses. Figure 3–2 depicts the analysis of a panel of purine-derived compounds binding to a macromolecular target, human liver glycogen phosphorylase (HLGP).[37] Based on the crystal structure, these compounds bind at a specific site in each subunit of the homodimeric enzyme (Figure 3–2A and 3–2B). Even though these compounds were small (130–790 Da) and displayed low affinities for HLGP, SPR signals were observable and reproducible (Figure 3–2C) and could be fit to a simple model to obtain the affinities for a range of chemical entities (Figure 3–2D).

Figure 3–3 demonstrates how the biosensor can be used to examine interactions of more complex systems. In this example, the sensor chip surface mimicked the *in vivo* environment of membrane-bound receptors. CXCR4, a chemokine-recognizing G-protein-coupled receptor (GPCR), was captured on an anti-CXCR4 surface and enveloped in a lipid bilayer (Figure 3–3A). The reconstituted receptor bound SDF-1α (a native chemokine ligand) in a concentration-dependent manner (Figure 3–3B) and with an affinity (K_D = 160 nM; Figure 3–3C),[38] similar to that determined for cell-associated receptor.[39]

Compared to more traditional methods to determine affinity constants, the biosensor possesses several technical advantages. For example, the ability to monitor interactions in real time permits detection of affinities ranging from $K_D \sim 1$ pM for tight interactions (e.g., ligand/receptor[40] and antibody/antigen pairs)[8,28] to $K_D \geq 1$ mM for weak interactions (e.g., small molecules binding to HLGP[37] or human serum albumin [HSA]).[34] In contrast, separation methods such as enzyme-linked immunosorbent assay (ELISA), pull-down, and filter-binding assays involve washing steps during which weak-affinity complexes are lost. Although other equilibrium methods such as nuclear magnetic resonance (NMR), analytical ultracentrifugation (AUC), fluorescence, and radio-ligand assays measure the amount of complex formed in the presence of free material and therefore could in principle be used to study transient interactions, these methods have their own drawbacks, including high sample consumption, low throughput, and/or the need to label one or both binding partners. SPR, on the other hand, requires no labeling, consumes little sample, and has comparatively high throughput. Additionally, the biosensor surface can be used to extract a binding partner from crude samples, which dramatically reduces sample preparation time.

People who are not familiar with SPR technology often worry that immobilizing a molecule on the surface affects its activity and therefore will give erroneous

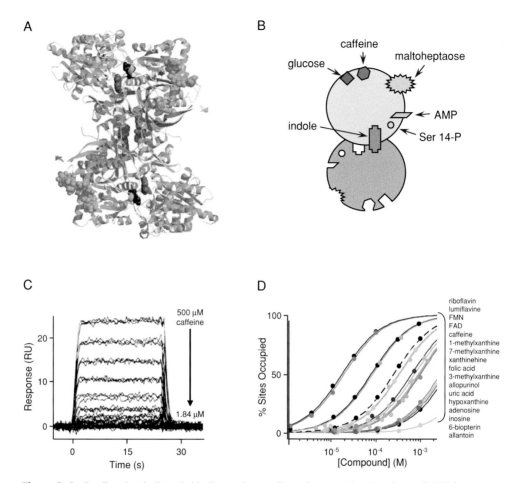

Figure 3–2: Small-molecule ligands binding to human liver glycogen phosphorylase a (HLGPa) homodimer. **(A)** Ribbon diagram of HLGPa showing the binding sites of different ligands. Helices and strands of one subunit are colored light and dark green; the helices and strands of the twofold symmetry related subunit are colored blue and purple. Ligands are shown as space-filling Corey-Pauling-Koltun (CPK) models; maltoheptaose in green, glucose in blue, adenosine 5′-monophosphate (AMP) in turquoise, caffeine in red, phosphorylated serine 14 in yellow and the indole inhibitor in pink at the dimer interface. **(B)** Cartoon showing the interactions targeted in this study. **(C)** Sensorgrams obtained for 0–500 μM caffeine binding to HLGPa. Responses at equilibrium were plotted against the tested caffeine concentrations and fit to a simple binding model to obtain an affinity of 108 ± 10 μM. **(D)** Isotherms generated from a simple model for compounds binding to HLGPa. The isotherm for caffeine is shown as a dashed line. Calculated enzyme affinities of these compounds ranged from 17 to 8000 μM. Panels A and B provided by VL Rath (unpublished results). Panels C and D were adapted from reference 37 with permission from Elsevier Science (copyright 2002). *See color plates.*

binding constants. First, it is important to remember that many biological interactions (e.g., ligands binding to cell membrane proteins or the extracellular matrix) occur at solution/surface interfaces. In the case of these systems, the biosensor is ideal for mimicking the receptors' native environment. Take the membrane receptor example described in Figure 3–3. The receptor is placed in a semirigid environment on chip surface, a membrane is built around the receptor, and analytes in solution are tested for binding.

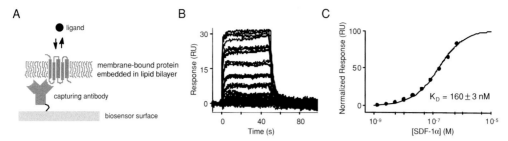

Figure 3–3: Biosensor-based affinity analysis of a membrane-bound receptor/ligand interaction. **(A)** Schematic for testing analyte binding to an antibody-captured/membrane-reconstituted GPCR. **(B)** Sensorgrams depicting 0–640 nM SDF-1α (a native CXCR4 ligand) binding to captured and reconstituted CXCR4. **(C)** Plotting the responses at t~40 s in panel B versus SDF1α concentration and fitting the data to a simple binding isotherm yields and affinity of 160 ± 3 nM. Panels B and C reproduced from reference 38 with permission from Academic Press (copyright 2003).

It is true that immobilization can chemically affect the activity of a molecule. However, out of the almost 10,000 articles describing SPR work published to date, there are only a few examples where direct immobilization affected the molecule's affinity. If a given immobilization method affects the molecule's activity, it is possible to choose alternative immobilization methods. The variety of available immobilization methods (via either covalent attachment or indirect capture) permits a range of molecules, from peptides and oligonucleotides to antibodies, receptors, and multi-protein complexes to be tethered to the sensor surface through different means.

The most common concern associated with the surface is that immobilization of the molecule will change its entropic properties, which would adversely affect affinity. It is important to remember that the vast majority of biosensor experiments are actually performed using a sensor chip coated with a dextran matrix. A molecule tethered to the non–cross-linked dextran layer (rather than a solid surface) retains much of its rotational entropic properties, as well as a degree of diffusional freedom.[41,42] In fact, affinities determined from carefully performed biosensor experiments of various biological systems (e.g., protein–protein,[43–45] antibody–antigen[8,46] protein–DNA,[47] and protein–small molecule)[41,48] match the values obtained using isothermal calorimetry, analytical ultracentrifugation, electromobility shift assay, KinExa (Sapidyne Instruments; Boise, Idaho, USA), and stopped-flow fluorescence. Also, using two protein–small-molecule model systems, we demonstrated the biosensor analysis is robust: Similar affinities were obtained by investigators having varying levels of expertise and using a variety of Biacore instrument platforms.[48,49] These comparative studies demonstrate the biosensor can yield reliable binding constants. Finally, when there is a credible concern that the surface may affect the affinity of the target molecule, the biosensor can be run in a solution competition model (described later in this chapter under the section heading *Indirect competition analysis*).

In this chapter we focus on using the biosensor to determine affinities for systems at equilibrium, which exists when equal numbers of complexes are forming and dissociating so the net change in response is zero (i.e., the response is flat).

We mathematically define the equilibrium dissociation constant (K_D), describe two complementary biosensor-based methods to measure K_D values, and provide worked examples of both methods.

DEFINING AFFINITY CONSTANTS

Throughout the scientific literature we find binding affinities reported using a variety of descriptors, including EC_{50}, IC_{50}, K_A, and K_D. EC_{50} (concentration at 50% efficacy) and IC_{50} (concentration at 50% inhibition) are parameters used most often in a broad physiological context and simply describe the molar concentration of a sample required to induce 50% of a maximal effect in a competition analysis. In contrast, K_D and K_A (the equilibrium association constant), which are related by the relationship $K_A = 1/K_D$, describe an interaction at the molecular level and are mathematically derived from first principles. We encourage investigators to report affinities in terms of K_D (rather than K_A) because the equilibrium dissociation constant has units of molarity and therefore correlates directly with the sample concentration range required to detect binding. Throughout this chapter, affinities are always expressed in terms of K_D.

For a simple 1:1 interaction, two binding partners, A and B, can form a complex ($A + B = AB$). Under steady-state conditions, the equilibrium dissociation constant, K_D, is defined as:

$$K_D = \frac{[A][B]}{[AB]}. \tag{3-1}$$

Information about the affinity can be obtained from two types of equilibrium-based biosensor experiments: direct equilibrium or competition (or inhibition) analyses. A direct equilibrium analysis examines the binding of two partners to yield an interaction affinity. The SPR biosensor injects A (analyte in solution) over a surface containing B (ligand immobilized on the sensor chip surface) and monitors AB complex formation in real time to produce a binding response, R. The fraction of ligand binding sites occupied corresponds to R/R_{max}, which is described in this single-site model as the average number of A bound [AB] per total number of B($[B] + [AB]$):

$$\frac{R}{R_{max}} = \frac{[AB]}{[B] + [AB]}, \tag{3-2}$$

where R_{max} is the response observed when all B binding sites are occupied. Substituting $[AB] = [A][B]/K_D$, Equation 3–2 can be rewritten as

$$R = \frac{R_{max}[A]}{K_D + [A]}. \tag{3-3}$$

(The explicit derivation of Equation 3–3 is found in reference 50). A plot of binding responses (R) versus injected analyte concentrations [A] is fit to an isotherm described by Equation 3–3 to yield R_{max} and K_D parameters. Example isotherms for an equilibrium analysis shown in Figure 3–4, panels A and B,

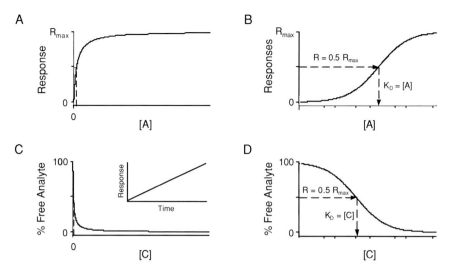

Figure 3–4: Typical binding isotherms obtained from affinity measurements of a simple single-site interaction. Binding isotherm generated from Equation 3–3 plotted on linear **(A)** and log **(B)** scales. Isotherm generated from Equation 3–6 plotted on linear **(C)** and log **(D)** scales, with a corresponding standard curve shown in the inset of panel C. Dotted lines indicate K_D's correspond to the concentration of analyte (A) or competitor (C) that produce half-saturation responses.

illustrate that as [A] increases R approaches R_{max} and the concentration of A that generates a half-saturation response (0.5 R_{max}) corresponds to the K_D.

A competition analysis, on the other hand, involves monitoring how two molecules ($B_{solution}$, the immobilized ligand, and $C_{solution}$, the soluble competitor) bind to a shared third partner (A): $A + B_{surface} = AB_{surface}$ and $A + C_{solution} = AC_{solution}$. In the most common cases, the competitor in solution is the same ligand immobilized on the surface. Determining an affinity constant from a competition analysis is more complex, for it involves first establishing a calibration curve for free analyte (A) available for binding to the ligand surface. The explicit derivation of K_D from a competition analysis is given in reference 22 and the experimental details and data fitting are described later in this chapter. Example isotherms obtained for competition analyses are shown in Figures 3–4C and 3–4D. Figure 3–4 also illustrates that the correlation between the half-saturation response and K_D is more apparent when concentration data are plotted on a log scale, as in panels B and D, rather than on a linear concentration scale shown in panels A and C.

METHODS FOR DETERMINING AFFINITY CONSTANTS USING SPR

Because direct equilibrium analysis is the most commonly used SPR-based method to measure K_D values,[7] we emphasize below the precautions necessary to obtain reliable equilibrium binding responses and interpret them correctly. We also highlight pertinent examples from the recent biosensor literature and provide two worked examples that demonstrate the application of this method. Competition analysis, although more laborious and therefore performed less

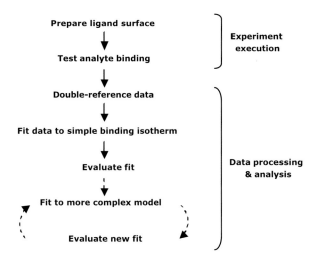

Figure 3–5: Flowchart of the steps involved in a direct equilibrium analysis. The dotted lines indicate steps used only when the data are not well described by a simple binding model.

frequently, provides a valuable alternative that can be employed when a direct equilibrium analysis might be unsuitable. We discuss when to apply a competition analysis, outline the assay requirements, and also provide a worked example of this method.

Direct equilibrium analysis

Using the biosensor, a direct equilibrium analysis is fairly straightforward to perform. The most important thing to remember is that this is an analysis of an interaction *at equilibrium* so you need to monitor the association phase until equilibrium is achieved. For weak interactions, the assay can often be completed rapidly, with data collection for an entire experiment requiring only a few hours of instrument time. This is because many of the systems studied by this method exhibit k_d's $> 10^{-2}s^{-1}$, which means the complex dissociates completely within a few seconds to minutes after the analyte injection ends, and the response returns to baseline so no regeneration step is required. In cases like these, each binding cycle is usually only a few minutes long.

Generating equilibrium data

Figure 3–5 outlines the steps required in an equilibrium analysis. The first three steps produce the binding responses that are then fit to a model that describes the interaction. Obtaining high-quality equilibrium data is critical for determining accurate affinity constants and requires careful experimental design and execution as well as appropriate data processing.

Ligand surface preparation. Figure 3–6A depicts the setup of an equilibrium analysis: analyte in solution interacts with immobilized ligand, which is either directly attached (left panel) or indirectly captured (right panel) to the chip surface. Direct immobilization utilizes amine, thiol, aldehyde, or carbohydrate

A

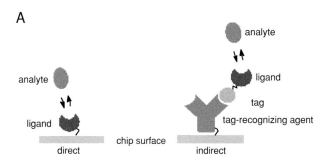

B

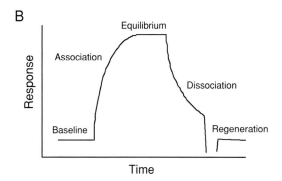

C

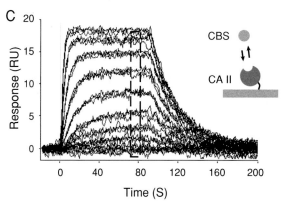

D

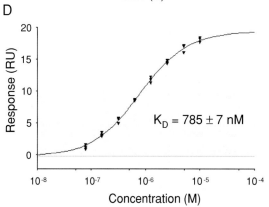

chemistry to form covalent bonds between the ligand and the dextran-coated sensor chip surface.[51] These chemistries produce stable surfaces because the ligand is permanently attached to the sensor chip and can yield the high-density ligand surfaces critical for examining small-molecule interactions. One significant drawback to this method, however, is that the ligand may be heterogeneously linked to the surface. Because the covalent bonds are random, reactive groups in or near the ligand's binding site might become tethered to the surface and thereby hinder analyte binding. In contrast, indirect immobilization produces an oriented ligand surface. Engineering the capturing tag distant from the binding site ensures the binding site is accessible to the analyte. Also, the opportunity to capture a ligand (via tags such as biotin, His_x, and glutathione-S-transferase, or by ligand-specific antibodies) means that ligand preparations need not be pure. Instead, the ligand can be selectively extracted from crude samples such as cell lysates or hybridoma supernatants. Because capturing involves noncovalent interactions, however, the ligand may dissociate from the surface over time (although this decay can usually be accounted for with careful data processing). Deciding which tethering methods to use depends on the interaction to be studied and most often must be determined empirically.

Analyte binding. Figure 3–6B depicts a typical sensorgram profile obtained from an equilibrium analysis. In this example, the SPR signal increases as the analyte begins to bind to the ligand, plateaus as equilibrium between complex formation and disassembly is achieved, and decreases as the analyte is washed from the surface. To obtain the K_D, a concentration series of the analyte must be tested. References 52 and 53 describe in detail how to prepare analyte samples and choose analysis conditions. Six experimental parameters that should be included to improve the reliability of a data set are emphasized below (and are also illustrated in Worked Example 1):

- *Collect equilibrium data.* The association phase in each binding cycle should be monitored long enough to observe a plateau in the binding response for each concentration of analyte. Achieving this plateau in the signal is critical because the responses at equilibrium are used to calculate interaction affinity.

←

Figure 3–6 *(facing page).* Equilibrium analysis. **(A)** Assay design. The ligand is tethered to the surface, either by direct immobilization or indirect capture by a high-affinity binding partner. **(B)** Sensorgram of a complete binding cycle. Binding responses are recorded as analyte is flowed across the ligand. Reaction equilibrium is demonstrated by the plateau in response at the end of the association phase. During the buffer wash of the dissociation phase, the signal decays as the ligand/analyte complex falls apart. A short regeneration pulse (usually of weak acid or base) strips residual analyte from the ligand surface and returns the signal to baseline. **(C)** Typical data set. In this example, a concentration series of CBS (0, 0.0781, 0.156, 0.313, 0.625, 1.25, 2.50, 5.00, and 10.0 μM) was injected in triplicate across immobilized CA II. A cartoon of this small-molecule inhibitor/enzyme interaction is shown in the inset. **(D)** The binding responses at equilibrium (highlighted by the dashed box in panel C) were plotted versus analyte concentration and fit to the simple binding isotherm described by Equation 3–3 to yield the affinity of the CBS/CA II interaction.

- *Test a wide range of analyte concentrations.* Because an equilibrium analysis uses only the plateau portion of the sensorgram, little information is extracted from individual sensorgrams. To obtain enough binding data to fit a particular model with confidence, a large number of analyte samples should be tested and they should span a wide concentration range (we recommend a >100-fold span in analyte concentration whenever possible). Also, if an estimate of the interaction affinity is known, the analyte concentrations should bracket this estimate by at least tenfold on each side. For example, if the estimated affinity is $1\mu M$, then the analyte would be tested from $<0.1\mu M$ to $>10\mu M$.
- *Test a fine dilution series of analyte.* Testing a wide range of analyte concentrations prepared in a fine dilution series (ideally, two- or threefold dilutions; Worked Example 1 describes an analyte tested in a twofold dilution series over a >100-fold concentration range) ensures that enough data will be collected to produce an accurate, precise K_D.
- *Test each analyte concentration in replicate.* Replicates reveal the reproducibility of the binding responses. Overlaying the data collected from replicate injections of the same analyte concentration demonstrates ligand and analyte stability over the time course of the analysis. And rather than testing the replicates sequentially, they should be randomly scattered throughout the analysis sequence to confirm the reagents' stability over the course of the entire experiment.
- *Include buffer blanks.* Samples of aliquotted running buffer should also be analyzed periodically throughout the experiment. These blanks reveal how instrument drift contributes to the apparent binding response and also confirm if an immobilized ligand is stable over time.
- *Collect data from a reference surface.* Each analyte and buffer sample should be flowed across both the immobilized ligand and a reference surface. Signals collected from the reference surface reveal how nonspecific binding, injection noise, and refractive index differences between the sample and running buffer contribute to the apparent binding response.

Data processing

Unlike binding profiles that display curvature and can therefore be fit to yield kinetic rates and affinity parameters, the sensorgrams obtained from an equilibrium analysis are often square-shaped (particularly those for weak-affinity interactions). But signal changes due to refractive index differences between the sample and running buffer are also square-shaped and can mask the actual binding events. Careful data processing that includes "double-referencing"[52–54] corrects for these bulk refractive index changes as well as for instrument drift and injection noise. An additional correction step should be included when using highly refractive buffer components (e.g., dimethylsulfoxide [DMSO] as a cosolvent).[34] Applying these data processing steps ensures that only real binding responses are used to determine the interaction affinity.

Figure 3–6C depicts the double-referenced data set obtained for a concentration series of a soluble small-molecule inhibitor (carboxybenzenesulfonamide, CBS; 201 Da) binding to an immobilized enzyme (carbonic anhydrase II, CA II; 30 kDa). In this example, each CBS concentration was tested in triplicate over both the CA II surface and an unmodified reference surface. The triplicates, interspersed with buffer blanks, were analyzed in random order. Association was monitored for 90 s, which was long enough to observe the equilibrium plateau for each analyte concentration. Dissociation was monitored for 120 s, during which the responses returned to baseline. Twofold dilutions of CBS spanned a ~125-fold concentration range, which proved sufficient to describe the interaction because the responses varied from almost no binding at 78.1 nM to approaching surface saturation at 5.00 and 10.0 μM (as evidenced by the responses for these two highest concentrations being spaced closely together).

The complete data set is information rich. For example, even without applying a binding isotherm we can evaluate the quality of the experiment, determine if the interaction is stoichiometric, and estimate its affinity. The overlay of binding responses for each CBS concentration demonstrates the analysis is reproducible. Because the responses for the highest CBS concentrations approach saturation, the CBS/CA II interaction displays a defined stoichiometry. In addition, we can visually estimate the affinity of this interaction because the analyte concentration that produces a response equal to half the saturation response corresponds to the K_D.

Applying a simple binding model

The remaining steps of the flowchart shown in Figure 3–5 involve fitting the response data with interaction models to obtain the affinity constant(s). The data set is first fit to the simple 1:1 model described by Equation 3–3 (unless corroborating evidence of complexity from complementary biophysical analyses exists). Because equilibrium was achieved at $t = 80$ s for each sensorgram in Figure 3–6C, the averaged responses highlighted by the dashed box can be plotted versus the injected CBS concentration and fit to a binding model to determine the K_D. Figure 3–6D shows the response versus concentration data points fit by the simple binding isotherm generated from Equation 3–3 to yield an affinity of 785 ± 7 nM. Because the isotherm intersects each set of data points (and the reported error is <1%), the CBS/CA II interaction is well described by this simple model. To illustrate in detail the experimental and data processing/analysis steps required to determine the affinity of the CBS/CAII interaction, we include it as Worked Example 1 at the end of this chapter.

The four examples shown in Figure 3–7 illustrate how the biosensor can be used to examine a range of interactions described by the simple 1:1 model.[55–58] These examples employ various ligand immobilization approaches and data fitting/presentation methods and also demonstrate the biosensor's ability to monitor multiple interactions at one time. For example, the data shown in Figures 3–7A and 3–7B were obtained from ligand (protein [7A] or peptide [7B]) directly immobilized onto the sensor chip surface using amine-coupling chemistry. In contrast,

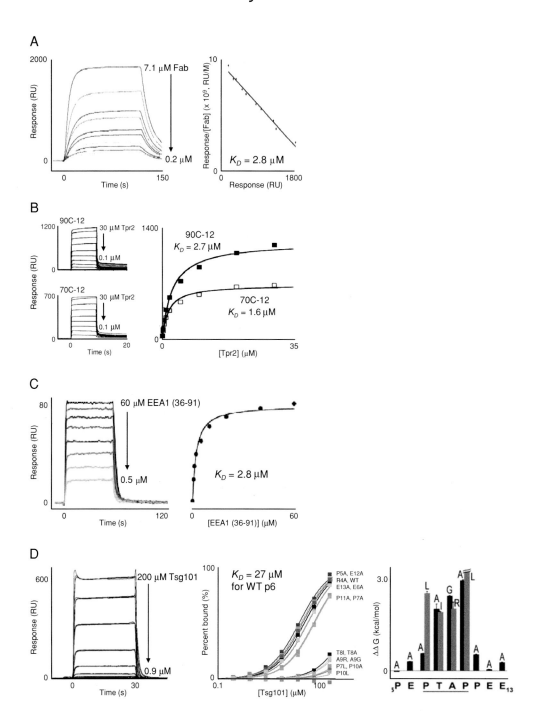

the data shown in Figures 3–7C and 3–7D were obtained from GST-tagged proteins captured on anti-GST antibody surfaces.

Figure 3–7A employs a Scatchard analysis to determine the equilibrium dissociation constant. Though this linear transformation makes it easy to tell if the data are well fit by a simple model, this analysis requires data manipulation. We prefer instead to fit the primary data directly to a binding isotherm. The equilibrium data in panels 7B and 7C are plotted on a linear scale and fit to binding isotherms, while the data shown in the center panel of Figure 3–7D are plotted on a log scale and fit to binding isotherms. The advantage of presenting data on a log scale as shown in Figure 3–7D (instead of a linear scale in which the data from low concentrations are bunched together) is that the data points are evenly distributed and it is easy to evaluate the model fit at all analyte concentrations.

The examples in Figures 3–7B and 3–7D illustrate another powerful feature of using many of the commercially available SPR biosensor platforms: the ability to analyze in parallel multiple ligands immobilized on different flow cell surfaces. For example, the two Hsp peptides examined in Figure 3–7B can be immobilized in separate flow cells and tested simultaneously for cochaperone binding. Taking full advantage of the biosensor's capabilities, the analysis shown in Figure 3–7D involved Tsg101 protein in solution binding to three different p6 protein surfaces at one time.

Figure 3–7D highlights SPR's utility in epitope mapping and thermodynamic studies. In this example, a protein, human tumor susceptibility gene protein 101 (Tsg101), was flowed over its binding partner, HIV-1 p6 (both wild-type p6 and a panel of site-specific mutants) to identify which residues in p6 were critical for Tsg101 recognition. $\Delta\Delta G$ values were determined from the calculated affinities and were plotted as shown in the right panel of Figure 3–7D. This analysis established that a four-amino acid motif, PTAP, in p6 PTAP motif is the energetically dominant binding epitope. We have included the steps used to obtain the data shown in Figure 3–7D as Worked Example 2 to illustrate in detail how this type of analysis is performed.

The data shown in Figure 3–7D also illustrate how the biosensor can provide affinity information about extremely weak interactions ($K_D > 100\mu M$).

Figure 3–7 *(facing page)*. Examples of direct equilibrium analyses. **(A)** *Left panel*: concentration series of soluble Fab HyHEL-63 V_LN32A injected across immobilized hen egg white lysozyme. *Right panel*: Scatchard analysis of the equilibrium binding responses. Adapted from reference 55 with permission from the American Chemical Society (copyright 2003). **(B)** *Left panel*: concentration series of soluble cochaperone Tpr2 injected across 12-mer C-terminal Hsp70 and Hsp90 peptides immobilized on the surfaces of two flow cells. *Right panel*: binding responses plotted on a linear scale versus injected Tpr2 concentration and fit to simple isotherms. Adapted from reference 56 with permission from the European Molecular Biology Organization (copyright 2003). **(C)** *Left panel*: concentration series of soluble early endosome antigen 1 (EEA1) injected across GST-Rab5c captured on an anti-GST antibody surface. *Right panel*: binding responses plotted on a linear scale versus injected EEA1 concentration and fit to a simple isotherm. Adapted from reference 57 with permission from the American Society for Biochemistry and Molecular Biology (copyright 2003). **(D)** *Left panel*: concentration series of soluble Tsg101 injected across GST-p6 captured on an anti-GST antibody surface. *Middle panel*: Epitope mapping of the Tsg101 binding site on p6. For a panel of p6 proteins, binding responses were plotted on a log scale versus injected Tsg101 concentration and fit to simple isotherms. *Right panel*: Changes in free energy of Tsg101 binding caused by mutations in p6. Dark bars represent the changes due to single alanine substitutions in residues 5–13 and light bars represent alternative substitutions. Left and right panels adapted from reference 58 with permission from Cell Press (copyright 2001). *See color plates.*

Although this interaction was limited by the availability of soluble Tsg101 at high concentrations, the analysis yielded reproducible responses (even for the lowest-affinity P_7–P_{10} mutants) that could be fit to obtain K_D values. Though we recommend testing analyte concentrations that approach ligand saturation, this may not always be feasible. Particularly for weak interactions, the analysis may be limited by analyte availability or solubility at concentrations much above the K_D.

Figure 3–8 further supports using SPR to examine weak-affinity interactions. If we obtain high-quality, reproducible responses that have little scatter and assume a simple binding model, we can determine interaction affinities using analyte concentrations that do not even begin to saturate all the ligand binding sites. An analyte concentration series that spans <0.01–$100 \times K_D$ and approaches ligand saturation produced the plot shown in Figure 3–8A. Fitting these responses to a simple binding isotherm yielded an affinity of $10.00 \pm 0.01\,\mu M$ and a maximum binding response (R_{max}) of 100.02 ± 0.03 RU (upper left panel in Figure 3–8B). Sequentially omitting data points for the highest analyte concentrations (as shown in the other panels of Figure 3–8B; omitted data points are indicated by "+") demonstrates how much information is required to obtain a reliable affinity constant. For example, when the highest included analyte concentration approximates the K_D (as shown in panels for 44% and 60% saturated), the reported affinity and maximum response differ by less than 2% from that obtained from the analysis of the full data set. In fact, in this analysis approximate K_D and R_{max} values could be determined with confidence using only analyte concentrations that produced responses that were $<20\%$ of the ligand saturation level.

Applying complex binding models

While the K_D determinations for the examples in Figure 3–7 all assume a 1:1 interaction, affinities can also be determined for more complex interactions. We recommend always fitting SPR data initially to the simplest relevant biological model, which most often is a simple 1:1 interaction. In the event this model insufficiently describes the data *and* complementary biophysical analyses (e.g., analytical ultracentrifugation, NMR, or crystallography) suggest the interaction is complex, a more elaborate model may be required. When determining the appropriate model to fit a data set, additional reaction steps or complexity should be added in a stepwise manner to keep the model as simple as possible. The following examples illustrate instances in which complex models can be appropriately applied to interpret SPR responses.

Independent, equivalent interactions. Equation 3–3 can also be applied to determine the affinity of *n:n* interactions if the binding events are both equivalent and independent (such as the 2:2 stoichiometry of the human liver glycogen phosphorylase (HGLP)/purine compound interactions).[37] Of course, the stoichiometry *n* of the interaction must first be determined by another biophysical method such as X-ray crystallography or analytical ultracentrifugation. For interactions in which an analyte binds at multiple equivalent and independent sites in the ligand (e.g., antitumor agents binding with $>1:1$ stoichiometry to DNA),[59]

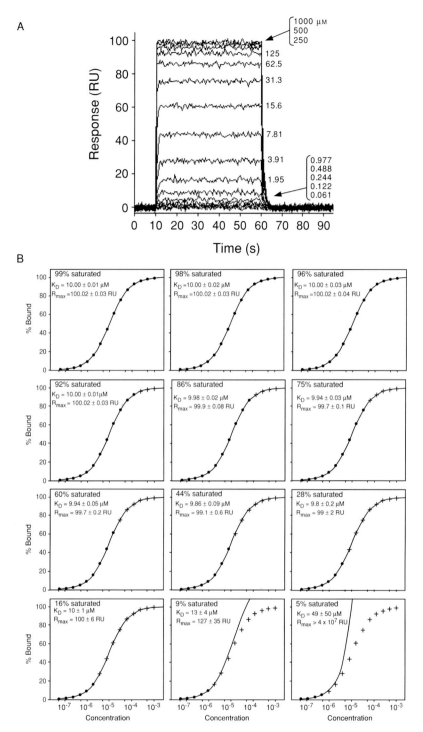

Figure 3–8: Determining interaction affinity from analyte concentrations that do not saturate binding sites on the ligand surface. **(A)** Responses for a rapidly equilibrating system tested at analyte concentrations of 0.061–1000 µM. **(B)** The upper left panel depicts the fit of the entire data set to a simple binding isotherm to yield $K_D = 10.00 \pm 0.01$ µM and $R_{max} = 100.02 \pm 0.03$ RU. The remaining panels illustrate the model fit, % saturation and reported parameters obtained as the responses for the highest analyte concentrations are omitted stepwise from the analysis. Circles represent data points included in each fit, and crosses indicate omitted data points.

the K_D value can be determined from Equation 3–4:

$$R = \frac{nR_{\max}[A]}{(K_D + [A])},$$ (3–4)

where n is the analyte per ligand stoichiometry.

Independent, nonequivalent interactions. For biological systems having independent binding sites of differing affinities, Equation 3–3 can be expanded to yield individual K_D and $R_{\max}$ values for each class of binding sites:

$$R = \sum_i \frac{R_{\max_i}[A]}{(K_{D_i} + [A])},$$ (3–5)

in which i is the number of binding classes. (This equation assumes only one site of each affinity class exists. Combining Equations 3–4 and 3–5 would produce a model appropriate for a multisite interaction in which several sites are of equivalent affinity and other sites are not.) Figure 3–9 illustrates how Equation 3–5 was applied to determine K_D's for drugs binding to HSA. HSA has four predominant binding sites (Figure 3–9A), which are targeted by distinctly different structural classes of drugs.[34]

(In fact, the number and independence of the drug binding sites in HSA has been well established by solution-based methods such as equilibrium dialysis and steady-state fluorescence.)[60,61] At low concentrations, a drug occupies its primary binding site in HSA. At higher concentrations, however, the drug also occupies additional, lower-affinity sites (Figure 3–9B).

Sensorgrams obtained for warfarin in solution binding to immobilized HSA are shown in Figure 3–9C. For each injected warfarin concentration, the equilibrium plateau is achieved within the first few seconds of the association phase ($t = 0$–30 s) and the complex readily dissociates in the washout phase ($t > 30$ s). From inspection of this data it is apparent, however, that this interaction does not obey a simple model: the responses (even for the highest warfarin concentrations) do not approach saturation, which indicates the warfarin/HSA interaction is complex. The plots in Figure 3–9D confirm this complex binding behavior. The inset in Figure 3–9D depicts the fit of a simple 1:1 model (from Equation 3–3) to the data points generated by plotting the averaged binding responses at 15–25 s versus injected warfarin concentration. The simple model poorly describes the data, particularly the shoulder observed at ~2–10 μM. Applying the isotherm generated by Equation 3–5 (with $i = 2$) yields the fit shown in the main panel of Figure 3–9D. Although assuming $i = 2$ is a simplification of the actual binding events, this approximation proved sufficient to characterize the highest-affinity, primary interaction.[34] Figure 3–9E illustrates that this equilibrium analysis can be used to characterize the high-affinity interactions of drugs that target different binding sites in HSA. For example, phenylbutazone and dicumarol bind at the warfarin site, while ibuprofen and naproxen bind at another site in HSA.[35] This assay is now routinely used to characterize compounds whose HSA binding properties are unknown.

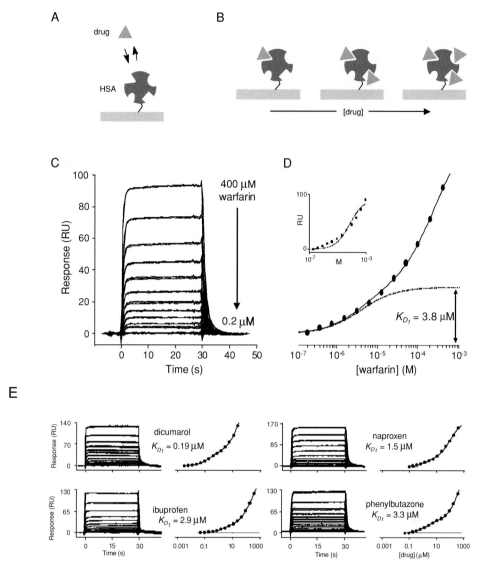

Figure 3−9: Affinity determination for drugs binding to HSA. **(A)** Assay design: a drug in solution is flowed across immobilized HSA. **(B)** At low concentrations, the drug binds at a single, high-affinity site in HSA, but as its concentration increases, the drug binds at other, lower-affinity sites. **(C)** Sensorgrams obtained from quadruplicate injections of warfarin (0.2−400 μM) over a surface of 11.0 kRU immobilized HSA. **(D)** Response data at equilibrium (t = 10−25 s) plotted versus warfarin concentration. Standard errors for the data points are less than the size of the symbols. The inset depicts the fit of the data to the simple model described by Equation 3−3. In the main panel, the solid line depicts the fit of the data to the model described by Equation 3−5 (i = 2) while the dashed line represents the saturation of the high-affinity site. This fit yields equilibrium dissociation constants of 3.8 μM and 273 μM for the high- (K_{D1}) and low-affinity (K_{D2}) sites, respectively. **(E)** Determining four drugs' affinity for their primary binding sites in HSA. Each was injected in triplicate in a twofold dilution series that spanned 0.1−500 μM (except for dicumarol, which was tested at 0.003−25 μM). Response data at equilibrium were plotted against drug concentration and fit to Equation 3−5 with i = 2 to obtain the affinity (K_{D1}) for the primary binding site. Panels C and D (main panel) adapted from reference 34 with permission from Elsevier Science; panel E adapted from reference 35 with permission from the American Pharmaceutical Association.

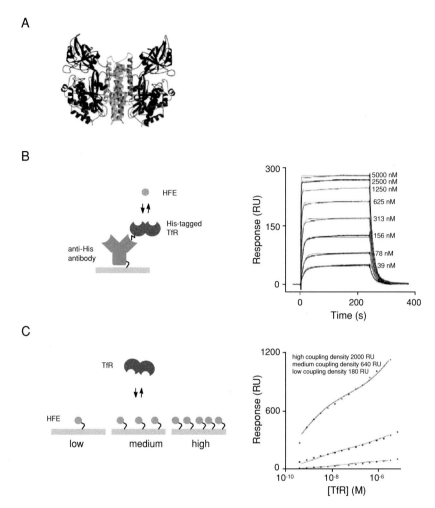

Figure 3–10: (A) Ribbon diagram of transferrin receptor. The HFE binding site on the helical domain closest to the viewer is highlighted in green. The HFE binding site on the other helical domain is omitted for clarity. **(B)** Concentration series of soluble HFE injected across His-tagged transferrin receptor captured on an anti-His antibody surface. Affinities determined from this orientation were $K_{D1} = 98$ nM and $K_{D2} = 520$ nM (gray lines represent the best fit to a bivalent ligand kinetic model). **(C)** Equilibrium binding responses plotted on a log scale versus TfR concentrations injected across HFE immobilized at three densities. HFE was immobilized at low, medium, and high densities to vary the avidity effect on apparent TfR binding. All three binding curves were globally fit to a bivalent analyte model, yielding affinities of $K_{D1} = 46$ nM and $K_{D2} = 210$ nM. Adapted from reference 62 with permission from Academic Press (copyright 2001). *See color plates.*

Interdependent interactions. Figure 3–10 highlights a particularly well-performed SPR study that yielded the affinities for a more complex interaction: a class I major histocompatibility complex (MHC)-related protein called HFE binding to two nonequivalent, interdependent sites in the homodimeric transferrin receptor (TfR).[62] Crystallographic analysis demonstrated the receptor possesses two potential HFE binding sites (Figure 3–10A). Equilibrium gel filtration and analytical ultracentrifugation measurements established HFE bound at both sites in the receptor with submicromolar affinities. Initial SPR studies, however,

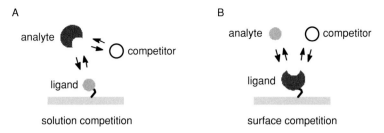

Figure 3–11: Formats of competition analyses. **(A)** In the solution-based analysis, the analyte can bind to either soluble competitor or immobilized ligand. **(B)** In the surface-based analysis, the immobilized ligand binds to either analyte or competitor, which are both in solution.

showed the interaction was poorly described by a simple 1:1 (or 2:2) interaction model. Panels B and C depict complementary SPR experiments that confirmed HFE binds at two sites in the dimeric receptor with affinities that differ by five-fold. In addition, the experiment performed using the orientation in panel C illustrates how avidity affected the apparent interaction affinity.

This example demonstrates the care required when fitting SPR responses to complex models. The binding model applied to the HFE/TfR biosensor data was supported by both crystallographic evidence and solution-based studies. In addition, the SPR studies involved (1) testing the interaction in both orientations, (2) measuring the responses obtained from multiple surface densities to evaluate the avidity effect, (3) demonstrating first the responses were poorly described by a simple interaction model, and (4) when possible, fitting the responses to both kinetic (Figure 3–10B) and equilibrium (Figure 3–10C) models of sequential binding.

More complex interactions. SPR-based affinity analyses can even be used to interpret complex interactions. Using equilibrium analysis to characterize a small-molecule/DNA interaction, Wilson and coworkers employed a cooperative binding model supported by NMR and DNase I footprinting studies.[63,64] Additional, increasingly elaborate models that can be applied when there is supporting biological evidence that the interaction is complex are found in reference 50.

Indirect affinity determination

As illustrated in Figure 3–11, an indirect affinity analysis can be measured in two formats: solution competition (the analyte binds to either an immobilized ligand or a soluble competitor) or surface competition (the immobilized ligand binds to either analyte or soluble competitor). In a competition analysis, the affinity of the ligand/analyte interaction is not determined. Instead, analyte binding to the ligand surface is used as an indicator to determine the affinity of competitor binding. Unfortunately, few examples of well-executed surface competition analysis are found in the biosensor literature.[65–67] Biosensor users' success with solution-based competition is more apparent: recent noteworthy examples are described in references 22 and 68 through 73.

In the early and mid-1990s, shortly after biosensors became commercially available, solution competition was most often used to quantitate small-molecule binding to macromolecular targets. Recent advancements in biosensor technology and methodology now permit small-molecule/target interactions to be monitored directly.[6,7] Today, solution competition analyses are primarily employed when immobilizing the ligand may affect its activity, when a capturing tag cannot be incorporated, or when an analyte may bind to more than one site on the ligand surface to produce an avidity-enhanced apparent affinity. Disadvantages of a solution competition analysis include: (1) the assay is more complicated than a direct equilibrium analysis, both in experimental design and data analysis, (2) the analyte/competitor interaction must be preincubated until equilibrium is achieved (which may require several hours (or even days for very for high-affinity interactions), (3) there is significant consumption of both analyte and competitor, and (4) a simple interaction model must be assumed.

An important aspect of a solution competition assay is that the concentration of analyte in solution available for ligand binding (free analyte, A_{free}) must be determined. This step is critical to obtain an accurate affinity of the analyte/competitor interaction, yet it is omitted from most biosensor-based solution competition analyses reported in the literature. To demonstrate how calculating A_{free} (rather than assuming A_{free} is equivalent to the total concentration of analyte, A_{total}) influences reported affinity, Karlsson showed K_D calculated for a peptide/antibody interaction increased almost fourfold when the A_{free} parameter was correctly applied. He also showed that "corrected" affinity matched that determined from the ratio of association and dissociation rate constants.[66]

Figure 3–12 outlines the steps required in a solution competition analysis. This analysis is performed in two stages: (1) constructing a standard curve of analyte binding to immobilized ligand in the absence of soluble competitor, and (2) examining analyte binding to soluble competitor in the presence of immobilized ligand to determine the K_D value of the analyte/competitor interaction. Both stages require careful experimental execution and thorough data processing and analysis.

Ligand surface preparation. The first step of a solution competition analysis involves immobilizing the ligand on the sensor chip surface at a high density, if possible. At high ligand densities, analyte binding is dominated by transport to the surface, not the inherent interaction kinetics. Under these conditions the binding response is linear, which makes the rate of binding constant over time and means the affinity can be determined with high precision. (This is unlike an equilibrium analysis, in which the ligand is immobilized at low density so that mass transport effects are minimized and the interaction kinetics predominate.)

Analyte binding (for standard curve). This step is similar to testing analyte binding in an equilibrium analysis: buffer samples and an analyte concentration series are injected over the ligand and reference surfaces. As described previously,

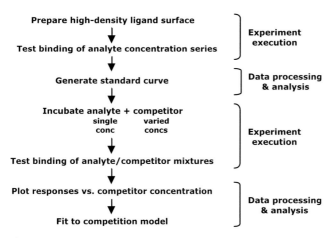

Figure 3–12: Flowchart of the steps involved in a solution competition analysis.

the samples are prepared in a fine dilution series over a wide concentration range and tested in replicate. In this analysis, it is essential to collect sufficient association-phase data to measure the slope of the response accurately and precisely. Oftentimes this may require monitoring analyte binding for several minutes.

Data fitting (for standard curve). Analyte binding responses are double-referenced and the binding rates, which correspond to the slopes ($\Delta RU/\Delta t$) of the linear association-phase data, are calculated. Typically, the slopes of the binding responses will increase in proportion to the injected analyte concentration, A_{total}. Plotting the binding rate of each analyte concentration and fitting these data points to a straight line produces the standard curve eventually used to determine A_{free} in a competition analysis.

Analyte binding (for competition analysis). The next steps in the flowchart (Figure 3–12) require analyzing the ligand/analyte interaction as competitor is added to the analyte solution. Aliquots of a single analyte concentration are incubated with varying concentrations of competitor, and these mixtures are then tested for the amount of analyte that remains available to bind the immobilized ligand. Again it is essential to include buffer blanks in the analysis and inject the samples (in replicate) over both the ligand and reference surfaces to reference the binding data properly. It is also important to collect sufficient association-phase data and test a wide range of competitor concentrations in a fine dilution series to ensure that the K_D can be accurately determined from the data. Competitor concentrations should span the range from no detectable effect to complete inhibition of the analyte/immobilized ligand interaction. In addition, analyte blanks (solutions containing analyte but no competitor) need to be included as controls to (1) show the analyte binding response does not change over the course of the experiments, and (2) determine a data point for analyte binding when competitor is not present.

A

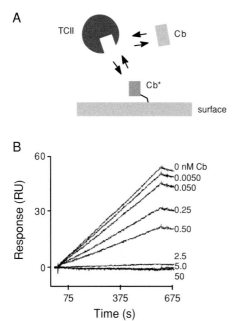

B

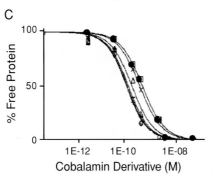

C

Figure 3–13: Solution competition analysis to determine the affinities of transcobalamin (TCII) for cobalamin (Cb) and Cb analogs. **(A)** Assay design. An immobilized cobalamin derivative (cyanocobalamin-*b*-aminopentylamide, Cb*) and Cb compounds in solution compete for the binding site in TCII. **(B)** TCII binding to immobilized Cb* in the presence of increasing concentrations of soluble Cb. Similar binding responses were obtained for other, related cobalamin compounds. **(C)** Competition isotherms were fit to a simple binding model to obtain affinities for each of the TCII/derivative interactions. The data points obtained by utilizing data from panel B are denoted with filled circles. The other symbols correspond to five other cobalamin derivatives. Panels B and C are adapted from reference 71 with permission from Elsevier Science (copyright 2002).

Data fitting (for competition analysis). To obtain the affinity of the competitor/analyte interaction, the percentages of analyte available to bind the immobilized ligand (calculated using the slopes of the association-phase data) are plotted versus competitor concentration. These data points are then fit to the isotherm described by Equation 3–6:

$$A_{free} = (A_{total} - C_{total} - K_D)/2 + [(A_{total} + C_{total} + K_D)^2/4 - C_{total}A_{total}]^{1/2}, \quad (3–6)$$

where A_{free} is the concentration of free analyte in solution (determined from the standard curve), A_{total} is the injected concentration of analyte, C_{total} is the injected concentration of the soluble competitor, and K_D is the equilibrium dissociation constant. Reference 22 explicitly describes the derivation of this equation.

Figures 3–13–3–15 depict examples of solution competition experiments. In the experiment shown in Figure 3–13, Cannon *et al.* took advantage of the mass transport-limiting conditions induced by a high-density (~600 RU) cobalamin analog surface to determine the affinities of cobalamin and its derivatives for a

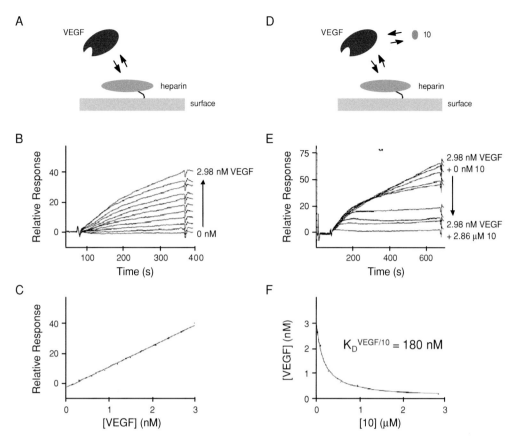

Figure 3–14: Solution competition analysis to determine the affinity of VEGF for oligosaccharide 10. **(A)** Assay design used to generate the standard curve for the VEGF/heparin interaction. **(B)** Sensorgrams obtained for a concentration series of VEGF binding to immobilized heparin. **(C)** Responses from panel B plotted versus VEGF concentration to yield the standard curve. **(D)** Assay design used to monitor the VEGF/heparin interaction in the presence of the soluble competitor, 10. **(E)** Sensorgrams obtained for VEGF binding to heparin as the concentration of 10 increases. **(F)** Free VEGF concentration plotted versus 10 concentration and fit to a simple binding isotherm described by Equation 3–6 to obtain the solution-phase affinity for the VEGF/10 interaction. Panels B, C, E, and F are adapted from reference 22 with permission from the American Chemical Society (copyright 2003).

cobalamin-binding protein, transcobalamin (TCII).[71] Once a standard binding curve was constructed for TCII binding to the immobilized cobalamin analog Cb*, other Cb derivatives were added to the TCII solution to determine their affinities for TCII (Figure 3–13A). As the concentration of soluble Cb derivative increased, TCII binding to the Cb* surface decreased (Figure 3–13B). A standard curve of TCII binding the surface Cb* was used to convert the binding responses in Figure 3–13B to the percentage of protein available to bind the surface Cb* (and not bound to the solution competitor). This percentage was plotted versus the concentration of soluble Cb derivative and fit to an isotherm described by Equation 3–6 to yield a K_D of 430 nM for Cb* in solution binding to TCII. Figure 3–13C illustrates how different compounds varied in their affinities for TCII. To demonstrate further the steps involved in the affinity determination outlined in

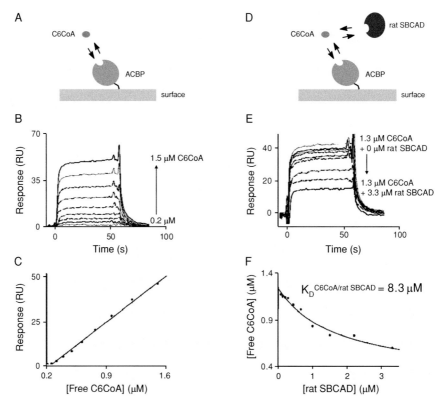

Figure 3–15: Solution competition analysis to determine the affinity of hexanoyl-CoA (C6CoA) for rat short/branched chain acyl-CoA dehydrogenase (SBCAD). **(A)** Assay design used to generate the standard curve for the C6CoA/Acyl-CoA binding protein (ACBP) interaction. **(B)** Sensorgrams obtained for a concentration series of C6CoA binding to immobilized ACBP. **(C)** Equilibrium responses from panel B plotted versus C6CoA concentration to yield the standard curve. **(D)** Assay design used to monitor the C6CoA/ACBP interaction in the presence of the soluble competitor, SBCAD. **(E)** Sensorgrams obtained for C6CoA binding to ACBP as the SBCAD concentration increases. **(F)** Concentration of free C6CoA plotted versus SBCAD concentration and fit to a simple binding isotherm described by Equation 3–6 to obtain the solution-phase affinity for the C6CoA/SBCAD interaction. Panels B, C, E, and F are adapted from reference 74 with permission from the American Society for Biochemistry and Molecular Biology (copyright 2003).

Figure 3–12 and illustrated in Figure 3–13, this experiment is described in detail as Worked Example 3.

The experiment shown in Figure 3–13 tracked the slope of the linear association-phase responses to monitor competition binding events. In contrast, the two examples in Figures 3–14 and 3–15 tracked the response intensities at the end of the association phase. The experiment described in Figure 3–14 employed a heparin surface to determine the affinities of oligosaccharide antitumor agents binding to the growth factors fibroblast growth factors 1 and 2 (FGF-1 and FGF-2), and vascular endothelial growth factor (VEGF).[22] Using a standard curve for the growth factor/heparin interaction, Cochran *et al.* determined the growth factor affinities of a panel of oligosaccharides to establish optimal chain length, as well as degrees of phosphorylation and sulfation. Using a similar assay design (Figure 3–15), He *et al.* determined the solution-based affinities constants for

CoA/acylCoA dehydrogenase interactions to identify the key enzyme residues responsible for substrate specificity.[74]

SUMMARY

In our review of the year 2003 optical biosensor literature,[7] we found that one-fourth of all SPR-based experiments were affinity measurements, either equilibrium or competition analyses (and this percentage has held steady over the past several years). The biosensor's versatility and reliability permit users to obtain K_Ds for diverse biological interactions that span the range of extremely weak (mM) to very tight (sub-pM) affinities. We are confident that as users apply attention to experimental details and appropriate data processing steps (detailed in the accompanying Worked Examples) SPR will increasingly impact various areas of life science research and drug discovery.

WORKED EXAMPLES

We describe in detail three examples of determining the affinity of an interaction using SPR: two direct equilibrium analyses (1: a small molecule in solution binding to target protein immobilized on the surface, and 2: a soluble protein binding to its protein partner that is captured on the surface) and a solution competition analysis of a protein/small-molecule interaction. The protocols in these examples are based on the operating software provided with Biacore 2000 and 3000 platforms, but they can be easily modified for application with other Biacore platforms or other manufacturers' instruments.

Example 1: Direct equilibrium analysis of a small-molecule/protein interaction

Background. Carbonic anhydrase II (CAII; 30 kDa) catalyzes the reversible conversion of carbon dioxide to bicarbonate.[75] Sulfonamide compounds, including carboxybenzenesulfonamide (CBS, 201 Da) competitively inhibit this reaction and bind the enzyme with 1:1 stoichiometry.

Assay design. CAII was immobilized on the sensor chip surface and a concentration series of CBS was injected across the protein (Figure 3–16A). Additional details of this analysis are described in references 41, 48, and 49.

Materials needed
- CM5 sensor chip (Biacore AB)
- PBS running buffer (20 mM NaH_2PO_4-Na_2HPO_4 • H_2O, 150 mM NaCl, pH 7.4) prepared, filtered, and degassed immediately prior to use
- amine-coupling kit (Biacore AB)
- carbonic anhydrase II from bovine erthyrocytes (Sigma Chemical Corp.)
- carboxybenzenesulfonamide (Sigma Chemical Corp.)

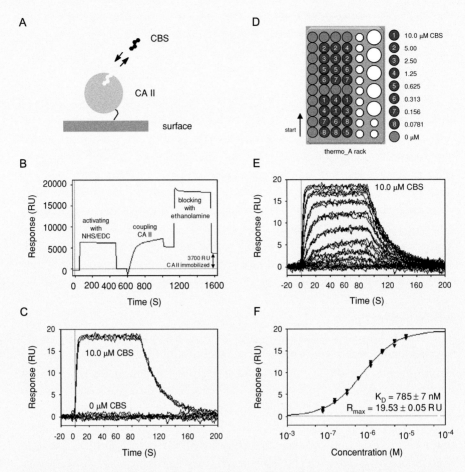

Figure 3–16: Steps involved in Worked Example 1. **(A)** Assay design. **(B)** CA II immobilization. **(C)** CBS binding test. **(D)** Arrangement of buffer blanks and analytes in a Biacore sample rack. **(E)** Double-referenced CBS binding responses. **(F)** Responses at equilibrium fit to a simple binding isotherm to obtain K_D and R_{max} values.

Ligand immobilization. CA II (15 μg/mL, dissolved in 10 mM sodium acetate, pH 5.0) was immobilized at a density of 3700 RU on one flow cell surface using a standard amine-coupling protocol (Figure 3–16B). The unmodified surface of another flow cell served as a reference.

Analyte preparation. CBS was dissolved in PBS running buffer to prepare a stock 2.00 mM solution. From this stock, a 2-mL working solution of 10.0 μM CBS in PBS running buffer was prepared. This working solution was used in both the analyte binding test and the binding assay.

Analyte binding test. Three 150-μL aliquots of 10.0 μM CBS in PBS running buffer and five 150-μL aliquots of PBS running buffer were injected over both the CAII and reference surfaces using a flow rate of 100μL/min. The reproducible, double-referenced[52] binding responses shown in Figure 3–16C demonstrate that (1) the ligand remained active when immobilized on the sensor chip surface over the course of the binding test, (2) the ligand immobilization density and the tested analyte concentration were both sufficient to obtain significant response signals, and (3) the interaction achieved equilibrium during the association phase of the binding cycle. And because the complex dissociated completely within minutes, no surface regeneration between binding cycles was required.

Analyte dilution series. From the 10.0 μM CBS working solution, triplicate 150-μL aliquots of a twofold serial dilution series (10.0, 5.00, 2.50, 1.25, 0.625, 0.313, 0.156, and 0.0781 μM) were prepared and arranged in a Biacore sample rack as shown in Figure 3–10D. To stabilize the system, the analysis began with nine injections of PBS running buffer, which were not included in the data analysis. (We find that the first few runs of an automated analysis are usually outliers from the remainder of the data set.) Other buffer blanks (to be used in double-referencing the binding responses) were included periodically throughout the analysis (Figure 3–16D).

Binding assay. The binding assay was performed at 25°C using a flow rate of 100 μL/min. Each sample was injected for 90 s (using "KINJECT" injection mode) and dissociation was monitored for 2 min. Although KINJECT requires a larger sample volume than the INJECT or QUICKINJECT options, this function provides a uniform analyte injection.

Data processing and analysis. All data were double-referenced[52] using SCRUB-BER (BioLogic Software Pty, Australia) to produce the profiles shown in Figure 3–16E. The average of the responses at $t = 75$–85 s were plotted versus injected CBS concentration (data points in Figure 3–10F). Fitting the data to the simple binding isotherm described by Equation 3–3 (automatically calculated in SCRUB-BER; alternatively, the isotherm can be constructed using Excel, Kaleidagraph, or a similar curve-fitting program) yielded the K_D and R_{max} values shown in Figure 3–16F. This data set can also be fit to obtain kinetic rate and affinity parameters because the sensorgrams have curvature and therefore contain kinetic information (see chapter entitled *Extracting Kinetic Rate Constants from Binding Responses*). Reference 41 demonstrates the affinities determined for this interaction using both equilibrium- and kinetic-based analysis methods are very similar.

Example 2: Direct equilibrium analysis of a protein–protein interaction with one binding partner captured from crude sample

Background. The protein–protein interaction of human Tsg101 and viral p6 is essential for the release of infectious HIV-1 from the host cell.[58,76–78] Characterizing this interaction is the first step in understanding, and perhaps arresting, the virus life cycle at this critical juncture.

In addition to the biological relevance of this interaction, the biosensor provides two tremendous technical advantages for this analysis: (1) capturing the ligand out of crude sample eliminates purification steps, and (2) measuring binding

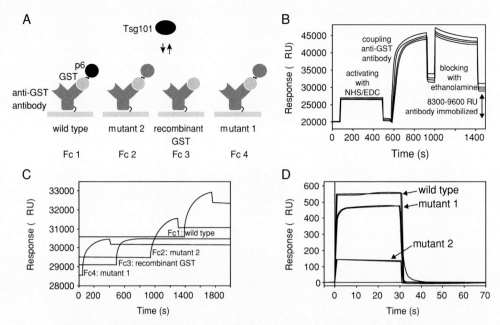

Figure 3–17: Initial steps involved in Worked Example 2. **(A)** Assay design. The three p6 proteins are individually captured on anti-GST antibody surfaces in three flow cells. Recombinant GST is captured on the remaining flow cell surface. Tsg101 is injected across all four surfaces. **(B)** Anti-GST antibody immobilization on the four flow cell surfaces using amine-coupling chemistry. **(C)** Capture of p6 proteins and recombinant GST at 1440–1840 RU. **(D)** Double-referenced responses obtained for the Tsg101 (200 μM) binding test.

events in real time permits the analysis of very weak interactions. (In traditional affinity analyses such as ELISA and pull-down assays, the Tsg101/p6 complex would be lost during the washing steps.)

Assay design. Figure 3–17A depicts the three main steps involved in this assay: (1) anti-GST antibody was immobilized on the surfaces of the four flow cells; (2) GST-tagged p6 proteins were individually captured out of crude cell lysate supernatants on the anti-GST antibody surfaces of flow cells 1, 2, and 4 (to produce a reference surface that mimics the reaction surfaces, recombinant GST was antibody-captured on flow cell 3); and (3) a concentration series of Tsg101 was injected across the four surfaces. Additional details of this analysis are described in reference 58.

Materials needed
- CM5 sensor chip (Biacore AB)
  for immobilization of anti-GST antibody
- amine-coupling kit (Biacore AB)
- HEPES-buffered saline (HBS) immobilization buffer (10 mM HEPES, 150 mM NaCl, 0.005% P20, pH 7.4) prepared, filtered, and degassed immediately prior to use

- 0.8 mg/mL anti-GST antibody (Biacore AB) for capture of GST-tagged p6 proteins
- HBS running buffer (10 mM HEPES, 150 mM NaCl, 0.005% P20, 100 μg/mL)
- BSA, pH 7.4, prepared, filtered, and degassed immediately prior to use
- 0.2 mg/mL recombinant GST (Biacore AB)
- lysate supernatants from cells overexpressing each GST-tagged p6 protein for Tsg101 binding studies
- HBS running buffer
- 200 μM stock solution of purified Tsg101 UEV domain in HBS running
- buffer

Anti-GST antibody immobilization. The antibody (diluted 1/10 in 10 mM sodium acetate, pH 5.0) was immobilized at densities of 8300–9600 RU on the four flow cell surfaces using a standard amine-coupling protocol (Figure 3–17B) and HBS as the immobilization buffer.

Capture of GST-tagged p6 proteins. Each p6-containing supernatant was centrifuged at 14 krpm for 5 min, diluted 1/10 in HBS running buffer, and flowed across an antibody surface for 5 min at 5 μL/min. To produce a reference surface that mimics the reaction surfaces, recombinant GST (diluted 1/10 in HBS running buffer) was similarly antibody-captured on flow cell 3. The GST-tagged proteins were stably captured at levels of 1440–1840 RU on the antibody surfaces (Figure 3–17C).

Tsg101 binding test. Based on the protein production level, 200 μM was the highest concentration of Tsg101 that could be studied in this analysis. Three 100-μL aliquots of 200 μM Tsg101 in HBS running buffer and three 100-μL aliquots of HBS running buffer alone were injected over the four surfaces using a flow rate of 100 μL/min. Figure 3–11D depicts the double-referenced binding responses generated from the three ligand surfaces. All three Tsg101/p6 interactions reached equilibrium within seconds, Tsg101 dissociated rapidly from each p6 surface (so no regeneration step was required), and the overlay of the three Tsg101 injections over each surface demonstrated the analysis was reproducible. Also, differences in Tsg101 binding to the three ligand surfaces is immediately apparent: the two mutants displayed decreased levels of binding relative to wild-type p6.

Tsg101 dilution series. From the 200 μM Tsg101 stock solution, triplicate 100-μL aliquots of a threefold serial dilution series (200, 66.7, 22.2, 7.41, 2.47, 0.823, 0.274, 0.0914, and 0 μM) were prepared and arranged in a Biacore sample rack as shown in Figure 3–18A. To stabilize the system, the analysis began with nine injections of PBS running buffer, which were not included in the data analysis. Other buffer blanks (used in double-referencing) were included periodically throughout the analysis (Figure 3–18A).

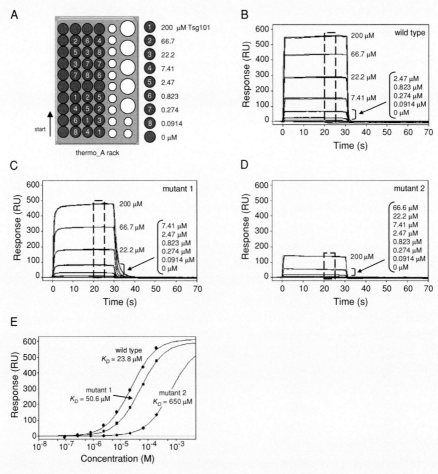

Figure 3–18: Final steps involved in Worked Example 2. **(A)** Arrangement of buffer blanks and analytes in a Biacore sample rack. Sensorgrams obtained for triplicate aliquots of 0–200 μM Tsg101 injected across wild-type **(B)**, mutant 1 **(C)**, and mutant 2 p6 **(D)** proteins. For comparison, panels B–D are plotted on the same axes. **(E)** Equilibrium responses (highlighted by the dashed boxes in panels B–D) plotted versus injected Tsg101 concentration and simultaneously fit to simple binding isotherms described by Equation 3–3. The reported affinities for each interaction are given in the panel.

Binding assay. The binding assay was performed at 20°C using a flow rate of 100 μL/min. Each sample was injected for 30 s using KINJECT and dissociation was monitored for 1 min.

Data processing and analysis. All data were double-referenced[52] using SCRUBBER to produce the profiles shown in Figure 3–18B–3–18D. In SCRUBBER, the average of the responses at $t = 20$–25 s generated from each protein surface was plotted versus injected Tsg101 concentration and fit simultaneously to the simple binding isotherm described by Equation 3–3 to obtain the affinities shown in Figure 3–18E. (In this fitting procedure, a single R_{max} was determined for the entire data set because the p6 proteins were captured at similar densities.)

Example 3: Indirect competition analysis of a small-molecule/protein interaction

Background. Cyanocobalamin (Cb, vitamin B_{12}; 1355 Da) is an essential human nutrient because it is a precursor of cofactors required for methionine synthase and methylmalonly CoA mutase. Cellular internalization of Cb depends on three transport proteins: haptocorrin (HC), transcobalamin (TCII), and intrinsic factor (IF). Each of these proteins has a single, high-affinity cobalamin-binding site.[79] In this example, we measured binding of these proteins to an immobilized Cb analog, Cb*, to determine indirectly each protein's affinity for cobalamin. We used a high-density Cb* surface to induce mass transport-limited binding. Under these conditions, responses during the association phase are linear, which makes the binding rate constant over time and thereby increases the precision of the analysis.

Assay design. The analysis was performed in two phases: (1) generating standard protein binding curves, and (2) determining the interaction affinity in solution. To generate the standard curves, individual concentration series of cobalamin-binding proteins (HC and TCII) were injected across cyanocobalamin-*b*-aminopentylamide (Cb*) immobilized on the sensor chip surface (Figure 3–19A). To determine the affinities of the HC/Cb and TCII/Cb interactions in solution, each protein was injected across the immobilized Cb* in the presence of soluble Cb (Figure 3–20A). The concentration of Cb in solution was increased until no protein binding to the immobilized Cb* was detected. Reference 71 describes additional details of this analysis.

Materials needed
- CM5 sensor chip (Biacore AB)
- HBS immobilization buffer (10 mM HEPES, 150 mM NaCl, 3.4 mM EDTA, 0.005% P20, pH 7.5, prepared, filtered, and degassed immediately prior to use)
- amine-coupling kit (Biacore AB)
- 10 mM sodium acetate, pH 4.5
- cyanocobalamin-*b*-5-aminopentylamide (Cb*, prepared as described in reference 54)
- HBS running buffer (10 mM HEPES, 150 mM NaCl, 3.4 mM EDTA, 1 mg/mL BSA, 0.005% P20, pH 7.5, prepared, filtered, and degassed immediately prior to use)
- haptocorrin (Sigma Chemical Co.)
- transcobalamin (produced and purified as described in reference 71)

Ligand immobilization. Cb* (dissolved in 10 mM sodium acetate, pH 4.5) was immobilized at a density of 650 RU on one flow cell surface at 37°C using a standard amine-coupling protocol (Figure 3–19B) and HBS immobilization buffer. The unmodified surface of another flow cell served as a reference.

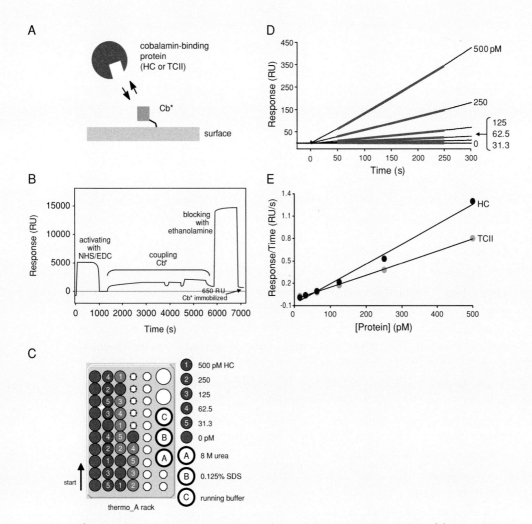

Figure 3–19: Generating standard curves for a solution competition analysis. **(A)** Assay design. In this example, HC and TCII were tested for binding to immobilized Cb* (cyanocobalamin-b-aminopentylamide). **(B)** Immobilization of Cb* using traditional amine-coupling chemistry to produce a 650 RU capacity surface. **(C)** Arrangement of buffer blanks and analytes in a Biacore sample rack, HC samples are denoted in dark gray with concentrations listed on the right. A similar concentration series was prepared for TCII in light gray. **(D)** Responses for HC binding to immobilized Cb* (for simplicity, only the association phase is shown). Similar binding profiles were obtained for TCII. **(E)** Protein binding standard curves. The slopes of the responses from 50–250 s were plotted versus the injected protein concentration and fit with a straight line to yield the standard curve for free protein (A_{free}) in solution. Panels D and E adapted from reference 71 with permission from Elsevier Science (copyright 2002).

Analyte preparation for generation of standard curves. HC and TCII were diluted in HBS running buffer to prepare stock 500-pM solutions. From these stock solutions, duplicate 200-μL aliquots of a twofold serial dilution series (500, 250, 125, 62.5, 31.3, and 0 pM) were prepared in duplicate and arranged in a Biacore sample rack as shown in Figure 3–19C. Nine aliquots of HBS running buffer, which were used to stabilize the system but were excluded from the data analysis, were also added to the sample rack.

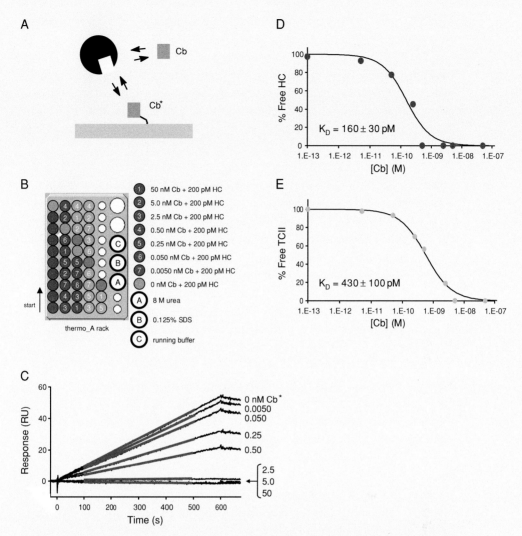

Figure 3–20: Determining K_D's from a solution competition analysis. **(A)** Assay design. **(B)** Arrangement of buffer blanks and analytes in a Biacore® sample rack, HC samples are denoted in dark gray with concentrations listed on the right. A similar concentration series was prepared for TCII (denoted in light gray). **(C)** Responses for TCII binding to immobilized Cb* in the presence of increasing concentrations of cobalamin (Cb). Similar binding profiles were obtained for HC. **(D)** The slopes of the responses (highlighted) in panel (C) plotted versus the soluble Cb concentration and fit to a simple binding isotherm described by Equation 3–6 to obtain the solution-phase affinity for the TCII/Cb interaction. **(E)** Responses obtained for HC binding to immobilized Cb* in the presence of soluble Cb were similarly fit to obtain the solution-phase affinity of the HC/Cb interaction. Panels C–E adapted from reference 71 with permission of Elsevier Science (copyright 2002).

Binding assay for generation of standard curves. The binding assay was performed at 30°C using a flow rate of 20 μL/min. (This slow flow rate was used so that binding of the analyte solution could be monitored over several minutes and to induce mass transport-limiting conditions.) Each sample was injected for 10 min and dissociation was monitored for 1 min. The Cb* surface was regenerated with 8 M urea and 0.125% SDS (followed by a buffer wash) between binding cycles.

Data processing and standard curve generation. All data were double-referenced.[52] Because the binding responses were linear during the association phase, the slope of each sensorgram (at $t = 50$–250 s, as illustrated in Figure 3–19D for HC) could be determined. These slopes were plotted versus the injected protein concentration and fit to straight lines (Figure 3–19E).

Analyte preparation for competition analyses. HC was diluted in HBS running buffer to prepare a stock solution of 200 pM. A concentration series of Cb ranging from 100–0.01 nM was prepared and then added to an equal volume of 200 pM HC. The final concentrations of Cb (50, 5.0, 2.5, 0.50, 0.25, 0.050, 0.0050, and 0 nM) were prepared in duplicate and allowed to equilibrate for 5 hr. The same Cb concentration series was also prepared using a 200 pM TCII stock solution. Aliquots of 200 μL of each protein/Cb mixture were arranged in a Biacore sample rack as shown in Figure 3–20B. Nine aliquots of HBS running buffer alone (used to stabilize the system but excluded from the data analysis) were also added to the sample rack.

Binding assay for competition analyses. The binding assay was performed at 30° C using a flow rate of 20 μL/min. Each sample was injected for 10 min and the dissociation was monitored for 1 min. The Cb* surface was regenerated with 8 M urea and 0.125% SDS (followed by a buffer wash) between binding cycles.

Data processing and analysis of competition analysis. All data were double-referenced (as shown in Figure 3–20C for TCII) and the slopes of the binding responses were used along with the standard curves to determine the percentage of free protein in solution available to bind to the immobilized Cb*. The percentage of free protein was then plotted versus the concentration of Cb in solution (Figures 3–20D and 3–20E). Fitting these data to simple binding isotherms described by Equation 3–6 yielded the two proteins' affinities for Cb. Reference 71 compares the TCII/Cb affinities from this competition assay and from direct kinetic analysis and also illustrates how this method was used to characterize a panel of cobalamin derivatives.

REFERENCES

1. J. Smal, J. Closset, G. Hennen, P. de Meyts, *Biochem J* **225**, 283 (1985).
2. S. Bass, M. Mulkerrin, J. Wells, *Proc Natl Acad Sci* USA **88**, 4498 (1991).
3. T. Clackson, J. Wells, *Science* **267**, 383 (1995).
4. G. Fuh *et al.*, *Science* **256**, 1677 (1992).
5. B. Cunningham *et al.*, *Science* **254**, 821 (1991).
6. R. L. Rich, D. G. Myszka, *J Mol Recognit* **16**, 351 (2003).
7. R. Rich, D. Myszka, *J Mol Recognit* **18**, 1 (2005).
8. A. Drake, D. Myszka, S. Klakamp, *Anal Biochem* **328**, 35 (2004).

9. R. Rauchenberger *et al., J Biol Chem* **278**, 38194 (2003).
10. B. Bernat, G. Pal, M. Sun, A. Kossiakoff, *Proc Natl Acad Sci USA* **100**, 952 (2003).
11. B. Seet *et al., Proc Natl Acad Sci USA* **100**, 15137 (2003).
12. Y. Huang, R. Rich, D. Myszka, H. Wu, *J Biol Chem* **278**, 49517 (2003).
13. G. Xu *et al., J Biol Chem* **278**, 5455 (2003).
14. R. Fisher *et al., J Biol Chem* **278**, 28976 (2003).
15. N. Papo, M. Shahar, L. Eisenbach, Y. Shai, *J Biol Chem* **278**, 21018 (2003).
16. P. S. Katsamba, M. Bayramyan, I. S. Haworth, D. G. Myszka, I. A. Laird-Offringa, *J Biol Chem* **277**, 33267 (2002).
17. M. Evans-Galea *et al., Biochemistry* **42**, 1053 (2003).
18. Y. Abdiche, D. Myszka, *Anal Biochem* **328**, 233 (2004).
19. I. Gustafson, *Colloids Surf B* **30**, 13 (2003).
20. A. Hall *et al., Infect Immun* **71**, 6864 (2003).
21. K. Kolev *et al., Blood* **101**, 4380 (2003).
22. S. Cochran *et al., J Med Chem* **46**, 4601 (2003).
23. K. Halkes, P. St Hilaire, P. Crocker, M. Meldal, *J Comb Chem* **5**, 18 (2003).
24. T. Gossas, U. Danielson, *J Mol Recognit* **16**, 203 (2003).
25. D. G. Myszka, *Anal Biochem* **329**, 316 (2004).
26. M. Donini *et al., J Mol Biol* **330**, 323 (2003).
27. J. Yi *et al., J Biol Chem* **277**, 12164 (2002).
28. P. Rathanaswami, S. Roalstad, D. Myszka, J. Babcock, *In preparation* (2005).
29. S. Swanson, *Dev Biol (Basel)* **112**, 127 (2003).
30. T. Chen, N. MacDonald, J. Wingard, *PharmaGenomic* **3**, 44 (2003).
31. D. Myszka, R. Rich, *Drug Discovery World* Spring, 49 (2003).
32. M. Hämäläinen *et al., J Biomol Screen* **5**, 353 (2000).
33. S. Löfås, *Modern Drug Discovery* May, 47 (2003).
34. R. Rich, Y. Day, T. Morton, D. Myszka, *Anal Biochem* **296**, 197 (2001).
35. Y. Day, D. Myszka, *J Pharm Sci* **92**, 333 (2003).
36. C. Baird, E. Courtenay, D. Myszka, *Anal Biochem* **310**, 93 (2002).
37. J. Ekstrom *et al., Chem Biol* **9**, 915 (2002).
38. P. Stenlund, G. Babcock, J. Sodroski, D. Myszka, *Anal Biochem* **316**, 243 (2003).
39. G. Babcock, T. Mirzabekov, W. Wojtowicz, J. Sodroski, *J Biol Chem* **276**, 38433 (2001).
40. T. Morton, D. Myszka, *Meth Enzymol* **295**, 268 (1998).
41. Y. S. Day, C. L. Baird, R. L. Rich, D. G. Myszka, *Protein Sci* **11**, 1017 (2002).
42. R. Karlsson, H. Roos, L. Fägerstam, B. Persson, *Methods* **6**, 99 (1994).
43. S. Yoo *et al., J Mol Biol* **269**, 780 (1997).
44. L. Leder *et al., J Exp Med* **187**, 823 (1998).
45. D. Myszka *et al., Protein Sci* **5**, 2468 (1996).
46. L. Joss, T. Morton, M. Doyle, D. Myszka, *Anal Biochem* **261**, 203 (1998).
47. D. Myszka, M. Jonsen, B. Graves, *Anal Biochem* **265**, 326 (1998).
48. D. G. Myszka *et al., J Biomol Techniq* **14**, 247 (2003).
49. M. J. Cannon *et al., Anal Biochem* **330**, 98 (2004).
50. K. van Holde, W. Johnson, P. Ho, *Principles of Physical Biochemistry*, Prentice Hall (Upper Saddle River, NJ, 1998).
51. *Biacore Applications Handbook*. Biacore AB: Uppsala, Sweden.
52. D. G. Myszka, *J Mol Recognit* **12**, 279 (1999).
53. R. Rich, D. Myszka, *J Mol Recognit* **15**, 352 (2002).
54. M. J. Cannon, D. G. Myszka, *Recent Res Devel Biophys Biochem* **3**, 333 (2003).
55. Y. Li, M. Urrutia, S. Smith-Gill, R. Mariuzza, *Biochemistry* **42**, 11 (2003).
56. A. Brychzy *et al., EMBO J* **22**, 3613 (2003).
57. E. Merithew, C. Stone, S. Eathiraj, D. Lambright, *J Biol Chem* **278**, 8494 (2003).
58. J. Garrus *et al., Cell* **107**, 55 (2001).
59. M. Facompré *et al., ChemBioChem* **4**, 386 (2003).
60. F. Larsen, C. Larsen, P. Jakobsen, R. Brodersen, *Mol Pharmacol* **27**, 263 (1985).
61. M. Dockal, M. Chang, D. Cater, F. Ruker, *Protein Sci* **9**, 1455 (2000).

62. A. West Jr *et al.*, *J Mol Biol* **313**, 385 (2001).
63. L. Wang *et al.*, *Proc Natl Acad Sci* USA **97**, 12 (2000).
64. F. Tanious *et al.*, *Biochemistry* **42**, 13576 (2003).
65. R. Karlsson *et al.*, *Anal Biochem* **278**, 1 (2000).
66. R. Karlsson, *Anal Biochem* **221**, 142 (1994).
67. M. Alterman *et al.*, *Eur J Pharm Sci* **13**, 203 (2001).
68. L. Nieba, A. Krebber, A. Plückthun, *Anal Biochem* **234**, 155 (1996).
69. M. Adamczyk, J. Moore, Z. Yu, *Methods* **20**, 319 (2000).
70. M. Montalto *et al.*, *J Immunol* **166**, 4148 (2001).
71. M. Cannon *et al.*, *Anal Biochem* **305**, 1 (2002).
72. A. Stephen *et al.*, *Biochem Biophys Res Commun* **296**, 1228 (2002).
73. C. Fong *et al.*, *Biochim Biophys Acta* **95**, 1596 (2002).
74. M. He, T. Burghardt, J. Vockley, *J Biol Chem* **278**, 37974 (2003).
75. S. K. Nair, J. F. Krebs, D. W. Christianson, C. A. Fierke, *Biochemistry* **34**, 3981 (1995).
76. O. Pornillos *et al.*, *EMBO J* **21**, 2137 (2002).
77. O. Pornillos, S. Alam, D. Davis, W. Sundquist, *Nat Struct Biol* **9**, 812 (2002).
78. O. Pornillos *et al.*, *J Cell Biol* **162**, 425 (2003).
79. R. Allen, P. Majerus, *J Biol Chem* **247**, 7702 (1972).

4 Extracting kinetic rate constants from binding responses

Rebecca L. Rich and David G. Myszka

INTRODUCTION	85
USING SPR TO MONITOR BINDING KINETICS	86
Correlating structure with function	87
Interpreting binding mechanisms	88
Engineering proteins	89
Discovering new drugs	90
Refining protein interaction maps	92
OBTAINING KINETIC RATE CONSTANTS	92
Interpreting kinetic data	95
Choosing an interaction model	96
Applying the fitting method	97
Evaluating fitting results	99
Applying complex models	101
SUMMARY	102
APPENDIX: WORKED EXAMPLES	102
Example 1: A small-molecule inhibitor/enzyme interaction fit to a mass transport model	103
Example 2: An antibody/antigen interaction having a slow dissociation rate	105
REFERENCES	108

INTRODUCTION

Many areas of biological research rely on kinetic analysis, the study of reaction rates, to give detailed information about chemical and biochemical processes. In a biochemical setting, the term "kinetics" most often refers to the rate at which an enzyme facilitates the conversion of substrate to product (Figure 4–1). Enzyme kinetics are studied by monitoring either substrate depletion or product formation as a reaction progresses. For example, carbonic anhydrase catalyzes the conversion of carbon dioxide to bicarbonate, and this catalysis can be tracked by the spectroscopic change of a pH indicator.[1] In contrast to its enzyme kinetics, the *binding kinetics* of carbonic anhydrase are the rates that molecules (such as

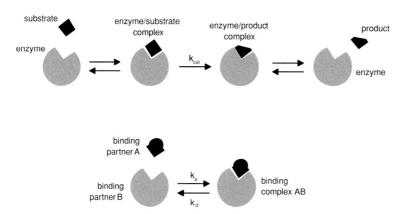

Figure 4–1: Kinetic schemes. In enzyme kinetics (*top*), k_{cat} is the rate that enzyme converts substrate (A) to product (C). In binding kinetics (*bottom*), k_a and k_d are the rates of reactant association and complex dissociation, respectively.

inhibitors) bind to and dissociate from the enzyme without being modified (Figure 4–1).[2,3]

Studies of binding kinetics are not limited to the analysis of enzyme/substrate interactions. Rather, binding kinetics can be measured for any interaction in which two or more partners collide with sufficient energy and in the appropriate orientation to form a complex. Kinetic output parameters obtained from a biosensor are the rate at which the free binding partners join to form a complex (association rate constant, k_a; measured in units of $M^{-1}\,s^{-1}$) and the rate at which the complex falls apart to yield free binding partners (dissociation rate constant, k_d; measured in units of s^{-1}). The ratio of these two rate constants (k_d/k_a) yields the affinity of the interaction (equilibrium dissociation constant, K_D; measured in units of M and described in more detail in the previous chapter entitled *Extracting Affinity Constants from Biosensor Binding Responses*). In this chapter we illustrate how kinetics contribute to understanding biological interactions and we outline the steps required to extract these parameters from surface plasmon resonance (SPR) binding responses.

USING SPR TO MONITOR BINDING KINETICS

Because SPR biosensors detect changes in mass in real time as complex forms, this technology is suited to studying the binding kinetics of a wide range of biological systems: from the very small (drug candidates) to the very large (viruses) and for interactions ranging from simple 1:1 complexes to complicated macromolecular assemblies. In addition, SPR can be used to study interactions that have rate constants spanning several orders of magnitude[4] (reported association rate constants of 10^2–$10^7\ M^{-1}\,s^{-1}$ and dissociation rate constants of 10^{-6}–$10^1\ s^{-1}$, which correspond to affinities of 10^{-1}–10^{-13} M). To illustrate this technology's capabilities in examining interaction kinetics, we highlight five examples (structure/function, binding mechanisms, protein engineering, drug discovery, and

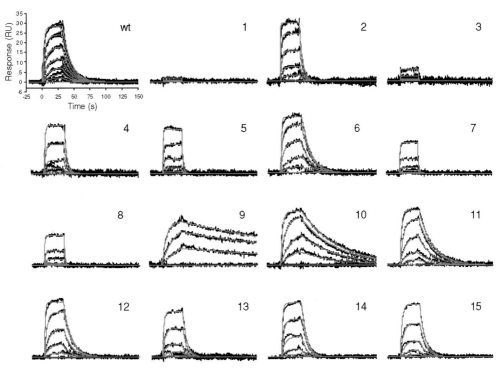

Figure 4–2: Kinetic analysis of p35/MC3 caspase interactions. Concentration series of p35 (wild-type or one of fifteen mutants) were flowed across immobilized MC3. All sensorgram data were fit to a simple 1:1 interaction model (shown as red lines) to obtain kinetic and affinity parameters. Table 4–1 summarizes the parameters obtained for each binding pair. Adapted from reference 5 with permission from the American Society of Biochemistry and Molecular Biology (copyright 2003). *See color plates.*

pathway mapping) in which SPR-determined rate constants impact basic and applied biological research.

Correlating structure with function

SPR can be used in structure/function analysis to obtain information about the roles individual amino acids play in the interaction interface. For example, Xu *et al.* systematically mutated 15 residues within the MC3 caspase binding site of p35 and measured the rate constants for each binding pair (Figure 4–2 and Table 4–1).[5] Mutants 1 and 3 displayed affinities >100-fold weaker than that of wild-type p35. The other mutants varied threefold in their association rates and >150-fold in their dissociation rates, resulting in >100-fold range in affinities. This rate information revealed which mutations predominately affected complex formation (e.g., faster association for mutant 2 and slower association for mutants 9, 10, and 13) or dissociation (e.g., faster for mutant 7 and slower for mutant 9). Other recent reports of SPR-based kinetic mapping studies include investigation of the U1A hairpin RNA-binding site in U1A protein[6] and the homodimerization interface of human growth hormone receptor.[7]

Table 4–1. Kinetic parameters for p35/caspase interactions

Interaction	k_a (M^{-1} s^{-1})	k_d (s^{-1})	K_D (μM)
p35-WT MC3	6.69(3) × 10^5*	0.0779(2)	0.116(1)
p35-1	0.37(2) × 10^5	1.27(5)	34(1)
p35-2	8.6(1) × 10^5	0.200(2)	0.233(5)
p35-3	0.94(2) × 10^5	1.0(1)	11(1)
p35-4	6.5(1) × 10^5	0.252(5)	0.39(1)
p35-5	5.07(9) × 10^5	0.491(8)	0.97(2)
p35-6	5.68(6) × 10^5	0.0719(5)	0.127(2)
p35-7	5.9(2) × 10^5	0.65(2)	1.11(5)
p35-8	4.9(2) × 10^5	0.56(1)	1.15(5)
p35-9	3.26(2) × 10^5	0.00366(5)	0.0112(2)
p35-10	3.40(3) × 10^5	0.01775(1)	0.0523(5)
p35-11	4.70(3) × 10^5	0.0466(2)	0.0990(8)
p35-12	6.31(2) × 10^5	0.0788(5)	0.125(1)
p35-13	3.81(4) × 10^5	0.157(3)	0.412(9)
p35-14	5.36(5) × 10^5	0.124(1)	0.231(3)
p35-15	5.08(7) × 10^5	0.111(1)	0.218(4)

* The numbers in parentheses indicate the errors in the last digit of the reported parameters.

Interpreting binding mechanisms

Analysis of SPR data to obtain the best fit kinetic parameters also provides mechanistic details about how complexes form and dissociate. For example, a kinetic analysis of estrogen receptor/ligand interactions demonstrated clear differences in how agonists and antagonists bind to estrogen receptor.[8] As illustrated by the examples in Figure 4–3, agonists (ligands that activate the receptor) bound with

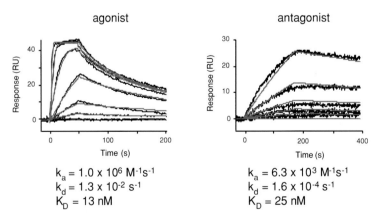

Figure 4–3: Representative data sets (black lines) for the SPR kinetic analysis of ligand/estrogen receptor interactions. Concentration series of the agonist ligand (estriol, left set of sensorgrams) and the antagonist ligand (nafoxidine, right set of sensorgrams) were flowed across the immobilized receptor. Red lines represent the global fits of the data to a 1:1 bimolecular interaction model. Adapted from reference 8 with permission from the National Academy of Sciences USA (copyright 2002). *See color plates.*

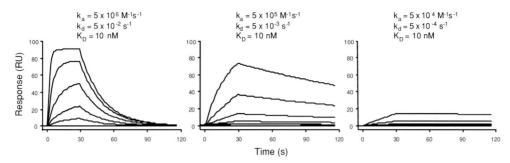

Figure 4–4: Simulated data sets for interactions displaying a range of kinetics but identical affinities. For each simulation, a maximum binding capacity of 100 RU and analyte concentrations of 0, 1.23, 3.70, 11.1, 33.3, and 100 nM were assumed.

relatively fast association rate constants, while antagonists (ligands that inhibit the receptor) bound with much slower association rates. The slower association rates observed for the antagonist are consistent with conformational changes accompanying inhibitor/estrogen receptor complex formation.

Because nuclear hormone receptors are structurally and functionally homologous, one may predict that progesterone and testosterone receptors would also bind antagonists with slow association rates. Members of another receptor family (e.g., G-protein coupled receptors), however, may rapidly bind antagonists but require a conformational change to bind agonists. Though there may not be a predictable, universal kinetic pattern to distinguish between activators and inhibitors of all ligand/receptor interactions, rate constants could provide the basis for deciphering binding mechanisms among protein families.

Engineering proteins

Kinetics can also play an important role in engineering, in which a protein is optimized to bind with high affinity through successive rounds of modification. Tracking which modifications increase the association rate or decrease the dissociation rate (or do both) allows for a more methodical approach to optimization. For example, promising antibodies are traditionally selected from a panel of hybridomas based only on antigen affinity. Figure 4–4 illustrates the value of examining kinetic parameters when choosing which antibodies to pursue as therapeutic candidates. The figure depicts simulated binding responses for three antibody/antigen interactions that have the same affinity but very different kinetics. In a simple equilibrium analysis, these three antibodies would be misleadingly considered equivalent when in fact they undergo very different antigen binding mechanisms. This example emphasizes the need to evaluate both complex association and dissociation when selecting antibodies of interest.

Kinetic analysis of antibody/antigen interactions has traditionally been sample- and time-consuming. Recently, Canziani *et al.* developed a higher-throughput screening method in which kinetic parameters could be determined

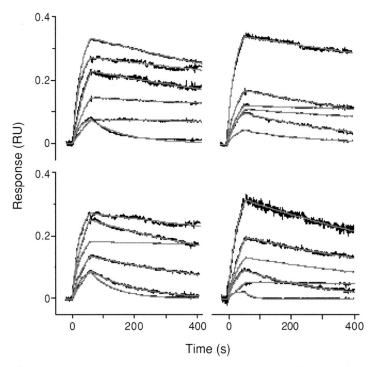

Figure 4–5: Kinetic screening of antibodies. Antigen responses (black lines) are shown in groups of six interactions per plot. The responses were all simultaneously fit with a 1:1 interaction model (red lines) using a single [B] (or R_{max}) for the entire data set. Reproduced from reference 9 with permission from Elsevier Science (copyright 2004). *See color plates.*

for approximately 200 antibodies per day.[9,10] In this analysis, antibodies are captured on immunoglobulin G (IgG) surfaces and a single antigen concentration is tested for antibody binding. The binding responses are normalized with respect to the density of antibody captured and the data are fit globally. Figure 4–5 shows an example data set for 24 antibodies. Constraining the data set to global fitting (in which the maximum binding response is common for the entire data set, but individual k_a's and k_d's are determined) produces reliable rate constants from injection of only one antigen concentration.

Discovering new drugs

Because of recent improvements in experimental design, instrument sensitivity, and throughput, SPR-based kinetic analysis of small molecules (<500 Da) is becoming routine. An example of how these studies are applied in drug development is illustrated by Markgren and coworkers' study of HIV inhibitors. As a first step in the engineering of next-generation therapies, these researchers examined four known inhibitors of HIV-1 proteinase (Figure 4–6A).[11] This work compared the drugs' relative association and dissociation rate constants and established baseline kinetic parameters against which drug candidates could be compared. Expanding on this initial study, this group kinetically screened 58 inhibitor analogs against HIV-1 proteinase and determined how structural

A.

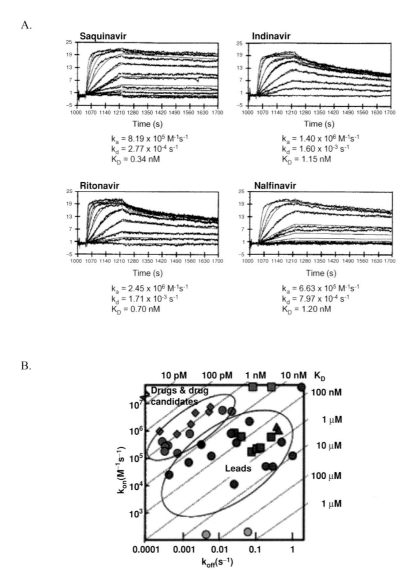

B.

Figure 4–6: (A) Kinetic analysis of four HIV-1 proteinase inhibitors. Concentration series of the inhibitors were flowed over the immobilized proteinase. Response data were fit to a 1:1 interaction model that included a mass transport parameter to obtain rate constants for each interaction. Adapted from reference 11 with permission from Elsevier Science (copyright 2001). **(B)** k_{on}-k_{off}-K_D map of representative HIV-1 proteinase inhibitors: drugs, drug candidates, and leads. Reproduced from reference 12 with permission from the American Chemical Society (copyright 2002).

modifications in the compounds affected the rate constants.[12] Plotting association rate constant versus dissociation rate constant yielded the k_{on}-k_{off}-K_D map shown in Figure 4–6B. This map illustrates the correlation between kinetics and drug efficacy: the most promising inhibitory compounds display fast association rates and slow dissociation rates. This work exemplifies how kinetic information can contribute in drug optimization.

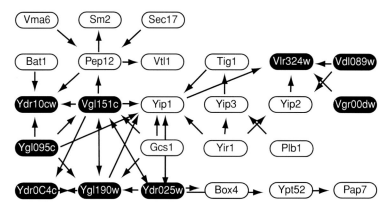

Figure 4–7: Example protein interaction map. Network of proteins involved in vesicular transport as determined from two-hybrid analysis. Arrows indicate the orientation of each two-hybrid interaction, from the bait to the prey. Hypothetical proteins of unknown functions are shown in black. Reproduced from reference 13 with permission from National Academy of Sciences USA (copyright 2001).

Refining protein interaction maps

With the advent of protein interaction maps, researchers can now begin to identify binding partners across an organism's proteome. The interaction map shown in Figure 4–7 illustrates the complexity of even a relatively simple network: proteins involved in vesicular transport.[13] To fully understand this or any other pathway or network, however, requires not only identifying all combinations of binding partners but also determining the kinetics of each interaction. Kinetic analysis reveals the time scale of interrelated binding events by establishing the stability of each complex formed, determining the likelihood that two (or more) species compete for a common binding partner and identifying the overall rate-determining step(s). Though the contribution of kinetics to interaction mapping is in its infancy, the combination of rate information with studies that reveal the spatial and temporal concentrations of each species will provide the basis for functional interpretation of the proteome.

As shown in Figures 4–2–4–7, a kinetic analysis can provide a wealth of information about interacting systems. Additional examples from a wide range of biological interactions that were readily amenable to kinetic analysis are highlighted in references 14 and 15. They illustrate that, when properly designed and executed, SPR-based kinetic analyses can be routinely executed. In this chapter we describe how to set up a kinetic analysis, how to choose a model, how the data-fitting routines are applied, and how to evaluate the reported parameters. To illustrate these points, we also include two worked examples of kinetic analyses.

OBTAINING KINETIC RATE CONSTANTS

The flowchart in Figure 4–8 outlines the steps required in a typical kinetic analysis. We find that with careful attention to each step, interpretation of the kinetics,

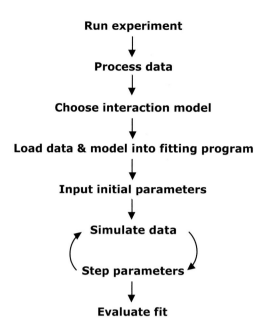

Run experiment

↓

Process data

↓

Choose interaction model

↓

Load data & model into fitting program

↓

Input initial parameters

↓

Simulate data

Step parameters

↓

Evaluate fit

Figure 4–8: Steps involved in a kinetic analysis.

which can most often be described by a simple interaction model, is actually straightforward.

Generating kinetic data. In a kinetic analysis, the analyte in solution is flowed over the ligand, which is either directly immobilized or indirectly captured on the sensor chip surface (Figure 4–9). A typical kinetic analysis involves testing a series of analyte concentrations in replicate. Ideally, these span a range from concentrations that saturate the ligand surface to dilutions that produce little or no binding response. For example, testing eight samples prepared in a threefold dilution series (e.g., 1000, 333, 111, 37.0, 12.3, 4.12, 1.37, 0.457, and 0 nM, where 0 = buffer alone) in triplicate permits the relatively rapid examination of >1000-fold span in analyte concentration and establishes the reproducibility of the analysis. This first step, collecting reproducible and information-rich data, is the cornerstone of a reliable kinetic analysis.

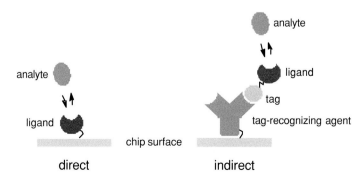

Figure 4–9: Assay design for a kinetic analysis.

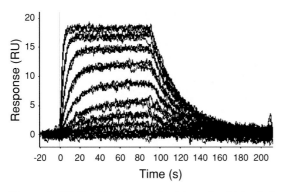

Figure 4–10: Double-referenced data set obtained for a small-molecule inhibitor (4-carboxybenzene-sulfonamide, CBS) binding to an enzyme (carbonic anhydrase II, CA II). Triplicates of 0, 0.078, 0.156, 0.313, 0.625, 1.25, 2.50, 5.00, and 10.0 μM CBS were injected across immobilized CA II.

Processing the sensorgrams is the second step. This requires double-referencing the binding data to account for systematic artifacts related to nonspecific binding, signal drift, and bulk refractive index changes and yields sensorgrams amenable to fitting simple kinetic models.[14,16,17] Fitting an unprocessed (or inadequately processed) data set to interaction models may yield erroneously complex binding kinetics.

Figure 4–10 is a double-referenced data set well suited to a kinetic analysis. In this example, triplicate injections of an analyte concentration series were flowed over a ligand surface for 90 s and dissociation of the ligand/analyte complex was monitored for 120 s. All of the responses returned to baseline within 2 min (so no surface regeneration was required) and the replicates of each analyte concentration overlaid well, which demonstrated the reproducibility of the analysis. By visually inspecting the binding responses it is apparent that this data set should be readily amenable to the fitting process because (1) the binding responses display significant curvature in both the association and dissociation phases, and (2) the sensorgrams vary in both intensity and shape.

Because binding events are described by exponentials, the curvature in the data provides information necessary to identify a unique solution for the equations used to model the interaction. When there is no curvature in the data (i.e., the responses are linear), it is difficult to extract meaningful kinetic information because a straight line can be described by a variety of exponential equations. Also, testing a range of analyte concentrations is critical for fitting the data robustly because the varying binding responses contribute differently to the fitting process. For the highest analyte concentrations (5.00 and 10.0 μM), the apparent binding rates (ΔResponse/ΔTime, which measures the number of collisions with respect to time and is dependent on analyte concentration) were very rapid and the interaction equilibrium was achieved within 30 s. These binding responses therefore help establish the active ligand concentration [B] but by themselves yield a poor estimate of the rate constants. Conversely, the binding responses generated by the lower analyte concentrations do not aid in estimating

[B] but instead provide information regarding the rate constants because they have slower apparent binding rates.

Interpreting kinetic data

The routines used to fit SPR data have evolved since commercial biosensors were first introduced in 1990. For the first few years, kinetic rate constants were determined using linear transformation methods that assume the binding events could be described by a simple model (A + B = AB). The responses recorded during the association and dissociation phases of a binding cycle were individually transformed to straight lines from which k_a and k_d could be determined. Though this approach was well suited for simple interactions or responses that had a high signal-to-noise ratio, it could not be applied to complex binding events (e.g., interactions exhibiting mass transport, conformational change, or surface heterogeneity). For example, transformed data obtained from a complex interaction would display curvature and be poorly described by a straight line. Though an investigator could conclude that this simple model did not mimic the observed binding events, more complex models could not be applied to interpret the interaction. In addition, systems that produce low binding responses were problematic to fit with this linear transformation approach because the noise in low binding responses becomes amplified when linearly transformed, creating plots with too much scatter. This made it extremely difficult to determine definitively the rate constants for small molecules, as well as for analyte preparations having a low percentage of active material.

To overcome some of these drawbacks, O'Shannessy *et al.* demonstrated how primary SPR data could fit to an integrated rate equation to obtain kinetic parameters.[18] Fitting the binding responses themselves (rather than the transformed data) to a model represented a significant development in the refinement of fitting methods for SPR-based data. Data fitting was still constrained, however, to analysis of individual binding curves and the assumption of a simple interaction model (or, at best, a sum of independent binding events). In addition, this approach was still restricted to fitting the association and dissociation phase data separately.

In the mid-1990s, kinetic data analysis was revolutionized by the introduction of numerical integration modeling methods coupled with global fitting routines tailored to SPR applications.[19] Numerical integration allows any interaction model to be simulated, so biomolecular complexes that undergo conformational change, that involve multiple partners binding sequentially, or that are influenced by mass transport can be described by the appropriate model. Global data fitting provides the ability to discriminate between models that appear equally valid when using earlier fitting methods. And in global fitting a complete data set is analyzed at once: the association- and dissociation-phase data collected for a range of analyte concentrations are fit simultaneously to yield one k_a and one k_d that describe the interaction. This comprehensive analysis can become even more rigorous if data obtained from an analyte concentration series tested over

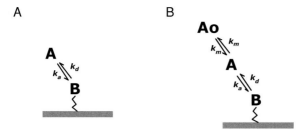

Figure 4–11: Interaction schemes, in which A is the analyte and B is the immobilized ligand. **(A)** 1:1 interaction described by the simplest model. **(B)** 1:1 interaction model that includes a parameter (k_m) to account for mass transport of the analyte from bulk solution to the unstirred solvent layer at the surface. Ao represents the analyte in bulk solution.

multiple ligand surfaces are fit globally, as illustrated in reference 20. The stringency of applying global fitting to obtain a single set of kinetic parameters from a collection of binding responses makes it easy to determine if the chosen model accurately describes the interaction.

Today, most of the SPR community has adopted global fitting methods to determine kinetic rate constants. With the availability of shareware data-fitting programs[21] and the advent of high-speed personal computers, these flexible yet thorough methods are now accessible for all biosensor users. The ease with which the analysis programs can be applied to large data sets allows rate constants to be obtained within a few seconds. Continuing with the example data set in Figure 4–10, we now describe the next steps in the flowchart (Figure 4–8): the process of globally fitting data to yield reliable kinetic parameters.

Choosing an interaction model

Although current fitting methods allow data to be simulated by a large variety of models, Occam's razor (the simplest answer is the best answer) should be applied. So, we recommend beginning with the simplest model that is consistent with what is known about the interaction. We find that the majority of SPR kinetic data, when properly collected and processed, can be described by one of two schemes shown in Figure 4–11.

Figure 4–11A depicts the simplest bimolecular interaction model: analyte A in solution binding reversibly to immobilized ligand B. Output parameters from this simple model are the rate constants (k_a and k_d) and the active concentration of ligand (B) immobilized on the surface (reported in response units proportional to the relative molecular masses of ligand and analyte; also referred to as R_{max}). For comparison, Figure 4–11B depicts a slightly more complex model frequently appropriate for interactions measured using the biosensor: a bimolecular interaction that includes mass transport events. Mass transport influences binding responses when diffusion of the analyte to the ligand surface is comparable to or slower than analyte binding to the ligand itself (this phenomenon is described in more detail in the chapter dedicated to mass transport). Output parameters from this mass transport-influenced model are the rate constants, the active ligand concentration, and the mass transport rate (k_m).[19,20] To begin fitting the example data set shown in Figure 4–10, we start by choosing the simple 1:1 interaction model shown in Figure 4–11A.

Applying the fitting method

This simple interaction model (A + B = AB) can be described mathematically by differential equations 1–3:

(1) $d[A]/dt = 0$
(2) $d[B]/dt = -k_a[A][B] + k_d[AB]$
(3) $d[AB]/dt = k_a[A][B] - k_d[AB]$

In Equation 1 there is no change in the analyte concentration, [A], because it is held constant through the interaction (during the association phase, A is continually replenished by sample flow over the surface, and during the dissociation phase there is no A present). Equations 2 and 3 track the concentrations of B and AB over the course of each binding cycle.

The fitting program employs numerical integration to obtain approximate solutions to these equations, yielding values for k_a, k_d, and [B]. This process involves three steps: (1) choosing initial estimates of the kinetic parameters, (2) generating a simulated set of response curves based on the parameters chosen in step 1, and (3) iteratively finding a set of rate constants that best mimics the experimental data. Though step 1 requires user effort, steps 2 and 3 are performed automatically by the fitting program.

1. Initial conditions. Once a model is chosen, the experimental data set and model are loaded into the fitting program (the two most commonly used programs are BiaEval 3.1 [www.biacore.com] and SCRUBBER (Biologic Software Pty, Australia). Next, values of the known and measured parameters, as well as initial guesses for the unknown parameters, are entered into the program.

Figure 4–12 illustrates the parameters involved in applying a simple 1:1 interaction model. The concentration of injected analyte A is known (10 μM was the highest concentration tested in the example shown in Figure 4–10) and the concentration of complex AB corresponds to the binding response signal, so it is a measured parameter (0 is entered for this parameter as a starting condition because no AB complex exists at time = 0). Though the concentration of immobilized ligand B remains constant throughout the interaction, the concentration of B *available for binding to A* changes as A binds to and dissociates from the surface. In the modeling algorithm, the parameter B is therefore an unknown and corresponds to the maximum binding response observable for this interaction. For the example data set in Figure 4–10, 20 RU is an appropriate estimate for B because the binding response plateaus (at ~18 RU) for the two highest analyte concentrations are closely spaced, which indicates the ligand surface is approaching saturation. Other unknowns are the association and dissociation rate constants. Suitable estimates of k_a and k_d are 1×10^4 M^{-1}s^{-1} and 1×10^{-2} s^{-1}, respectively. These rate estimates are chosen based on our experience fitting similarly shaped

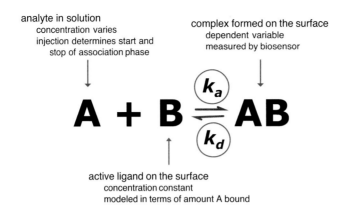

Figure 4–12: The five parameters used to determine kinetic rate constants (circled) from a simple 1:1 interaction model.

binding profiles. (Also, these initial estimates are intermediate in the range of rates measured by SPR.)

2. Data simulation. For simplicity, we initially describe how to apply the fitting method to only the response generated for the highest analyte concentration (10 µM). First, the fitting program incorporates the inputted parameters for k_a, k_d, and [B] into the set of differential equations (Figure 4–13A). Next, the fitting program numerically approximates a solution to the differential equations to produce a simulated binding curve. (This numerical integration is necessary because the output by the biosensor is not an exact function but rather a set of data points [response intensity over each time interval] collected during each binding cycle.) The poor overlay of the simulated and experimental data indicates that one or more of the initial modeling parameters does not adequately describe the interaction.

Though we can visually inspect this overlay to establish the quality of the fit, the program determines the quality of the fit by calculating the residuals. We obtain residuals by plotting the difference between experimental and simulated data versus time. Figure 4–13D illustrates that the residuals for the simulation in panel B are not randomly distributed about zero. Instead, there are systematic deviations, which demonstrate the simulated curve is a poor approximation of the actual binding response.

3. Fit iterations. There are a variety of nonlinear least squares fitting methods (triangulation, steepest descent, bootstrapping) that can be applied to SPR data. A fairly standard fitting method that works well for the time scales involved in SPR analyses is the Levenburg-Marquardt minimization algorithm.[22] During the fitting procedure, the simulated data are iteratively subtracted from experimental data to calculate the residuals. If the residuals are larger after a round of fitting, the algorithm steps back to generate a new set of simulated data. If, instead, the residuals are smaller after a round of fitting, this new set of parameters is kept and the algorithm again steps forward to further reduce the residuals. After

A $\quad$ **d[A]/dt = 0**

$\quad$ **d[B]/dt = -1e4[10 uM][B] + 1e-2[AB]**

$\quad$ **d[AB]/dt = 1e4[10 uM][B] − 1e-2[AB]**

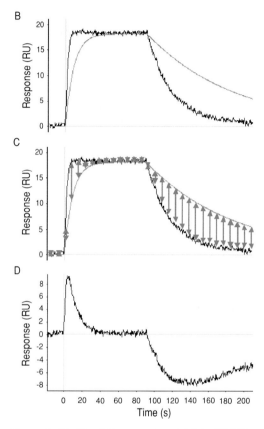

Figure 4–13: Simulating experimental data. **(A)** Differential equations used to simulate the highest binding response shown in Figure 4–10. The initial estimates of the kinetic rate constants are highlighted. **(B)** Overlay of experimental data with simulated data generated from the equations shown in **(A)**. **(C)** Arrows indicate the difference between experimental and simulated data throughout the binding cycle. **(D)** Residuals plotted for the fit shown in **(B)**.

several iterations, the simulated parameters should converge to a unique solution that closely approximates the data and outputs a unique set of rate constants. Figure 4–14 shows initial and final fits of the example data set shown in Figure 4–10. Though the inputted initial estimates do not describe the experimental data well, the final fit overlays the binding responses and the residuals are small and randomly distributed about zero.

Evaluating fitting results

Though we can visually examine the fit and residuals shown in Figure 4–14, we can use the binding responses generated by replicate injections of each analyte concentration to judge the quality of a fit even more rigorously. Averaging the replicates yields the experimental noise within the analysis. The experimental

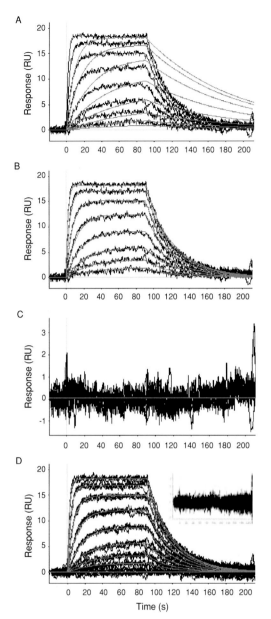

Figure 4–14: Fitting an experimental data set to a 1:1 interaction model. For illustrative purposes, sensorgrams of the buffer blank and replicate analyte injections are omitted from panels A–C. **(A)** Simulation of the data based on initial guess parameters of $k_a = 1 \times 10^4$ M^{-1}s^{-1}, $k_d = 1 \times 10^{-2}$ s^{-1}, and [B] = 20 RU. **(B)** Data set overlaid with the final fit of the model. **(C)** Residuals of the fit shown in **(B)**. **(D)** The complete data set (blanks and replicates included) fit to a 1:1 bimolecular interaction model (red lines) to yield $k_a = (4.47 \pm 0.01) \times 10^4$ M^{-1}s^{-1}, $k_d = (3.416 \pm 0.006) \times 10^{-2}$ s^{-1}, $K_D = 764 \pm 2$ nM, and [B] = 19.54 RU. The inset shows the fit residuals for the complete data set. *See color plates.*

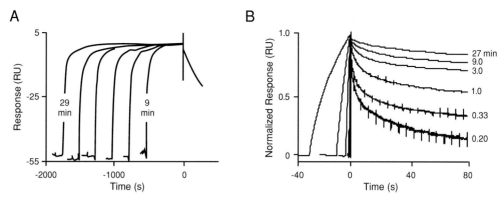

Figure 4–15: Varying the association phase time. **(A)** A gp120 mutant protein binding to immobilized CD4, with the association time varied from 9 to 29 min. The overlay of the dissociation phase responses (t > 0) indicates the interaction is independent of contact time. **(B)** Aβ peptide binding to immobilized Aβ fibril. Data were normalized by setting the response at the start of the dissociation phase equal to 1. The decrease in dissociation rate with increased association time implies that more peptide is incorporated into the fibril the longer it is exposed to the fibril surface. In both panels, the start of the dissociation phase was set to $t = 0$. Adapted from references 26 and 27 with permission from the American Chemical Society and Academic Press, respectively (copyrights 2001 and 1999).

noise level is the minimization endgoal of the fitting algorithm. So, in an optimal fit the level of noise and randomness within the residuals would closely approximate those of the experimental noise.

Applying complex models
In the event the simple 1:1 interaction model shown in Figure 4–11A inadequately describes a data set, a more complicated model (e.g., to account for mass transport, multistate binding, heterogeneity, conformational change) may be required. These complex models should be applied with care and, ideally, supported by a biological justification (e.g., the analyte is a dimer in the case of bivalent interactions) or additional experimentation.

Figure 4–15 illustrates how the biosensor can be used to confirm if an interaction is complex or not: analyte is flowed over the ligand surface at a range of injection times and the obtained sensorgrams are overlaid with $t = 0$ at the beginning of the dissociation phase. For a simple interaction, responses in the dissociation phase are independent of the analyte/ligand contact time and the sensorgrams overlay at $t > 0$ (Figure 4–15A). For a complex interaction such as the polymerization example in Figure 4–15B, the dissociation phase data do not overlay. Instead, the dissociation rate decreases as the association time increases.

Figure 4–16 demonstrates how models can be used to disprove a binding mechanism but not to identify the mechanism. The data set in this figure is poorly described by the simple 1:1 and mass transport models, indicating these models are inappropriate. Each of the more complex models, however, fits the data set well, demonstrating that a good fit does not mean that the model is correct. In fact, from this analysis we cannot tell if the interaction is driven by avidity or if

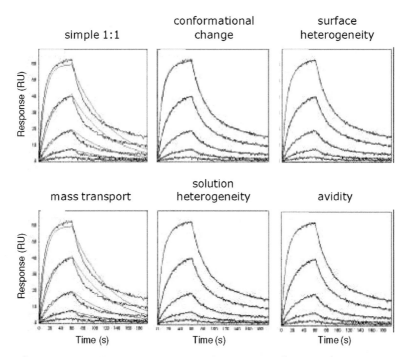

Figure 4–16: Binding responses for a complex interaction (black lines) overlaid with six different models (red lines). *See color plates.*

the ligand or analyte (or both) is heterogeneous or undergoes a conformational change upon binding. This example illustrates an important point: you cannot justify a particular complex binding mechanism by fitting alone. Instead, additional analyses would be required to identify which of the models used in Figure 4–16 is appropriate.

SUMMARY

With attention to experimental details and thoughtful application of fitting models, SPR data can yield rate constants useful for interpreting an interaction's binding events. As the scientific community at large recognizes the power of SPR as a biophysical tool and as expertise in the biosensor user community continues to improve, we are seeing high-quality SPR kinetic studies applied to an ever-increasing range of biological applications.

APPENDIX: WORKED EXAMPLES

We describe in detail two examples of determining kinetic rate constants. These examples focus on two very different biological systems and address some challenges that we regularly encounter. Obtaining the high-quality data in these examples requires a well-maintained and well-equilibrated instrument. In both instances, the biosensor was thoroughly cleaned, the Biacore "desorb" procedure was performed twice, and the instrument was primed five times with running buffer prior to the experiment.

Worked Example 1 outlines the steps required to fit a data set to obtain kinetic parameters for a mass transport-influenced small-molecule/target interaction. The data set was initially fit to a simple interaction model, but this model insufficiently described the interaction. The data set was then fit to a model that includes a parameter to account for the transport rate of analyte diffusion from the bulk solution to the ligand surface. Inspection of the fit and residuals illustrates how the mass transport model more appropriately describes the interaction.

Worked Example 2 outlines the kinetic study of an interaction having an exceptionally slow dissociation rate. This two-phase analysis involved (1) testing a concentration series of the analyte to collect association and estimated dissociation information, (2) retesting the highest analyte concentration to collect more accurate dissociation information, and (3) fitting both data sets simultaneously to obtain k_a and k_d parameters.

Example 1: A small-molecule inhibitor/enzyme interaction fit to a mass transport model

Assay design. A concentration series of acetazolamide (222 Da) was injected across carbonic anhydrase II immobilized on the sensor chip surface (Figure 4–17A). Because the complex dissociated completely within minutes, no surface regeneration between binding cycles was required. Additional details of this analysis are described in reference 23.

Materials needed
- CM5 sensor chip (Biacore AB)
- PBS running buffer (20 mM NaH_2PO_4-Na_2HPO_4 • H_2O, 150 mM NaCl, pH 7.4, prepared, filtered, and degassed immediately prior to use)
- acetazolamide (Sigma Chemical Corp.)
- carbonic anhydrase II from bovine erthyrocytes (Sigma Chemical Corp.)
- amine-coupling kit (Biacore AB)

Ligand immobilization. Carbonic anhydrase II (dissolved in 10 mM sodium acetate, pH 5.0) was immobilized at a density of 7100 RU on one flow cell surface using a standard amine-coupling protocol. The unmodified surface of another flow cell served as a reference.

Analyte preparation. Acetazolamide was dissolved in PBS running buffer to prepare a stock 1 µM solution. From this stock, triplicate 150-µL aliquots of a three-fold serial dilution series (333, 111, 37.0, 12.3, 4.12, 1.37, and 0 nM) were prepared and arranged in a Biacore sample rack as shown in Figure 4–17B.

A

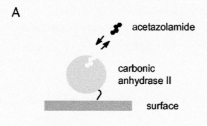

B

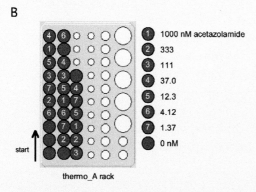

thermo_A rack

Figure 4–17: (A) Assay design for Worked Example 1. **(B)** Arrangement of acetazolamide aliquots in the sample rack, emphasizing the inclusion of blanks, as well as the triplicate and random analysis of each analyte concentration.

Binding assay. The binding assay was performed at 25°C using a flow rate of 100 μL/min. Each sample was injected for 1 min (using "KINJECT" injection mode) and dissociation was monitored for 5 min. To stabilize the system, the analysis began with two injections of PBS buffer not included in the data analysis.

Data processing and analysis. All data were double-referenced prior to model fitting. Initially, the data set was fit to the simple 1:1 interaction model described in Figure 4–11A. The fit of this model to the acetazolamide/carbonic anhydrase II interaction is shown in Figure 4–18A. From visual inspection it is obvious that this model did not describe the interaction well. This was confirmed by the nonrandom trends in the fit residuals (Figure 4–18B). Figure 4–18C depicts the fit of a 1:1 interaction model that included a parameter to account for mass transport (shown in Figure 4–11B). This fit yielded $k_a = (1.704 \pm 0.006) \times 10^6 \ M^{-1}s^{-1}$, $k_d = (5.64 \pm 0.02) \times 10^{-2} \ s^{-1}$, $k_t = (1.097 \pm 0.003) \times 10^7 \ s^{-1}$, and $K_D = 33.1 \pm 0.2$ nM. The randomness of the residuals about zero (Figure 4–11D) confirms that the incorporation of a mass transport parameter improved the fit of the model.

To demonstrate that we can obtain this quality of data by the general biosensor operator, we invited a panel of biosensor users to replicate the analysis described in Worked Example 1.[24] These users varied in their expertise level and the instrument platform they used. Overall, the compiled data sets were very encouraging:

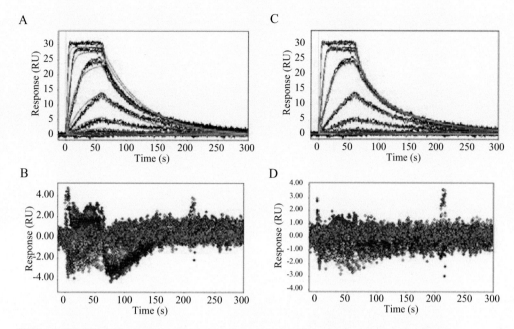

Figure 4–18: Fitting kinetic data obtained for the acetazolamide/carbonic anhydrase interaction to a 1:1 mass transport. **(A)** Final fit of the data set to a simple 1:1 interaction model. **(B)** Residuals of the fit shown in panel A. **(C)** Final fit of the data set to a 1:1 interaction model that includes a parameter to account for mass transport. **(D)** Residuals of the fit shown in panel C. *See color plates.*

the participants obtained similar results, and the standard errors in the rate constants were ~20%.

Example 2: An antibody/antigen interaction having a slow dissociation rate

Assay design. In this analysis, antigen in solution was injected across immobilized antibody (Figure 4–19A). Between binding cycles, the antibody surface was regenerated with phosphoric acid. To collect association rate information, the binding of an antigen concentration series was monitored for a few minutes; to collect dissociation rate information, the binding of a single analyte concentration was monitored for several hours. Simultaneously fitting the responses obtained from these two experiments yielded definitive kinetic and affinity parameters for this unusually tight binding pair. Additional details of this analysis are described in reference 25.

Materials needed
- CM5 sensor chip (Biacore AB)
- HBS-P + BSA running buffer (10 mM HEPES, 150 mM NaCl, 0.005% P20, 100 μg/mL BSA, pH 7.4, prepared, filtered, and degassed immediately prior to use)
- antibody/antigen pair
- amine-coupling kit (Biacore AB)
- 15 mM H_3PO_4 (Sigma Chemical Corp.)

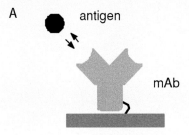

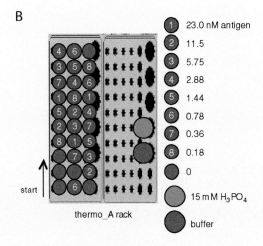

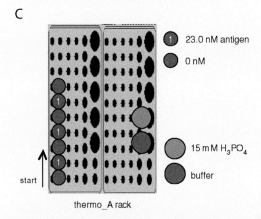

Figure 4–19: **(A)** Assay design for Worked Example 2. **(B)** Sample arrangement for the first phase of the analysis, in which a concentration series of antigen was tested in triplicate. **(C)** Sample arrangement for the second phase of the analysis, in which triplicate antigen and buffer blank injections were tested. In both phases, the antibody surface regenerated with phosphoric acid at the end of each binding cycle.

Ligand immobilization. Antibody (dissolved in 10 mM sodium acetate, pH 5.0) was immobilized at a density of 2300 RU on one flow cell surface using a standard amine-coupling protocol. The unmodified surface of another flow cell served as a reference.

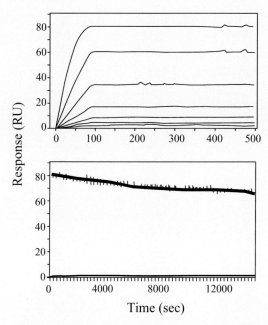

Figure 4–20: Two-phase kinetic analysis of the slowly dissociating antigen/antibody complex. The top panel shows the data collected in phase 1: the binding responses for triplicate injections of antigen (0, 0.18, 0.36, 0.72, 1.44, 2.88, 5.75, 11.5, and 23.0 nM) flowed across immobilized antibody. In each binding cycle, the association and dissociation phases were monitored for 90 and 450 s, respectively. The bottom panel shows the data collected in phase 2: triplicate binding responses of 23.0 nM antigen having a dissociation phase of 4 hr. The experimental data (black lines) from both panels were globally fit (red lines) to determine the kinetic rates, $k_a = 2.7 \times 10^6$ $M^{-1}s^{-1}$; $k_d = 1.6 \times 10^{-5}$ s^{-1}; $K_D = 6.1$ pM. Adapted from reference 5 with permission from Elsevier Science (copyright 2004). *See color plates.*

Analyte preparation for Phase I. Antigen was diluted in running buffer to prepare a stock solution of 23.0 nM. From this stock, triplicate 150-μL aliquots of a twofold serial dilution series were prepared and arranged in a Biacore sample rack as shown in Figure 4–19B.

Binding assay for Phase I. The binding assay was performed at 25°C using a flow rate of 100 μL/min. Each sample was injected for 90 s (using "KINJECT" injection mode) and dissociation was monitored for 450 s. To fully equilibrate the system, the experiment began with two buffer blank injections not included in the data analysis. At the end of each binding cycle, 10 μL 15 mM H_3PO_4 was injected to regenerate the surface and then 20 μL buffer was injected to wash the regeneration solution from the system.

Because no decay was evident in the binding response during the few minutes the dissociation phase was monitored, it would be difficult to accurately determine k_d from this data set alone. A complementary analysis that tracks complex dissociation over several hours can be coupled to this data set to provide both association and dissociation rate information.

Analyte preparation for Phase II. Three 150-μL aliquots of the highest antigen concentration (23.0 nM) were prepared and arranged in the sample block as shown in Figure 4–19C. A blank buffer was tested between each antigen sample to monitor systematic drift that may become apparent when monitoring an interaction over several hours.

Binding assay for Phase II. The binding assay was performed at 25°C using a flow rate of 100 μL/min. Each sample was injected for 90 s (using "KINJECT" injection mode) and dissociation was monitored for 4 hr (4 hr was a sufficiently long dissociation phase because significant decay was observed within this time). At the end of each binding cycle, 10 μL 15 mM H_3PO_4 was injected to regenerate the surface and then 20 μL buffer was injected to wash the regeneration solution from the system.

Data processing and analysis. Data were double-referenced and globally fit to a 1:1 interaction model that included a mass transport parameter. The responses generated from both phases of the analysis were constrained to a single k_a and k_d and, because the same antibody surface was used in both phases, fitting was also constrained to a global [B]. The fit of the model and the corresponding kinetic constants are shown in Figure 4–20. Similar kinetic rate constants were obtained for this interaction using a KinExA flow spectrofluorimeter.[25]

REFERENCES

1. S. Huang, B. Sjoblom, A. E. Sauer-Erkisson, B.-H. Jonsson, *Biochemistry* **41**, 7628 (2002).
2. Y. S. Day, C. L. Baird, R. L. Rich, D. G. Myszka, *Protein Sci* **11**, 1017 (2002).
3. S. K. Nair, J. F. Krebs, D. W. Christianson, C. A. Fierke, *Biochemistry* **34**, 3981 (1995).
4. D. G. Myszka, *Curr Opin Biotechnol* **8**, 50 (1997).
5. G. Xu *et al.*, *J Biol Chem* **278**, 5455 (2003).
6. P. S. Katsamba, M. Bayramyan, I. S. Haworth, D. G. Myszka, I. A. Laird-Offringa, *J Biol Chem* **277**, 33267 (2002).
7. S. Walsh, L. Jevitts, J. Sylvester, A. Kossiakoff, *Protein Sci* **12**, 1960 (2003).
8. R. L. Rich *et al.*, *Proc Natl Acad Sci* USA **99**, 8562 (2002).
9. G. A. Canziani, S. Klakamp, D. G. Myszka, *Anal Biochem* **325**, 301 (2004).
10. M. J. Cannon, D. G. Myszka, *Methods for Structural Analysis of Protein Pharmaceuticals.* W. Jiskoot and D. Crommelin, Eds., AAPS Press (2005). pp. 527–544.
11. P.-O. Markgren *et al.*, *Anal Biochem* **291**, 207 (2001).
12. P.-O. Markgren *et al.*, *J Med Chem* **45**, 5430 (2002).
13. T. Ito *et al.*, *Proc Natl Acad Sci* USA **98**, 4569 (2001).
14. R. L. Rich, D. G. Myszka, *J Mol Recognit* **15**, 352 (2002).
15. R. L. Rich, D. G. Myszka, *J Mol Recognit* **16**, 351 (2003).
16. D. G. Myszka, *J Mol Recognit* **12**, 279 (1999).
17. M. J. Cannon, D. G. Myszka, *Recent Res Devel Biophys Biochem* **3**, 333 (2003).
18. D. J. O'Shannessy, M. Brigham-Burke, K. K. Soneson, P. Hensley, I. Brooks, *Anal Biochem* **212**, 457 (1993).
19. T. A. Morton, D. G. Myszka, I. M. Chaiken, *Anal Biochem* **277**, 176 (1995).
20. D. G. Myszka, T. A. Morton, M. L. Doyle, I. M. Chaiken, *Biophys Chem* **64**, 127 (1997).
21. D. G. Myszka, T. A. Morton, *Trends Biochem Sci* **23**, 149 (1998).

22. W. Press, S. Teukolshy, W. Vetterling, B. Flannery, Numerical Recipes In C *Cambridge University Press* (1992).

23. D. G. Myszka, *Anal Biochem* **329**, 316 (2004).

24. M. J. Cannon *et al.*, *Anal Biochem* **330**, 98 (2004).

25. A. W. Drake, D. G. Myszka, S. Klakamp, *Anal Biochem* **328**, 25 (2004).

26. W. Zhang, A. P. Godillot, R. Wyatt, J. Sodroski, I. Chaiken, *Biochemistry* **40**, 1662 (2001).

27. D. G. Myszka, S. J. Wood, A. L. Biere, *Methods Enzymol* **309**, 386 (1999).

5 Sensor surfaces and receptor deposition

Matthew A. Cooper

INTRODUCTION	111
WHAT CHEMISTRY TO USE?	112
When to use amine coupling	113
When to use thiol coupling	113
When to use aldehyde coupling	114
When to use capture reagents	115
Covalent coupling chemistry summary	116
Noncovalent coupling chemistries	117
Biotin	117
Epitope tags	117
NTA-nickel	118
Heterodimers	119
SENSOR SURFACES	119
Planar surfaces	119
Biotin SAMs	119
Hydrophobic and amphipathic SAMs	119
Carboxylic acid SAMs	120
Dendrimers	120
Polymer surfaces	120
Dextran	121
Polyethyleneimine	125
Hyaluronic acid	125
Graft polymers	126
Gels	126
Molecularly imprinted polymers	126
WORKED EXAMPLES	127
Amine coupling	127
Thiol coupling	129
Aldehyde coupling.	136
Photo-activated coupling	137
Noncovalent capture	137
Cross-linkage	138
STOCK SOLUTIONS	139
EDC and NHS	139
REFERENCES	140

INTRODUCTION

The interface between a sensor surface and the chemical or biological systems to be studied is a vital component of all sensor systems, including label-free biosensors. With the exception of solution-phase systems such as calorimetry or analytical ultracentrifugation, receptors must be attached to some form of solid support to transduce a binding event to the sensor. During this process, receptors must retain their native conformation and binding activity, attachment to the sensor must be stable over the course of the assay, and binding sites must be presented to the solution phase to interact with the analyte to generate a detectable signal. Most importantly, the support must be resistant to nonspecific binding of analyte and other sample components that could mask a specific binding signal.

Many coupling strategies utilize a bespoke chemical linker layer between the sensor and the biological component to achieve these ends. Functionalized alkane thiols[1] and alkoxy silanes,[2] which form stable monolayers on planar surfaces act as ideal linkers. The alkyl termini of these molecules can be derivatized with ethyleneglycol subunits to produce a protein-resistant planar surface[3] or can be mixed with molecules that possess suitable reactivity for receptor capture, for example, −epoxy, −carboxyl, −amino, −biotinyl, −nitrilotriacetic acid.[3-5] The larger binding partner (e.g., a protein target) is normally immobilized on the surface, and the smaller binding partner (e.g., a drug candidate) is allowed to bind to this surface from free solution. However, in some cases drug-like molecules have been attached directly to the chemical linker layer and receptors passed over the surface.[6,7] Of note, a biotechnology company (Graffinity GmbH) has pushed this approach still further to challenge the dogma that small molecules must always be screening from free solution. In this case, chemical compounds from a library fitted with a molecular tag are captured onto a gold surface coated with a protein-resistant linker layer. The resultant chemical microarray is then exposed to the target protein and the extent of binding to each spot in the array is determined simultaneously using surface plasmon microscopy imaging.

A short chemical linker layer can also be used as a substrate for attachment of a polymer coat or hydrogel that renders the surface highly resistant to nonspecific adsorption of proteins, nucleotides, and drugs. The polymer also provides a three-dimensional scaffold for receptor immobilization. The most widely employed biosensor polymer coat is carboxymethyldextran,[8] although other materials that produce a protein-resistant hydrogel can also be used (e.g., hyaluronic acid, polyvinyl alcohol, nitrocellulose, sepharose). There are many strategies for covalent or noncovalent attachment of receptors to either planar self-assembled surfaces or polymer coats. Selection of the correct coupling chemistry requires careful consideration of (a) the resultant orientation of receptor, (b) its local environment on the surface, (c) the stability of the linkage under the conditions used to regenerate the surface, and (d) possible effects of the coupling chemistry on components of the binding interaction. This chapter will cover the most commonly employed sensor surfaces and coupling chemistries. Cells, cell fragments, and integral and peripheral membrane proteins present their own unique challenges and are dealt with separately in the Membrane Chapter in this volume.

WHAT CHEMISTRY TO USE?

The choice of immobilization chemistry is of central importance to the design of a successful biosensor assay.[9] The coupling method must be efficient, must produce a highly stable association (to prevent signal drift), and must allow control of the amount of receptor immobilized. Ideally the surface should:

- adhere irreversibly to the biosensor transducer substrate and not interfere with transduction of the sensor-binding event to the readout interface
- be compatible with standard fabrication techniques for volume production of sensors
- be resistant to degradation from the process of fabrication to the final use in an assay
- allow maximal retained activity of a receptor
- allow maximal transduction of binding event signal through to the biosensor transducer
- be highly resistant to nonspecific binding (especially with complex samples such as serum)
- be insensitive to bulk matrix effects (temperature, pressure, pH, ionic strength solvent changes)
- allow suitable attachment chemistry for as many receptor classes as possible
- not leach away during conditions of operation (typically flow of a biological buffer, buffer/dimethyl sulfoxide (DMSO); biological media or clinical sample)
- allow reproducible levels of receptor coupling coefficients of variation for protein coupling levels of less than 10%)

In addition to the above considerations, it is important to consider what effect the surface and method of coupling has on receptor orientation and homogeneity.[10] For example, amine coupling (e.g., to surface –Lys and –Arg residues on a protein) will lead to a heterogeneous population of receptors with random orientation on the surface, whereas affinity capture or sulfhydryl couplings can be used to produce a more homogenous population of oriented receptors on the surface.[11–13] Both heterogeneity and the decreased binding activities that accompany immobilization can be detrimental to robust biosensor analysis and can also limit the sensitivity and reproducibility when using surface-immobilized receptor films. It has also been suggested that for sensor-based diagnostic devices, random antibody spatial orientation limits device sensitivity, whilst other studies have reported that it is difficult to accurately predict biosensor detection kinetics using a heterogeneous immobilized antibody population. Heterogeneity can thereby

Table 5–1: Recommended coupling chemistries for different receptors and functional groups*

Biomolecules	Amine	Thiol	Aldehyde	Affinity capture
Neutral proteins	+	+	±	•
Acidic proteins	±	+	+	•
Basic proteins	+	±	±	•
Nucleic acids	−	±	−	•
Polysaccharides	−	−	±	•
Functional groups				
Peptides/proteins				
−NH$_2$	+	•	±	•
−SH	−	+	−	•
−COOH	−	•	−	•
−CHO	−	−	•	•
Polysaccharides				
−CHO	−	−	+	•
−COOH	−	•	•	•

* (+) recommended; (±) acceptable; (−) unsuitable; (•) requires additional modification.

complicate the design of biosensor devices as the kinetic and thermodynamic parameters that describe a molecular recognition event can be characterized by an average of a distribution of values rather than discrete constants.

When to use amine coupling

In general, it is often convenient to determine approximate specificity or affinity for a molecular interaction using simple amine coupling chemistry. This is by far the most widely adopted approach for protein analysis using commonly available biosensor systems. The coupling procedure is simple and straightforward, and results obtained are normally sufficient for a first-pass analysis of a molecular recognition event. Amine coupling is not suited to small peptides, small proteins, and when using negatively charged sensor surfaces such as carboxymethyldextran, as these receptors may possess very few or no basic residues. In addition, it is often found that conditions that favor effective preconcentration of receptor in the dextran matrix (i.e., low pH) are not conducive for efficient carbodiimide coupling (i.e., neutral or slightly basic pH) (Table 5–1).

When to use thiol coupling

Thiol coupling provides an alternative to amine coupling for a wide range of receptors. The procedure can use either intrinsic thiol groups in the receptor or reactive groups introduced by modification of carboxyl- or amine-containing residues. The choice of several functional groups for thiol coupling gives greater flexibility in designing immobilization conditions. Common receptors for which the thiol reaction are valuable include:

- acidic proteins, where the low pH required for electrostatic preconcentration in the surface matrix seriously impairs the chemical efficiency of coupling via amino groups
- peptides and other small receptors where amino groups are lacking or where the number of amino groups is limited and they may be required for biological function
- receptors for which a specific coupling site is important
- other biomolecules where amine coupling is unsatisfactory

Disulfide-coupled receptors cannot be used together with strong reducing agents such as dithiothreitol (DTT) or mercaptoethanol, nor with strongly (> pH 9) basic reagents because the disulphide bridge linking the receptor to the surface is unstable under such conditions. If reducing conditions are needed for the interaction being studied, or if basic conditions are needed for sensor regeneration, the stability of thiol-immobilized receptors should be checked first.

There are several variants of thiol-based coupling approaches available. Each relies on the presence of a thiol group, either specifically engineered into the protein or sensor surface or naturally present in the protein to be immobilized. Amine groups or carboxyl groups may be modified to allow coupling, a feature that may be of particular use in cases where amine coupling is unsuccessful. Thiol coupling has been used to couple a wide range of biomolecules including acidic proteins and peptides. Using this coupling approach allows design of efficient site-specific immobilizations and in most cases produces homogeneously oriented surfaces. In addition, thiol coupling may generate greater coupling yields than those generated through amine coupling, because hydrolysis of N-hydroxysuccinaminyl (NHS) groups is not as prevalent. Because these immobilizations are normally more reactive than amine coupling methods, creation of lower-density surfaces requires greater care taken in assay design, and lower receptor concentrations and shorter protein contact times are useful parameters to vary in this instance.

When to use aldehyde coupling

Aldehyde coupling is most often used for acidic proteins that cannot be successfully coupled using carbodiimide chemistry or with carbohydrate-containing proteins in which binding activity is better retained via linkage to the saccharide component of the receptor.[12-14] For example, in a study of carcinoembryonic antigen (CEA), periodate-mediated aldehyde coupling was more efficient than amine coupling methods due to the low isoelectric point (pI) of CEA. High-density CEA surfaces were used to observe mass transport limited binding of anti-CEA antibody fragments in a concentration quality control assay. These high-density surfaces also permitted screening of unpurified samples such as bacterial cell culture supernatants for CEA-specific molecules.

When to use capture reagents

By far the most widely adopted approach to circumvent the limitations of covalent coupling as outlined above is the capture of a receptor via an intermediate binding partner. This process involves attaching the receptor molecule indirectly to a surface through binding to another immobilized molecule. The covalently coupled protein in this case is referred to as the "capture molecule." This approach has several advantages including:

- generation of orientation-specific surfaces
- higher receptor homogeneity
- epitope tags in the receptor molecule can enable specific questions about analyte specificity or binding mode to be addressed
- purification requirements of the receptor molecule are minimized because the receptor can be "pulled down" *in situ* from impure mixtures
- avoidance of modification of the receptor molecule through direct chemical immobilization
- achieving consistent presentation of the receptor for interaction with the analyte over multiple binding cycles through use of appropriate regeneration conditions

Several factors are involved in designing an apposite capture surface:

- The covalently immobilized capture molecule should be as pure as possible (e.g., more than 95% affinity chromatography purified).
- It should be specific for the intended receptor molecule.
- It should not bind to other sample constituents (e.g., do not use a protein A as a capture molecule in an assay where significant amounts of nonspecific immunoglobulin may be present).
- It should bind the receptor molecule with a low off rate ($k_{diss} < 10^{-5}$ s^{-1} preferred).
- The capacity to bind receptor must be high enough to allow studies over a range of conditions (e.g., varying analyte or varying analyte binding affinity).
- It should not significantly alter the activity of the "capture molecule" and should not be altered significantly following regeneration procedures.

Given the above, some of the most versatile capture molecules are antibodies. Immunoglobulins such as rabbit antimouse Fcγ are excellent capturing agents and normally perform well in biosensor assays. These capture molecules can form stable surfaces, which are conveniently created through standard amine coupling procedures. Regeneration conditions required for these surfaces typically involve milder acidic solutions. There is also an extensive range of suitable capture antibodies currently available commercially.

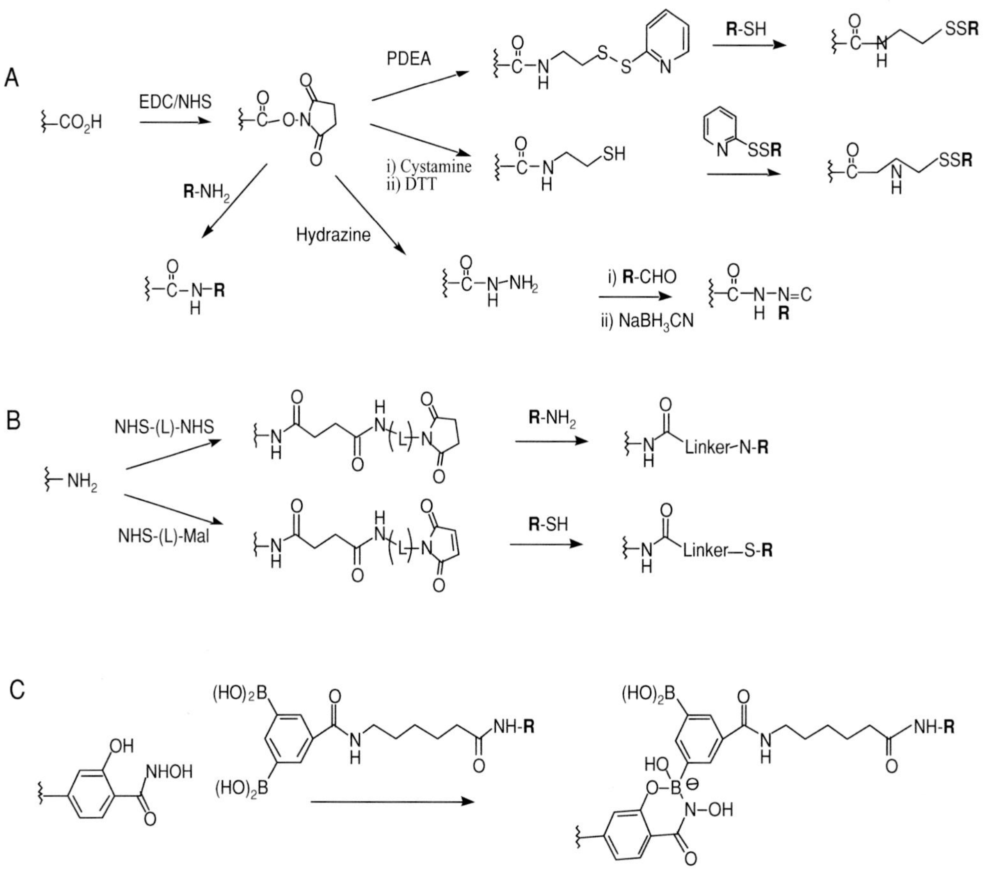

Figure 5–1: Commonly employed coupling methods for receptor immobilization. R = Receptor, (L) = Linker, Mal = maleimide, other abbreviations defined in the main text.

Covalent coupling chemistry summary

Water-soluble carbodiimides such as N-ethyl-N′-(dimethyl-aminopropyl)-carbodiimide (EDC) in combination with N-hydroxysuccinimide (NHS) are most often used for activation of carboxymethylated supports such as dextran or hyaluronic acid. The resultant reactive NHS ester can then be coupled directly with available amino moieties of a receptor to form a stable amide linkage (Figure 5–1). Acidic receptors (pI < 3.5) are difficult to immobilize by amine coupling because the low pH required for electrostatic preconcentration to the sensor surface protonates the primary amino groups and reduces coupling efficiency. Further derivitization with sulfydryl-reactive reagents (e.g., pyridinyldithioethanamine [PDEA] or 3-[2-pyridinyldithio] propioic acid N-hydroxysuccinimide ester [SPDP]) allows reaction with free surface thiols (e.g., Cys, Met) to form a reversible disulfide linkage.[11–13] Similarly, stable thioether bonds may be formed using coupling reagents such as sulfosuccinimidyl-4-(N-maleimidomethyl) cyclohexanecarboxylate (Sulfo-SMCC) and N-(γ-maleimidobutyrloxy)sulfosuccinimide ester (GMBS).[15,16] The surface may also

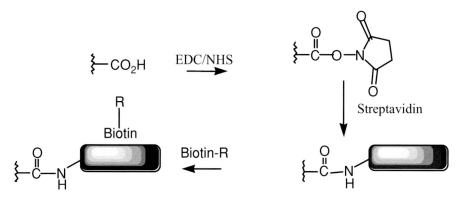

Figure 5–2: Capture of biotinylated ligands via immobilized streptavidin.

be derivatized with cystamine to effect coupling with disulfide-activated receptor. Finally, treatment with hydrazine followed by reductive amination enables coupling with aldehydes. The aldehyde groups may be native to the receptor or formed by mild oxidation of any *cis*-diols present.[14]

Amino-presenting surfaces[5,17] can be treated with commercially available bifunctional linking reagents to effect coupling with free amino or sulfhydryl groups on the receptor. Surfaces derivatized with salicylhydroxamic acid (SHA) can be used to produce reversible complexes with receptors activated with phenyldiboronic acid (PDBA).[18]

Noncovalent coupling chemistries

Biotin

Immobilization of a receptor molecule can be readily achieved through use of neutravidin- or streptavidin-presenting surfaces specific for biotinylated receptors. The multiple biotin binding sites of streptavidin on each side of the molecule allow biotinylated receptors to be cross-linked by the streptavidin "double adaptor" (Figure 5–2, 5–3). Streptavidin is commonly employed to immobilize 3′ or 5′-biotinylated oligonucleotides,[39–42] antibodies,[11,56] and biotinylated vesicles.[43,44]

This highly efficient method leads to very stable complexes but is effectively irreversible, thus rendering receptor exchange impossible. In these instances an alternative approach is to immobilize an antibiotin antibody to capture biotinylated receptor. This surface can be regenerated through use of mildly acidic or medium basic solutions.

Epitope tags

Epitope tags are now widely employed in both bacterial and eukaryotic expression systems to allow efficient protein purification and subsequent study. Commonly employed tags include glutathione S-transferase (GST), QPELAPEDPED (HSV), DYKDDDDK (FLAG), 6 × His (Figure 5–3), 10 × His, green fluorescent protein (GFP) and c-myc. Many commercially available proteins available with

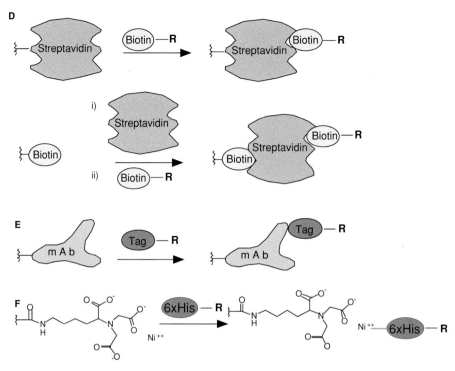

Figure 5–3: Capture of receptors via selected noncovalent affinity interactions.

these tags can be readily captured onto a biosensor surface with appropriate antibodies.[9,45,46] For example, FLAG fusion proteins (Sigma-Aldrich) can be captured through creation of an anti-FLAG antibody surface through standard amine coupling procedures of the anti-FLAG antibody or through biotinylation of the anti-FLAG antibody and subsequent capture on a streptavidin surface.[47] We typically achieve regeneration of these surfaces via use of mild acidic conditions. Note that care should be taken when using GST-fusions in biosensor analysis due to the propensity for the GST moiety to dimerise.

NTA-nickel

Metal coordinating groups such as iminodiacetic acid (IDA) and nitrilotri-acetic acid (NTA) have been widely employed for direct immobilization of $6 \times$ His and $10 \times$ His-tagged receptors.[4,48–50] His-tags are commonly used in immuno-affinity chromatography and metal–ion affinity chromatography. The His-tag generally comprises six to ten histidine residues and may be conveniently captured through use of a nitrilotriacetic acid (NTA) or an anti-hexahistidine antibody-coated sensor surface (Figure 5–3). However, the moderate affinity of the chelate-Ni^{2+}-histidine ternary interaction means there is sometimes considerable decay in the level of immobilized receptor. For this reason anti-$6 \times$ His mAbs are often employed to effect stable, oriented immobilization of His-tagged receptors[51]. In instances where EDTA is a necessary component of the buffer systems employed, capture of His-tagged proteins via an anti-His antibody immobilized on the sensor surface should be utilized rather than NTA. Regeneration of the anti-His antibody surface is generally achieved by way of mild to medium acidic solutions.[51]

Heterodimers

Heterodimeric coiled-coil systems offer a novel approach to creating capture surfaces. The heterodimeric E/K coiled-coil peptides have been used successfully as an alternate noncovalent capture system.[52] This strategy involves immobilizing one of the peptides (e.g., K peptide) followed by injection of the other peptide (e.g., protein E), which had been previously conjugated with the receptor molecule of interest. Rapid association rates and slow dissociation rates of the conjugated peptides render this system a suitable capturing regime for tagged proteins and antibodies in particular.

SENSOR SURFACES

Planar surfaces

Biotin SAMs

The use of biotinylated alkane thiol SAMs on gold was first exploited by Knoll and coworkers[53–55] to create streptavidin and antibiotin antibody layers on gold surfaces. Surface plasmon resonance (SPR) was used to directly and sensitively detect the formation of protein monolayers and in some cases multilayers.[57] More recently, the kinetics of adsorption and competitive desorption of wild-type and mutant streptavidins to biotinylated SAMs have been investigated by SPR.[58] Here an ethylene/glycol terminating thiol was used to create a protein-resistant mixed biotin SAM that specifically bound only to streptavidin. The composition, orientation, order, and thickness of these mixed SAMs was subsequently investigated using a variety of surface physical techniques.

Hydrophobic and amphipathic SAMs

These surfaces for capture of lipids and membrane receptors are dealt with in great detail in Chapter 7 in this volume. Briefly, surfaces composed of long-chain alkanethiol molecules form a flat, quasi-crystalline hydrophobic layer. They are designed to facilitate liposome-mediated hydrophobic adsorption of a user-defined polar lipid monolayer. In addition to the components essential for the formation of the lipid layer, other membrane-bound molecules can be incorporated. These molecules remain embedded in the lipid monolayer after the liposome-mediated hydrophobic adsorption process is completed and impart a particular binding specificity to the surface. A complete coverage of the surface with lipids or membranes helps reduce nonspecific binding of proteins and provides a surface suitable for interaction studies performed in an aqueous, detergent-free environment.

Carboxylic acid SAMs

Planar carboxylic acid-presenting monolayers possess lower binding capacities than polymer-based surfaces but are often employed to study interactions in which it is desirable to immobilize extremely large molecules, molecular complexes, or particles (e.g., phage, viruses, bacteria, and cells) as close as possible to the surface. In addition, this approach can help maintain sensitivity in evanescent-based detection systems in which field strength decays exponentially away from the system and can also reduce avidity effects with multivalent interaction partners compared to higher-capacity, more flexible polymer supports that assist in cross-linking of analytes by multiple receptors. Finally, such surfaces are also useful for interactions that may be affected by the presence of polysaccharides such as dextran.

Dendrimers

Dendrimers are large, branching molecules formed by repeated addition of serious functional groups to a specified core. The most common are based on polyamidoamine (PAMAM) synthesized from an ethylenediamine core and are available in different sizes (generations of growth layers) with a variety of amino, carboxylate, and hydroxyl surface groups. Smaller (generation 0.5–2) dendrimers have open, flat structures, and larger (generation >4) dendrimers are spheroidal, mimicking a globular protein in shape and dimension. As such, they have found utility in immunodiagnostic and gene expression profiling applications.[59,60] Dendrimers can be simply physisorbed to base materials[61] or more often chemically coupled[62] to self-assembled monolayers on a metal surface (Figure 5–4), silanated glass, quartz, or silica.

Of note is a detailed study of dendrimer-activated solid supports for nucleic acid and protein microarrays based on aminosilylated glass.[63] Here the surface amino groups were activated with a homobifunctional linker, disuccinimidylglutarate (DSG) or 1,4-phenylenediisothiocyanate (PDITC), and then allowed to react with a generation 4 amino dendrimer. Subsequently, the dendritic monomer layer was activated and cross-linked with a homobifunctional spacer, either DSG or PDITC. This led to the formation of a thin, chemically reactive polymer film, covalently affixed to the glass substrate, which was used for the covalent attachment of proteins or amino-modified oligonucleotides. It was claimed that the resulting surfaces displayed a surface capacity twofold greater than that found with conventional microarrays containing linear chemical linkers. In addition, the surfaces were found to be resistant to repeated alkaline regeneration procedures.

Polymer surfaces

Soft biocompatible interfaces on solid surfaces consisting of ultrathin soft polymer films have found widespread utility in label-free biosensors. They have been developed not only for label-free biosensor use but also for microarrays, chromatography, protein purification, diagnostic devices, and microtitre plates. Most

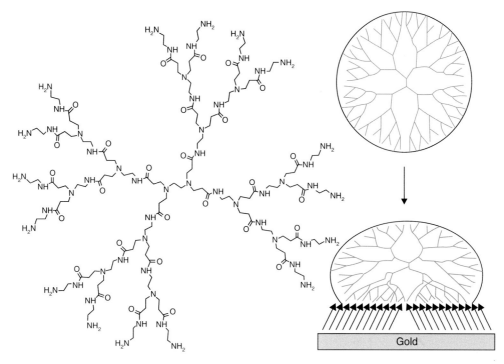

Figure 5–4: Immobilized amino dendrimer (*left*) on a gold surface decorated with a self-assembled monolayer (*right*).

importantly, polymeric interfaces can provide multiple functions essential for biosensor assays: (1) they serve to passivate or mask the inherent chemical properties of the underlying sensor material (e.g., plastic, glass, silica, quartz, metal, or metal oxide); (2) they can provide a three-dimensional scaffold that can increase the receptor-binding capacity, and hence the sensitivity of the biosensor; (3) they are often conformationally mobile, and hence are often termed "self-healing," because point defects in the underlying support are readily masked by the overlying mobile brush structure; (4) they can be readily functionalized with a wide variety of pendant chemical groups for coupling to moieties often found on the surface of biological molecules; (Table 5–2) and (5) they are readily fabricated by common manufacturing techniques (e.g., spin coating, photolithography, dip coating, immersion) that can be scaled to high volumes of product. In general polymers are characterized by their monomer identity, chain linearity, chain length, and monomer arrangement (in copolymers). This is an extremely wide field, and only the most commonly employed structures in biosensor detection are reviewed here.

Dextran

Dextran is a complex, branched polysaccharide made of glucose molecules joined into chains of varying length (from 5 to 150 kDa) often used as an antithrombotic (antiplatelet) and for the reduction of blood viscosity. By far the most widely used and flexible surface chemistries today are based on brush polymers,

Table 5–2. Selected chemistries for covalent immobilization of soluble receptors on polymeric supports

Activation reagent	Matrix	Functional group	Activation conditions	Activated structure	Receptor	Coupling pH range	Ref.
CNBr (Cyanogen bromide)	Polyol (esp. polysacch.)	–OH	aq./buffer		-NH^2	7–8.5	(19–21) (19, 20, 22)
CDAP (1-Cyano-[4-dimethylamino]pyridinium tetrafluoro borate)	Polyol (esp. polysacch.)	–OH	aq./organic		–NH^2	7–8.5	(19, 20)
DSC (Disuccinimidyl carbonate)	Polyol (esp. polysacch.)	–OH	Organic		-NH^2	6–8	(23)
CDI (Carbonyldiimidazole)	Polyol (esp. agarose)	–OH	Organic	Various	–NH^2	8–10	(24)
Tosyl chloride	Polyol (esp. agarose)	–OH	Organic		–NH^2 -SH	9–10	(25)
Tresyl chloride	Polyol (esp. agarose)	–OH	Organic		–NH^2 -SH	8–9	(25)

122

Method	Support	Matrix group	Conditions	Structure	Reactive groups	pH	Ref.
Bisoxiranes	Polyol	−OH	aq. pH 13–14 aq. pH 8–10	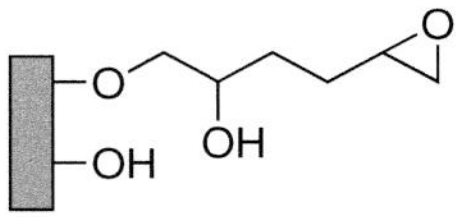	−OH −NH₂, SH	11.5–13 8–11	(26)
Epichlorohydrin	Polyol	−OH	aq. pH 13–14	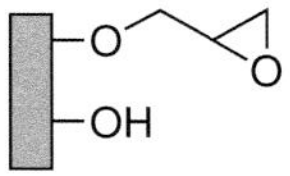	−OH, −NH₂, SH	11.5–13 8–11	(27, 28)
Divinylsulfone	Polyol	−OH	aq. pH 13–14	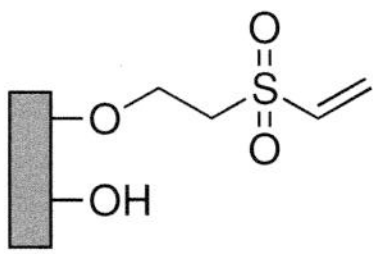	−OH, −NH₂, SH	10.5–12 8–11	(29)
Carbodimides	Polyol	−COOH	aq.	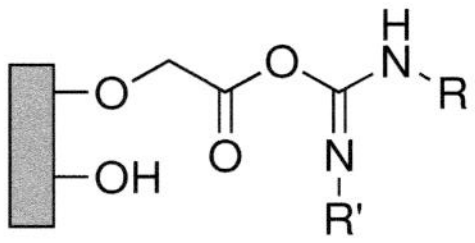	−NH₂	5	(30–32)
NHS/EDC	Polyol	−COOH	aq.	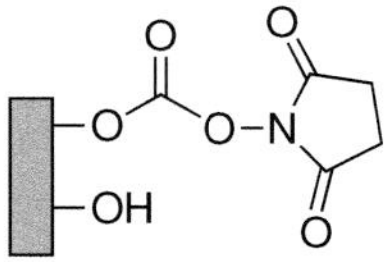	−NH₂	5–9	(33)
Matrix thiol	Polyol	−SH	aq.		−CH=CH−, =CO, −CNH	8–10	(30)

(continued)

123

Table 5-2 *(Continued)*

Thiol–disulfide exchange	Polyol	–SH	aq.	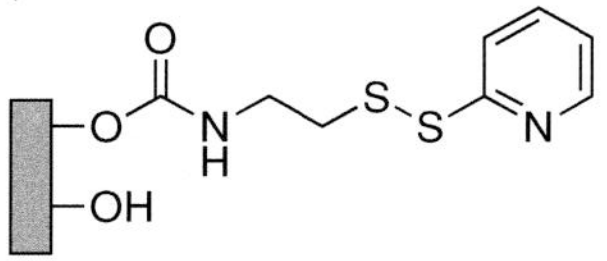	–SH	2–9	(34)
Glutaraldehyde	Polyamide	–CONH$_2$	aq.	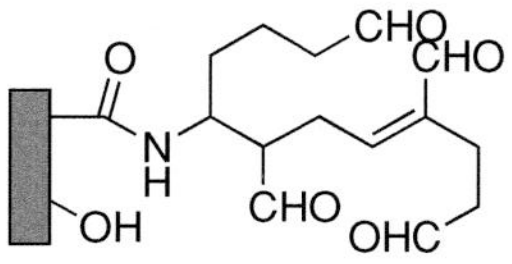	–NH$_2$	7	(35)
Hydrazine	Polyamide	–CONH$_2$	NaNO$_2$/HCl		–CHO, –CO	7–9	(36)
Silyl oxirane	Silica	–SiOH		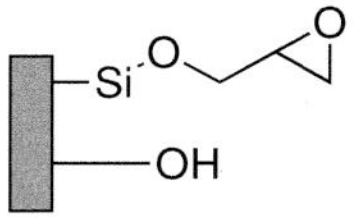	–NH^2, –SH	8	(37)
Isocyanide	Various	–COOH, –NH$_2$, =CO, –NC	aq. pH 6.5	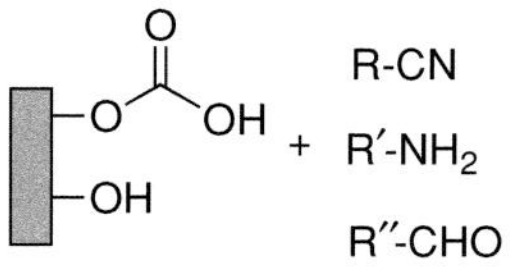	–NH^2,-COOH, -COH, =CO	6.5	(38)

124

principally due to the commercialization of carboxymethyldextran by Biacore/GE Healthcare in their SPR biosensor systems. These polymers are usually prepared via chemical coupling to underlying hydroxy- or carboxy-terminating SAMs via treatment with basic epichlorohydrin or hydrazine, respectively, then coupling to a basic dextran solution, and finally derivitization with basic bromoacetic acid to form carboxymethyldextran.[8] Carboxymethyldextran sensor surfaces with varying degrees of charge and polymer length are commercially available. For example, surfaces with lower negative charge density (such as the Biacore CM4 sensor chip) can be useful when using basic proteins or highly charged molecules such as nucleic acids. Surfaces with a short dextran brush length (such as the Biacore CM3 sensor chip) allow an interaction to take place closer to the sensor surface, which can improve sensitivity when working with evanescent detection instrumentation and large molecules, molecular complexes, viruses, or whole cells.

In addition to carboxymethyldextran, mixtures of dextran, aldehyde dextran, and aldehyde dextran sulfonate have been prepared to control contributions from covalent and electrostatic association of receptor.[64] Others have produced preactivated dextran surfaces that can be used for direct thiol coupling to sulfhydryl-containing receptors, obviating the need for complex preparation of thiol-reactive intermediates.[65] Such an approach is useful for proteins such as human serum albumin, which possesses a single cysteine for formation of a highly homogeneous population of oriented receptor and in which amine coupling is inefficient due to the low pI of the protein. Note that there have been numerous studies on the effect of carboxymethyldextran on interaction affinities and kinetics, particularly with regard to mass transport limitation of binding.[66,67] These issues are dealt with in detail in the kinetics and experimental design chapters in this volume.

Polyethyleneimine

As an alternative to negatively charged carboxymethyldextran, it is possible to couple proteins and other receptors to polyethyleneimine, normally via intermediate cross-linking to Schiff bases formed via treatment with glutaraldehyde. This approach has been used to couple protein A for antibodies specific to atrazine,[68] bacteria and bacterial enterotoxins,[69] complement C4,[70] alpha-fetoprotein,[71] hepatitis B virus,[72] and many others.

Hyaluronic acid

Hyaluronic acid, or hyaluronan, is a nonsulfated glycosaminoglycan $(-4GlcUA\beta1-3GlcNAc\beta1)_n$ distributed widely throughout connective, epithelial, and neural tissues. It is one of the chief components of the extracellular matrix, contributes significantly to cell proliferation and migration, and may also be involved in the progression of some malignant tumors. The mechanical and chemical properties of hyaluronic acid and mixed polymer layers for use in biocompatible implantation have been studied by both SPR and quartz crystal microbalance (QCM).[73,74] As the material is of human origin and nonantigenic, there is interest in the use of this material in biocompatible coatings for implantable biosensors, such as glucose sensors.[75]

Graft polymers

Graft copolymer systems based on polylysine have long been used for immobilization of proteins[76] and have been well characterized on a variety of negatively charged surfaces (mainly metal oxides such as Ta_2O_5, Nb_2O_5, TiO_2, SiO_2). Most are based on a composite layer of polylysine that supports a derivatized polyethylene glycol layer.[76–78] Other groups have synthesized block terpolymer with disulfide, PEG-OMe, and PEG-NHS side chains on a gold substrate,[79] poly(vinyl)alcohol and poly(acrylic acid) graft polymers for immunosensing applications.[80] These surfaces generally possess low nonspecific binding characteristics but have only one terminal reactive group and hence a lower receptor binding capacity than polymers such as carboxymethyldextran.

Gels

A few select papers utilize cross-linked matrices such as cellulose, polyacrylamide, and agarose to immobilize receptors for use with label-free biosensing. For instance, disulfide-modified polyacrylamides have been coated on gold and silver surfaces as cushions for polymer-supported lipid bilayers to provide a soft, deformable layer that will allow for transmembrane protein insertion and mobility.[81] Soluble enzymes such as hexokinase have been physically immobilized by inclusion in polyacrylamide on a piezoelectric quartz crystal microbalance.[82] Using a dual-channel active and reference channel QCM system with polyacrylimide-entrapped bovine serum albumin (BSA) or hexokinase, a large specific shift for glucose binding hexokinase could be observed. Similarly, glucose binding was also detected using acrylamidophenylboronic acid–acrylamide hydrogels.[83]

Agarose has also been extensively used for the immobilization and stabilization of proteins on solid supports, principally for immunofluorescence assays in diagnostic antibody arrays. In a detailed comparison of planar (3-mercaptopropyltriethoxysilane, 3-glycidyloxypropyltrimethoxysilane, 3-aminopropyltrimethoxysilane, octadecyltrichlorosilane, and 2-methoxy[polyethylenoxy]propyl] trimethoxysilane) with polymer thin films (polystyrene, polyimide, poly[dimethylsiloxane] and agarose), agarose was found to be far superior in resisting protein binding of Cy5-labeled bovine serum albumin, fibrinogen, and lysozyme.[84] Finally, there have been a number of studies carried out with biosensors employing polysiloxanes,[85] polystyrene,[86,87] poly(ethyleneglycol) diacrylate,[88] and polymethylmethacrylate.[89,90] However, these polymers tend to form impenetrable or low-capacity sol–gel layers and often possess poor protein resistance properties.

Molecularly imprinted polymers

Biomimetic recognition elements employed in biosensors for detection of analytes are normally based on proteinaceous affibodies, immunoglobulins, single-chain or single-domain antibody fragments, nucleic acids, or aptamers. An alternative supramolecular approach is to employ a molecularly imprinted polymer (MIP), in which the polymer scaffold itself is the molecular recognition element. Here the imprint of a template molecule is formed on a synthetic polymer that has cavities resembling the geometric shape of the template and binding sites for template recognition.[91] MIPs as synthetic receptors have several advantages over biological receptors; principally their stability under challenging conditions, in contrast to natural biomolecules, which are sensitive to environmental conditions and can denature readily.[92] Template molecules can be imprinted to a polymer with covalent, noncovalent, or ion-mediated interactions, followed by appropriate cross-linking agents.[93] Imprinting consists of three steps: (1) functional monomers are mixed with the template molecule in a solvent, (2) polymerization of monomers occurs in the presence of a cross-linker, and (3) the template is removed from the polymer using basic, acidic, or detergent solutions. The selectivity of MIPs toward a target molecule depends on many factors including the molecular diversity of the monomer units employed, the geometry of the imprinted cavity, the rigidity of the cavity, and associated implications for enthalpy/entropy compensation.[92] Most applications of MIPs reported to date involve the use of acoustic (QCM) sensors for small-molecule detection; these are reviewed extensively in reference[94]. However, there are also examples in which protein[95] or viruses[96,97] have been specifically detected.

WORKED EXAMPLES

There are a large number of coupling chemistries that can be employed in many different label-free biosensor platforms. Note that the majority of examples shown below refer to flow rates and immobilization levels specific to the flow-based Biacore SPR systems commonly used at the time of writing. Nevertheless, much of the content below is readily transferable to other biosensor systems,[56] in particular flow-based systems that allow efficient delivery and removal of coupling reagents and receptors.

Amine coupling

Amine coupling is the most generally applicable coupling chemistry used in biosensor assays. Immobilization is usually facile and can be employed for a wide range of macromolecules. Amine coupling introduces N-hydroxysuccinimide esters into the surface matrix by modification of the carboxylic acid groups with a mixture of NHS and EDC. These reactive intermediates can then form covalent bonds with amine groups present in a receptor molecule (Figure 5–5). We present a general procedure for amine coupling below, together with a discussion of several factors that should be considered for a successful immobilization.

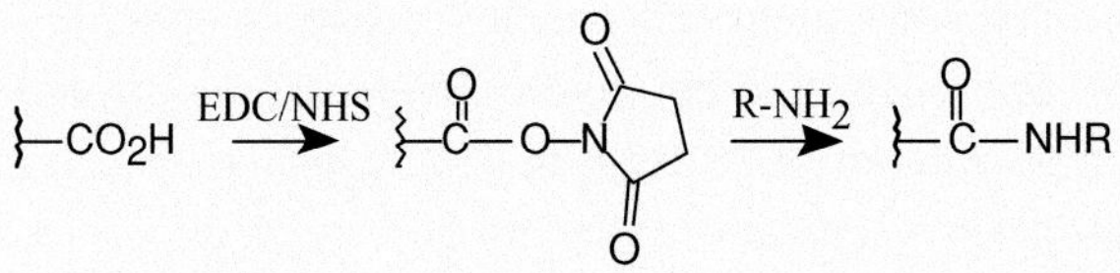

Figure 5–5: Capture of receptor via amine coupling to a reactive NHS-ester.

Preparation of coupling solutions:

Dissolve 115 mg NHS in 10 ml of filtered, deionized water.

Dissolve 750 mg EDC in 10 ml of filtered, deionized water.

It is recommended that each of these solutions, NHS, EDC, 10.5 ml 1 M ethanolamine hydrochloride-NaOH pH 8.5, respectively, be dispensed in 200 μL aliquots for storage (NHS and EDC store below –18° C; ethanolamine stores at +4–8° C).

Dilute receptor molecule to a suitable concentration, for example, 50 μg/ml in a low pH, low salt buffer (see notes below), for example, 10 mM sodium acetate, pH 5.0.

Prepare 1:1 mixture of EDC/NHS *in situ* for immediate use.

Preparation of sensor surface:

Running buffer = HBS, Flow rate = 5 μL/min

1. 35 μL NHS/EDC mixture
2. 35 μL receptor solution (step 3 above)
3. 35 μL 1 M ethanolamine pH 8.5

Several parameters in the above procedure can be modified to achieve optimal immobilization levels relevant to the specific experimental objective (see also the experimental design chapter). The following factors should be considered when using this type of coupling chemistry:

• Preconcentration of the receptor in the surface matrix is important for efficient immobilization of macromolecules. This is achieved by electrostatic attraction between negative charges on the surface matrix and positive charges on the macromolecule. Preconcentration is susceptible to the pH and ionic strength of the coupling buffer. The positively charged macromolecule is obtained by placing it in a buffer that is below its pI and has a low ionic strength (50 mM or less). In this manner, we can accomplish efficient immobilizations from relatively dilute solutions. One of the most common reasons for non-ideal preconcentration is overly high salt levels in receptor samples. Therefore, caution should be used when working with low receptor stock concentrations. Electrostatic preconcentration to charged carboxyl surfaces is less significant with low molecular weight receptors, which possess fewer ionizable groups and can also diffuse more rapidly to the sensor surface.

• Primary amine buffers such as Tris must be avoided, and carrier proteins such as BSA should be excluded from receptor preparations because those impurities having primary amine groups will be immobilized. Therefore, it is recommended that receptor molecules be affinity chromatography-purified to at least 95%.

- The protein contact time may also vary to adjust the immobilization level, particularly when trying to create low-density surfaces.
- The activation time (i.e., NHS/EDC exposure time) can be varied to increase or decrease immobilization levels.
- Adjusting the receptor concentration will also allow alterations in the immobilization level. However, the ideal receptor concentration will be the lowest value that gives maximal preconcentration, as there is no direct relationship between receptor concentration and amount immobilized.

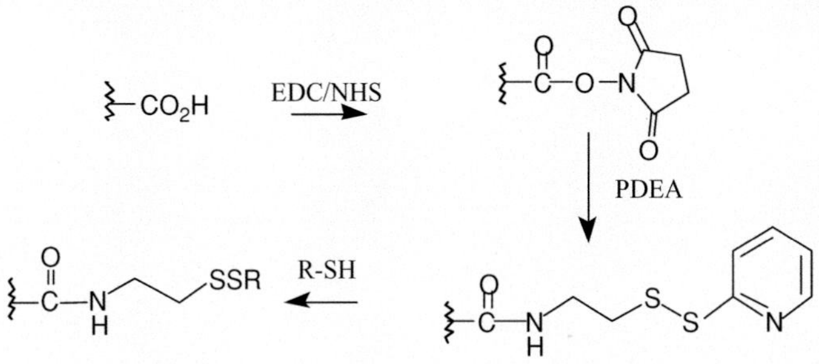

Figure 5–6: Capture of receptor via the receptor thiol method.

Thiol coupling

Three thiol-based approaches, (a) surface thiol, (b) receptor thiol, and (c) maleimide coupling, are outlined in the following section. Other reagents may be used to introduce reactive disulphide groups onto amino groups where this may be more appropriate, for example, SPDP (Amersham Pharmacia Biotech) and Sulfo-GMBS, SIA, SMPT (Pierce). The surface thiol method can also be used to immobilize receptors derivatized with other thiol-reactive reagents such as Sulfo-SIAB and Sulfo-SMCC (Pierce).

Receptor thiol. Receptor thiol coupling involves immobilizing receptor molecules through either intrinsic thiol groups or site-engineered cysteine moieties. This allows controlled coupling at a specific site in the receptor molecule. Thiol groups are coupled to reactive disulphide groups introduced onto the sensor surface (Figure 5–6, 5–7). This method may also be used for those receptors that have been modified as described above by PDEA, SPDP, Trauts reagent (2-iminothiolane) by reduction with dithioerythritol (DTE), followed by gel filtration.
Preparation of sensor surface:

Running buffer = HBS, Flow rate = 5 μL/min

1. 10 μL NHS/EDC mixture
2. 20 μL 80 mM PDEA (4.5 mg PDEA in 250 μl 0.1 M formate buffer, pH 8.5)

3. 35 μL receptor molecule (25–50 μg/ml) in low pH/low salt buffer
4. 20 μL 50 mM *l*-cysteine in 0.1 M sodium formate, pH 4.3 buffer, containing 1M NaCl, pH 4.3

Coupling will be optimized in both of these methods through optimal preconcentration conditions of the receptor. Care should be taken during regeneration to avoid use of basic solutions (pH 9.0) and reducing agents such as DTT, as these may result in loss of receptor from the surface. Cysteine and PDEA solutions should be used within 1 hr of preparation.

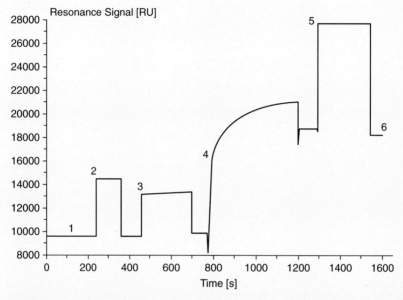

Figure 5–7: SPR sensorgram showing a typical immobilization sequence for the receptor thiol method. Staphylococcal protein A (SpA) derivatized with Traut's reagent is coupled to a PDEA-modified sensor chip. Numbers refer to the steps as follows:

1. Baseline signal for the unmodified sensor chip.
2. Injection of NHS/EDC to activate the sensor chip.
3. Injection of PDEA to derivatize the sensor chip.
4. Injection of 100 μg/ml Traut-SpA in 10 mM formate buffer pH 4.0.
5. Injection of cysteine/NaCl to deactivate excess reactive disulphides and remove noncovalently-bound protein.
6. Immobilized receptor corresponding to about 8600 RU.

Surface thiol. Surface thiol coupling involves modification of the receptor to include reactive disulphide groups (Figure 5–8, 5–9). Amine groups may be modified through use of *N*-succinimidyl 3-(2-pyridyldithio)-propionate (SPDP) or 4-succinimidyloxycarbonyl-methyla-[2-pyridyldithio]toluene (SMPT) and carboxyl groups by reaction with 2-(pyridinyldithio)ethane amine (PDEA). These reagents introduce pyridyldisulphide groups into the receptor, which are then coupled to thiol-derivatized sensor surfaces. Modification of carboxyl groups results in an increase in the isoelectronic point of the proteins, which is of additional benefit in the case of acidic proteins.

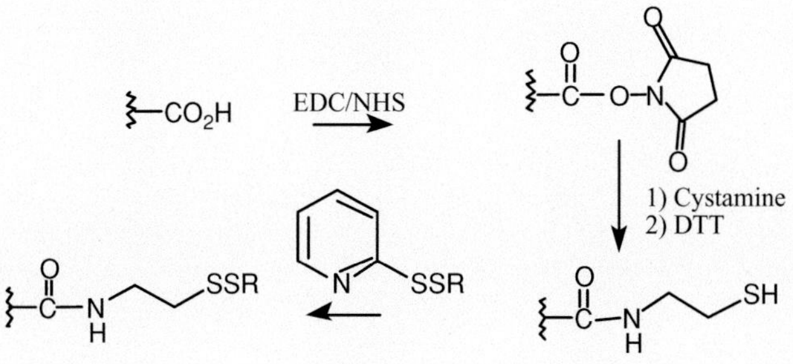

Figure 5–8: Covalent coupling of receptor via the surface thiol method.

Preparation of receptor molecules through PDEA modification:

1. Dissolve 1 mg receptor and 5.5 mg PDEA in 1 ml 0.1 M MES (2-morpholinoethanesulphonic acid) buffer pH 5.0. Cool on an ice bath.
2. Add 50 µl 0.4 M EDC in H_2O. Mix and allow to react for 1 hr on ice.
3. Remove low molecular weight reagents by gel filtration, using a desalting columns such as NAP-10 with 0.15 M NaCl as eluent.

Preparation of receptor molecules through SPDP modification:

1. Dissolve 1 mg receptor molecule in 1 ml 0.15 M NaCl.
2. Add 0.5 ml 0.3 M phosphate buffer pH 7.5 and 17 µl 2.5 mM SPDP in ethanol.
3. Mix and allow to react for 2 hr at room temperature.
4. Remove low molecular weight reagents by gel filtration, using an NAP-10 column with 0.15 M NaCl as eluent.

Preparation of sensor surface:

Running Buffer = HBS, Flow rate = 5 µL/min

1. 10 µL NHS/EDC mixture
2. 15 µL 40 mM cystamine dihydrochloride in 0.1 M borate buffer pH 8.5
3. 15 µL 0.1 M DTE in 0.1 M borate buffer pH 8.5
4. 35 µL receptor molecule (25–50 µg/ml) in low pH/low salt buffer
5. 20 µL 20 mM PDEA in 0.1 M formate buffer pH 4.3, containing 1 M NaCl

Derivatization of carboxyl groups with PDEA.
1. Dissolve 1 mg protein and 5.5 mg PDEA in 1 ml 0.1 M MES buffer pH 5.0. Cool in an ice bath. (A large molar excess of PDEA is used to minimize dimerization through EDC-induced polymerization of receptor molecules.)
2. Add 50 µl 0.4 M EDC in water. Mix and allow to react for 1 hr on ice.
3. Remove the reagents by gel filtration on an NAP-10 column with a suitable buffer.

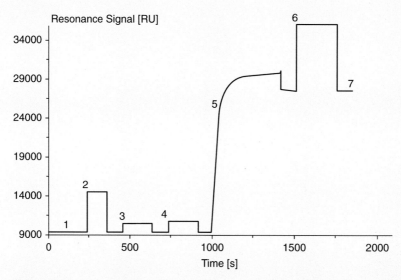

Figure 5–9: Sensorgram showing a typical immobilization sequence for the surface thiol method. SPDP-modified transferrin is coupled to a thiol-modified sensor chip. Numbers refer to the steps as follows:

1. Baseline signal for the unmodified sensor chip.
2. Injection of NHS/EDC to activate the sensor chip.
3. Injection of cystamine to derivatize the sensor chip.
4. Injection of DTE to reduce cystamine disulphides to thiols.
5. Injection of 50 μg/ml SPDP-transferrin in 100 mM acetate buffer pH 4.5.
6. Injection of PDEA/NaCl to deactivate excess thiol groups and remove noncovalently-bound protein.
7. Immobilized receptor corresponding to about 18 000 RU.

The reactive disulphide group introduced by SPDP or PDEA modification of proteins has limited stability in solution at 4–8° C. Stability is improved at low pH (4.5), and freezing is recommended if the receptor withstands storage in the frozen state.

Determining the degree of modification For proteins modified with PDEA, SPDP, or SMPT, the degree of modification can be determined approximately by reduction of the disulphide bond and spectrophotometric estimation of the thiopyridone produced (absorbance maximum 343 nm).

1. Measure the absorbance of the modified protein at 280 nm (A_{280}) and 343 nm ($A1_{343}$).
2. To 1 ml protein solution, add 50 μl 100 mM DTE in water. Mix and allow to react for a few minutes at room temperature.
3. Measure the absorbance again at 343 nm ($A2_{343}$). Calculate the degree of modification as follows:

$$C_{TP} = \frac{A2_{343} - A1_{343}}{l.e_{343,TP}}$$

$$A_{280,\text{prot}} = A_{280} - (l.C_{\text{TP}}.\varepsilon_{280,\text{TP}})$$

$$C_{\text{prot}} = \frac{A_{280,\text{prot}}}{l.\varepsilon_{280,\text{prot}}}$$

$$\text{molar degree of modicication} = \frac{C_{\text{TP}}}{C_{\text{prot}}},$$

where:

C_{TP}	the molar concentration of thiopyridone
C_{prot}	the molar concentration of protein
$A_{280,\text{prot}}$	the contribution of the protein to the absorbance at 280 nm
$\varepsilon_{343,\text{TP}}$	the molar extinction coefficient for thiopyridone at 343 nm (8.08×10^3 M^{-1}cm^{-1})
$\varepsilon_{280,\text{TP}}$	the molar extinction coefficient for thiopyridone at 280 nm (5.1×10^3 M^{-1}cm^{-1})
$\varepsilon_{280,\text{prot}}$	the molar extinction coefficient for the protein at 280 nm (determined separately or available from the literature)
l	the path length of the spectrophotometer cell in cm

Adjusting the degree of modification. A molar degree of receptor modification up to three is generally recommended. For applications where modification interferes with the biological activity of the receptor, the degree of modification can be reduced. When reactive disulphides are introduced to highly acidic receptors with PDEA, the modification serves the additional purpose of increasing the pI of the receptor to enhance electrostatic preconcentration. In such situations this effect may be the primary factor determining the optimal degree of modification. The amount of reactive disulphide groups introduced into the receptor depends on the nature of the protein and the reaction conditions. Table 5–3 summarizes the results of immobilization of a range of proteins with different reagents following the basic protocol described above. The degree of modification with PDEA can be adjusted either by varying the reaction time or by varying the amount of EDC added. For some proteins, reducing the reaction time may not have a marked effect on the degree of modification. In such cases, reduce the concentration of EDC to control the degree of modification.

Maleimide coupling provides an alternative form of thiol coupling, which involves formation of a stable thioether bond between reactive maleimido groups on the sensor surface and the thiol groups of the receptor molecule. Maleimide coupling offers all the advantages of the previous approaches to thiol-based coupling together with the capacity to generate surfaces that can withstand basic pHs (>9.5) as well as reducing agents such as β-mercaptoethanol and DTT. Several heterobifunctional reagents are available for introduction of reactive maleimido groups to the sensor surface, including sulfo-MBS (m-maleimidobenzoyl-N-hydroxysulfosuccinimide ester), sulfo-SMCC (sulfosuccinimidyl-4-[N-maleimidomethyl]cyclohexane-l-carboxylate]), GMBS (N-[γ-maleimidobutyrloxy]sulfosuccinimide ester]).

Preparation of sensor surface:

Running Buffer $=$ HBS, Flow rate $=$ 5 μL/min

Table 5–3: Representative immobilization data for selected proteins using the surface thiol method*

Protein	pI	Mwt (Da)	Modification		Buffer	pH	Immobilization	
			Reagent	Extent (molar)			[Protein] (μg/ml)	Amount coupled (kRU)
Chymotrypsinogen A	9.6	25700	SPDP	2.5	65 mM Acetate	5.0	50	1.7
RNase A	9.3	13700	SPDP	0.7	200 mM Acetate	4.5	50	3.5
Polyclonal IgG (RAMFc)	5.0–8.0	150000	SPDP	7.4	110 mM Acetate	5.0	50	11.3
IgG1 (anti-β2μ-globulin)	6.5–6.9	150000	PDEA	2.6	100 mM Acetate	4.5	50	18.7
			SPDP	0.9	120 mM Acetate	4.5	50	10.2
			SMPT	2.8	120 mM Acetate	4.5	50	7.2
IgG1 (anti-IgE)	5.8–6.2	150000	SPDP	4.3	110 mM Acetate	4.5	50	10.7
Transferrin	5.3–5.6	80000	SPDP	3.4	100 mM Acetate	3.5	50	12.2
Catalase	5.8	232000	SPDP	4.7	20 mM Formate	4.0	50	14.7
Protein A	5.1	41000	PDEA	3.5	10 mM Acetate	4.5	50	16.0
			SPDP	4.7	100 mM Formate	3.5	50	6.7
			SMPT	1.2	10 mM Acetate	4.5	50	8.5
Soybean trypsin inhibitor	4.5	20000	PDEA	4.0	10 mM Citrate	3.3	100	9.9
			SPDP	1.8	130 mM Formate	4.0	100	1.3
α1-antitrypsin	4.5–4.5	58000	SPDP	3.1	120 mM Citrate	3.0	50	3.9
Ovomucoid	3.8–4.5	28000	PDEA	0.9	10 mM Citrate	3.0	100	3.7
			SPDP	2.6	25 mM Citrate	2.5	50	1.9
Alkaline phosphatase	4.0	120000	SPDP	2.6	10 mM Formate	3.5	100	5.8
Carcinoembryonic antigen	<3.0	180000	PDEA	2.8	10 mM Citrate	3.0	200	7.3
Pepsin	2.2–3.0	34700	PDEA	3.2	10 mM Citrate	3.0	100	2.3

* Protein modification and immobilization data were obtained using a Biacore Sensor. CM5 performed following the recommendations described in the text.

1) 10 μL NHS/EDC mixture
2) 20 μL 1M ethylendiamine, pH 8.5
3) 40 μL 50mM sulfo-MBS in 25 mM sodium bicarbonate, pH 8.5
4) 30 μL receptor molecule (25–50 μg/ml) bin coupling buffer
5) 50 μL 100 mM cysteine in 10 mM sodium acetate, pH 4.0

Coupling buffers with pHs above 7.5 should be avoided due to the potential for hydrolysis of the maleimide group to non-reactive maleamic acid and the potential for other competing reactions. Activate the sensor chip first with a 2-min pulse of 0.05 M NHS/0.2 M EDC. This is the same basic activation procedure as for amine coupling, with the time of activation reduced to 2 min. The efficiency of the thiol coupling reaction is high, and longer activation times do not significantly increase the yield of immobilized receptor.

For the *receptor thiol method* (surface-bound disulphides), inject a 4-min pulse of 80 mM PDEA in 0.1 M borate buffer pH 8.5. For the *surface thiol method* (surface-bound thiol groups), inject a 3-min pulse of 40 mM cystamine dihydrochloride in 0.1 M borate buffer pH 8.5, followed by a 3-min pulse of 0.1 M DTE in 0.1 M borate buffer pH 8.5. DTT can be used in place of DTE.

Immobilizing the receptor. Inject a 7-min pulse of receptor in buffer at a pH below the isoelectric point of the receptor. The receptor may be native or modified, depending on the method used. Electrostatic preconcentration is required for efficient immobilization of macromolecular receptors.

Adjusting the immobilization level. Major factors affecting the amount of immobilized receptor are ionic strength, pH, and receptor concentration. The primary effect is on the extent of electrostatic preconcentration. Thiol coupling is relatively insensitive to the time of activation of the sensor chip. Varying the reagent concentration may in many cases be directly unsuitable. For example, low concentrations of cystamine in the surface thiol method can result in residual NHS-esters or unwanted coupling of DTE or DTT to the matrix.

- *Ionic strength.* Changing the ionic strength is usually the simplest and most effective way to regulate the amount of immobilized receptor. Increase the ionic strength to reduce the amount and *vice versa*.
- *pH.* Variations in pH can only be made over a relatively narrow interval for acidic proteins if an electrostatic preconcentration effect is to be retained.
- *Receptor concentration.* Typical concentrations for protein receptors are in the range 10–200 μg/ml. The use of excessively low concentrations may have an adverse effect on reproducibility.

When optimizing immobilization conditions, aim for electrostatic preconcentration levels corresponding to about 150% of the desired immobilization level.

Deactivation or capping. Excess reactive groups remaining on the surface after immobilization are deactivated by a 4-min pulse of a suitable reagent in buffer containing 1 M NaCl. The high ionic strength of this solution removes noncovalently bound material from the surface. For the *surface thiol method*, use PDEA for deactivation. For the *receptor thiol method*, use cysteine for deactivation.

Regeneration. Avoid using basic solutions (pH > 9) for regeneration, because the disulphide bond is unstable under these conditions and receptor may be lost from the surface.

Aldehyde coupling

Aldehyde coupling involves formation of a hydrazone bond via condensation of hydrazide groups on the sensor surface with aldehyde groups on the receptor molecule. These aldehyde moieties may be native to the protein or introduced through mild oxidation of cis-diols present in the receptor molecule. Aldehyde coupling is particularly useful for site-directed immobilization of glyco-conjugates, glyco-proteins, and polysaccharides and may also be useful for orientation-specific immobilization of proteins containing functional groups that may be converted to aldehyde moieties.

Preparation of receptor molecules containing cis-diols:

1. Prepare a cold solution of receptor in 100 mM sodium acetate pH 4.0 at 1 mg/ml.
2. Add freshly prepared 50 mM sodium metaperiodate in 100 mM sodium acetate pH 4.0 to a final concentration of 1 mM.
3. Allow the mixture to react for 20 min on ice.
4. Desalt the mixture on an NAP-5 column with 10 mM acetate pH 4.0 as eluent.
5. Refrigerate oxidized sample at 4°C.

Preparation of sensor surface:

Running Buffer = HBS, Flowrate = 5 μL/min

1. 15 μL NHS/EDC mixture
2. 35 μL 1 mM hydrazine or carbohydrazide
3. 35 μL 1 M ethanolamine pH 8.5
4. 35 μL of receptor sample in low pH buffer
5. 40 μL 0.1 M sodium cyanoborohydride in 100 mM sodium acetate pH 4.0
6. 3 × 4 μL 10 mM HCL or 0.1 M glycine

The above protocol is intended as an initial guideline to performing aldehyde coupling to carboxylic acid sensor surfaces. Care should be taken during chemical conversion of functional groups to aldehyde moieties to avoid loss of receptor activity. Several parameters can be varied depending on the specific application. Effective coupling will be governed by factors such as isoelectronic point (pI) of the protein, coupling buffer pH and ionic strength, and protein concentration. The reaction between the hydrazide surface and aldehyde groups is acid catalyzed. Therefore, immobilization will be most effective in buffers in the range pH 4–5. Variations in the buffer salt concentration may provide the most efficient mode of controlling immobilization levels, the upper limit being approximately 0.3 M salt. NHS/EDC activation times longer than 3 min do not have an appreciable effect on immobilized receptor levels. Indeed, higher hydrazide densities may result in surfaces that tend toward nonspecific adsorption of unoxidized proteins. Similarly, increased concentrations of hydrazine have relatively little impact on the immobilization level. Due to the toxicity of hydrazine it is not recommended to use concentrations above 5 mM. Carbohydrazide may be used as a less toxic alternative. The final step in the injection sequence given above involves reduction of the hydrazone bond with sodium cyanoborohydride that stabilizes the receptor surface toward acidic regeneration conditions.

Photo-activated coupling

Photo-activated cross-linkage is a novel approach to sensor surface derivatization that has been widely employed in the fabrication of protein and oligonucleotide microarrays. This approach was applied to the preparation of a biosensor surface by Stein and coworkers.[99] They employed a photoactive cross-linker, perfluorophenylazide hexylamine (PFPA), to functionalize a Biacore CM5 Sensor Chip under low light conditions outside of the biosensor instrument. Normal activation procedures were followed by pipetting 50 μL of 1:1 EDC/NHS onto the surface. The surface was then exposed for 7 min to a 27 mM solution of the photoactivatable cross-linker (5 μL of 300 mM PFPA diluted in DMSO with 10 μL methanol and 40 μL H_2O). Remaining active ester groups were deactivated with 1 M aqueous heptylammonium hydrochloride for 7 min. It was then possible to capture their receptor molecule, cell adhesion molecule, csA, via CH insertion of the cross-linker agent at the alkyl chains of the lipid anchor of the csA molecule. This was performed either through their biosensor instrument or externally, taking care to maintain low-light conditions. The gold-derivatized surface of the sensor chip was then covered with a layer of HEPES buffers saline (HBS) and a glass coverslip placed on top. This assembly was then placed dextran surface downward and exposed to ultraviolet (UV) radiation of 294 to 365 nm for 5 min.

They employed this immobilization procedure primarily to avoid denaturation of their receptor molecule by the usual coupling conditions (i.e., low salt, low pH coupling buffers). The authors demonstrated that their receptor molecule, csA protein, remained active following photo-activated cross-linkage, from analysis of the interaction of anti-csA antibodies with the csA surface. This strategy, though specific to the application presented by Stein and coworkers, may be of further use in generating surfaces toward understanding other membrane proteins and glycolipids via hydrophobic interaction with a selected photoactivatable cross-linker.

Noncovalent capture

Commonly used capturing molecules include protein A, streptavidin, nickel-chelating surfaces (NTA), and a range of antibodies generated against specific receptor molecules or epitope tags. However, users should not limit themselves to these capturing molecules. Rather, the molecules represent current design examples of capture-type surfaces. The flexibility and versatility of biosensor surface design is amenable to many more user-defined capture systems. A description of several capture systems follows.

Fusion protein capture. Capture of recombinant proteins fused to tags such as GST, the FLAG octapeptide (Sigma-Aldrich), and chelating polyhistidine tags provides an opportunity to employ a capturing approach to generating a biospecific sensor surface. For example, Biacore supplies a GST capture kit for site-directed affinity capture of GST fusion proteins that contains an affinity-purified goat anti-GST mAb at 0.8 mg/ml in a 75 μl coupling solution, regeneration solutions, and 0.2 mg/ml in 100 μl recombinant GST from *Schistosoma japonicum* for checking the activity of the immobilized anti-GST. The antibody exhibits high-affinity binding of GST fusion products and, in general, 10 mM glycine, pH 2.2, is the regeneration agent of choice.

Protein A capture. Protein A, Protein G, Protein L, and their hybrids, Protein LA and Protein A/G, are native or recombinant proteins of microbial origin with the ability to bind mammalian immunoglobulin molecules. The interaction between the various proteins and immunoglobulin isotypes (IgA, IgD, IgE, IgG and IgM) are not equivalent for all species or all antibody subclasses, hence further research is required prior to employing these molecules as capture molecules. A general protocol[11,98] for their use follows standard amine coupling procedures for protein immobilization followed by capture of the antibody of interest and is as follows:

Running buffer = HBS, Flowrate = 5 μL/min

1. Dilute protein to 400 μg/ml in 10 mM sodium acetate pH 4.0 and allow to stand at room temperature for 30 min.
2. Activate sensor surface (NHS/EDC) for 10 min.
3. Inject protein for 10 min.
4. Inject ethanolamine 1 M pH 8.5 for 10 min.

Expected immobilization level on an SPR biosensor system is 7–8000 RU. Regenerating these protein capture surfaces is usually feasible with 10 mM glycine pH 1.7, but they are also stable to regeneration with 100 mM HCl.

Cross-linkage

Capture surfaces may be further modified through cross-linkage of the receptor molecule to the capture molecule. The procedure provides a stable permanent receptor surface via chemical or photo-activated cross-linking (see earlier).

Chemical cross-linkage. Chemical cross-linkage involves a chemical agent to cross-link the receptor molecule to that capture surface. This strategy has been employed to cross-link IgG and Fab' fragments to protein A.[11]

Immobilization of protein
Running buffer = HBS, Flowrate = 5 mL/min

1. Dilute protein to 400 μg/ml in 10 mM sodium acetate pH 4.0 and allow to stand at room temperature for 30 min.
2. Activate sensor surface (NHS/EDC) for 10 min.
3. Inject protein for 10 min.
4. Inject ethanolamine 1 M pH 8.5 for 10 min.

Test binding of IgG of interest
1. Inject IgG, 100 μg/ml in 100 mM sodium borate pH 9.2 for 2 min.
2. Regenerate surface with 10 mM glycine pH 2.0 for 1 min.

Cross-linkage
1. Inject IgG (100 μg/ml in 100 mM sodium borate pH 9.2) for 2 min.
2. Inject 50 mM dimethyl pimelidate dihydrochloride in 0.1 M sodium borate buffer pH 9.2 for 20 min.
3. Inject 25 mM glycine pH 2.2 (removal of non–cross-linked IgG) for 1 min.
4. Inject 0.1 M ethanolamine pH 8.5 for 1 min to deactivate cross-linker.

Care should be taken to design a strategy so that the receptor molecule and/or capture molecule are not denatured or deactivated by the cross-linking agent.

STOCK SOLUTIONS

EDC and NHS

Dissolve 115 mg NHS in 10 ml filtered, deionized water. Dissolve 750 mg EDC in 10 ml filtered, deionized water. Filter each solution separately through a 0.22 μm hydrophilic, chemically resistant filter, dispense into separate 200 μL aliquots and store below –18° C. Rapidly thaw and mix equal volumes of EDC and NHS immediately before use (resulting in 200 mM EDC/50 mM NHS) and do not use if the mixed solution is kept for longer than 15 min at room temperature.

Ethanolamine. Dissolve 0.611 g of ethanolamine hydrochloride in 10 ml filtered, deionized water. Adjust to pH 8.5 with drops of 1 M NaOH or 1 M HCl as necessary, filter each solution separately through a 0.22 μm hydrophilic, chemically resistant filter, dispense into separate 200 μL aliquots and store at +4–8°C.

80 mM PDEA activation solution. Prepare 0.1 M borate buffer pH 8.5 by dissolving 0.31 g boric acid in 25 ml water, adjusting to pH 8.5 with 1 M NaOH and making up to 50 ml with water. Dissolve 4.5 mg PDEA in 250 μl 0.1 M borate buffer pH 8.5. Use within 1 hr of preparation.

50 mM l-cysteine–1 M NaCl deactivation solution. Prepare 100 mM sodium formate buffer pH 4.3 by dissolving 0.35 g sodium formate in 25 ml water, adjusting to pH 4.3 with formic acid and making up to 50 ml with water. Dissolve 1.5 mg *l*-cysteine and 14 mg NaCl in 250 μl 0.1 M formate buffer pH 4.3. Use within 1 hr of preparation.

40 mM cystamine activation solution. Dissolve 0.45 g cystamine dihydrochloride and 0.31 g boric acid in 25 ml water. Adjust to pH 8.5 with 1 M NaOH. Make up to 50 ml with water and store frozen at –20°C. This solution can be stored frozen for several months. The unfrozen solution is stable for three days.

0.1 M DTE or DTT reducing solution. Dissolve 0.77 g DTE or DTT and 0.31 g boric acid in 25 ml water. Adjust to pH 8.5 with 1 M NaOH. Make up to 50 ml with water. Store frozen at –20°C. This solution can be stored frozen for several months. The unfrozen solution is stable for three days.

20 mM PDEA–1 M NaCl deactivation solution. Prepare 0.1 M sodium formate buffer pH 4.3 by dissolving 0.35 g sodium formate in 25 ml water, adjusting to pH 4.3 with formic acid and making up to 50 ml with water.

Dissolve 1.2 mg PDEA and 14 mg NaCl in 250 μl 0.1 M formate buffer pH 4.3. This solution is stable for three days at 4–8°C. Preparation of stock solutions is not recommended.

REFERENCES

1. A. Ulman, *Chem Rev* **96**, 1533 (1996).
2. P. Silberzan, L. Leger, D. Ausserre, J. J. Benattar, *Langmuir* **7**, 1647 (1991).
3. J. Lahiri, L. Isaacs, J. Tien, G. M. Whitesides, *Anal Chem* **71**, 777 (1999).
4. G. B. Sigal, C. Bamdad, A. Barberis, J. Strominger, G. M. Whitesides, *Anal Chem* **68**, 490 (1996).
5. O. P. Ernst, C. Bieri, H. Vogel, K. P. Hofmann, *Meth Enzymol* **315**, 471 (2000).
6. M. A. Cooper, *Bioorg Med Chem* **8**, 2609 (2000).
7. M. Mrksich, J. R. Grunwell, G. M. Whitesides, *J Am Chem Soc* **117**, 12009 (1995).
8. S. Löfås, B. Johnsson, *J Chem Soc Chem Commun*, 1526 (1990).
9. R. Karlsson, A. Falt, *J Immunol Methods* **200**, 121 (1997).
10. R. A. Vijayendran, D. E. Leckband, *Anal Chem* **73**, 471 (2001).
11. B. Catimel *et al.*, *J Chromatogr* **776**, 15 (1997).
12. B. Johnsson *et al.*, *J Mol Recognit* **8**, 125 (1995).
13. S. Löfås *et al.*, *Biosens Bioelectron* **10**, 813 (1995).
14. R. Abraham *et al.*, *J Immunol Methods* **183**, 119 (1995).
15. B. Misselwitz, O. Staeck, T. A. Rapoport, *Mol Cell* **2**, 593 (1998).
16. D. Toroser, G. S. Athwal, S. C. Huber, *Febs Letters* **435**, 110 (1998).
17. W. Nunomura, Y. Takakuwa, M. Parra, J. G. Conboy, N. Mohandas, *J Biol Chem* **275**, 24540 (2000).
18. M. Stolowitz *et al.*, *Bioconjug Chem* **12**, 229 (2001).
19. J. Kohn, M. Wilchek, *Febs Letters* **154**, 209 (1983).
20. J. Kohn, M. Wilchek, *Appl Biochem Biotechnol* **9**, 285 (1984).

21. J. Porath, K. Aspberg, H. Drevin, R. Axén, *J Chromatogr* **86**, 53 (1973).
22. J. Porath, K. Asperg, H. Drevin, R. Axén, *J. Chromatogr* **86**, 53 (1973).
23. M. Wilchek, T. Miron, *Appl Biochem Biotechnol* **11**, 191 (1985).
24. M. T. W. Hearn, *Methods Enzymol.* K. Mosbach, Ed., (Academic Press, New York, 1987), pp. 102–117.
25. K. Nilsson, K. Mosbach, *Methods Enzymol.* K. Mosbach, Ed., (Academic Press, New York, 1987), pp. 65–78.
26. L. Sunberg, J. Porath, *J Chromatogr* **90**, 87 (1974).
27. J. Porath, N. Fornstedt, *J Chromatogr* **51**, 479 (1970).
28. R. Axen, J. Carlsson, J. C. Janson, J. Porath, *Enzymologia* **41**, 359 (1971).
29. J. Porath, *Methods Enzymol.* B. Jakoby, M. Wilchek, Eds., (Academic Press, New York, 1974), pp. 13–30.
30. D. Robinson, N. C. Phillips, B. Winchester, *FEBS Lett* **53**, 110 (1975).
31. H. Anttinen, K. I. Kivirikko, *Biochim Biophys Acta* **429**, 750 (1976).
32. S. L. Marcus, E. Balbinder, *Anal Biochem* **48**, 448 (1972).
33. P. Cuatrecasas, I. Parikh, *Biochemistry* **11**, 2291 (1972).
34. K. Brocklehurst, J. Carlsson, M. P. Kierstan, E. M. Crook, *Biochem J* **133**, 573 (1973).
35. J. L. Guesdon, S. Avrameas, *J Immunol Methods* **11**, 129 (1976).
36. J. K. Inman, *Methods Enzymol.* B. Jakoby, M. Wilchek, Eds., (Academic Press, New York, 1974), p. 30.
37. S. Ohlson, L. Hansson, P. O. Larsson, K. Mosbach, *FEBS Lett* **93**, 5 (1978).
38. L. Goldstein, *Methods Enzymol.* K. Mosbach, Ed., (Academic Press, New York, 1987), pp. 90–102.
39. P. Nilsson, B. Persson, M. Uhlén, P.-Å. Nygren, *Anal Biochem* **224**, 400 (1995).
40. K. K. Jensen, H. Orum, P. E. Nielsen, B. Norden, *Biochemistry* **36**, 5072 (1997).
41. D. J. Hart, R. E. Speight, M. A. Cooper, J. D. Sutherland, J. M. Blackburn **27**, 1063 (1999).
42. A. M. Grunden, W. T. Self, M. Villain, J. E. Blalock, K. T. Shanmugam, *J Biol Chem* **274**, 24308 (1999).
43. L. Masson, A. Mazza, R. Brousseau, *Anal Biochem* **218**, 405 (1994).
44. L. Jin *et al.*, *Biochemistry* **38**, 15659 (1999).
45. B. Kazemier, H. de Haard, P. Boender, B. van Gemen, H. Hoogenboom, *J Immunol Methods* **194**, 201 (1996).
46. E. Nice *et al.*, *J Chromatogr* **646**, 159 (1993).
47. C.-S. Suen *et al.*, *J Biol Chem* **273**, 27645 (1998).
48. D. J. O'Shannessy, K. C. O'Donnell, J. Martin, M. Brigham-Burke, *Anal Biochem* **229**, 119 (1995).
49. L. Nieba *et al.*, *Anal Biochem* **252**, 217 (1997).
50. U. Radler, J. Mack, N. Persike, G. Jung, R. Tampe, *Biophys J* **79**, 3144 (2000).
51. K. M. Müller, K. M. Arndt, K. Bauer, A. Plückthun, *Anal Biochem* **259**, 54 (1998).
52. H. Chao, D. L. Bautista, J. Litowski, R. T. Irvin, R. S. Hodges, *J Chromatogr B Biomed Sci Appl* **715**, 307 (1998).
53. L. Haussling, H. Ringsdorf, F. J. Schmitt, W. Knoll, *Langmuir* **7**, 1837 (1991).
54. W. Muller *et al.*, *Science* **262**, 1706 (1993).
55. F. J. Schmitt, L. Haussling, H. Ringsdorf, W. Knoll, *Thin Solid Films* **210**, 815 (1992).
56. S. Storri, T. Santoni, M. Minunni, M. Mascini, *Biosens Bioelectron* **13**, 347 (1998).
57. M. M. Stevens *et al.*, *Analyst* **125**, 245 (2000).
58. L. S. Jung, K. E. Nelson, P. S. Stayton, C. T. Campbell, *Langmuir* **16**, 9421 (2000).
59. R. Esfand, *Drug Discov Today* **6**, 427 (2001).
60. A. U. Bielinska, J. Johnson, J. R. Baker Jr., *Bioconjug Chem* **10**, 843 (1999).
61. H. C. Yoon, M. Y. Hong, H. S. Kim, *Anal Biochem* **282**, 121 (2000).
62. R. M. Crooks, A. J. Ricco, *Accounts of Chemical Research*, (1998), pp. 219–227.
63. R. Benters, C. M. Niemeyer, D. Wohrle, *Chembiochem* **2**, 686 (2001).
64. V. Chegel, Y. Shirshov, S. Avilov, M. Demchenko, M. Mustafaev, *J Biochem Biophys Methods* **50**, 201 (2002).

65. X. Li, C. Abell, M. A. Cooper, *Colloids Surf B Biointerfaces* **61**, 113 (2008).
66. R. W. Glaser, *Anal Biochem* **213**, 152 (1993).
67. P. Schuck, *Biophys J* **70**, 1230 (1996).
68. G. G. Guilbault, *Anal Chem* **55**, 1682 (1983).
69. S. H. Si, X. Li, Y. S. Fung, D. R. Zhu, *Microchem J* **68**, 21 (2001).
70. J. M. Hu, R. J. Pei, Y. Hu, Y. E. Zeng, *Chin J Chem* **16**, 219 (1998).
71. W. C. Tsai, I. C. Lin, *Sens Actuators B Chem* **106**, 455 (2005).
72. X. D. Zhou, L. J. Liu, M. Hu, L. L. Wang, J. M. Hu, *J Pharm Biomed Anal* **27**, 341 (2002).
73. P. Kujawa, G. Schmauch, T. Viitala, A. Badia, F. M. Winnik, *Biomacromolecules* **8**, 3169 (2007).
74. C. Picart *et al.*, *Langmuir* **17**, 7414 (2001).
75. S. S. Praveen, R. Hanumantha, J. M. Belovich, B. L. Davis, *Diabetes Technol Ther* **5**, 393 (2003).
76. L. A. Ruiz-Taylor *et al.*, *Proc Natl Acad Sci U S A* **98**, 852 (2001).
77. N.-P. Huang, K. E. Gabor Csucs, Y. Nagasaki, K. Kataoka, M. Textor, N. D. Spencer, *Langmuir* **18**, 252 (2001).
78. G. L. Kenausis *et al.*, *J Phys Chem B* **104**, 3298 (2000).
79. N. Xia, Y. Hu, D. W. Grainger, D. G. Castner, *Langmuir* **18**, 3255 (2002).
80. D. M. Disley *et al.*, *Biosens Bioelectron* **13**, 383 (1998).
81. J. C. Munro, C. W. Frank, *Langmuir* **20**, 10567 (2004).
82. S. Lasky, D. A. Buttry, *Abstracts of Papers of the American Chemical Society* **196**, 45 (1988).
83. R. Gabai *et al.*, *J Phys Chem B* **105**, 8196 (2001).
84. K. E. Sapsford, F. S. Ligler, *Biosens Bioelectron* **19**, 1045 (2004).
85. B. A. Cavic, M. Thompson, *Analyst* **123**, 2191 (1998).
86. C. C. Wang, H. Wang, Z. Y. Wu, G. L. Shen, R. Q. Yu, *Anal Bioanal Chem* **373**, 803 (2002).
87. S. P. Sakti *et al.*, *Sens Actuators A Phys* **76**, 98 (1999).
88. N. Y. Lee, J. R. Lim, M. J. Lee, S. Park, Y. S. Kim, *Langmuir* **22**, 7689 (2006).
89. S. Rauf *et al.*, *J Biotechnol* **121**, 351 (2006).
90. T. Serizawa, T. Sawada, H. Matsuno, *Langmuir* **23**, 11127 (2007).
91. G. Wulff, K. Knorr, *Bioseparation* **10**, 257 (2001).
92. S. A. Piletsky, N. W. Turner, P. Laitenberger, *Med Eng Phys* **28**, 971 (2006).
93. A. Bossi, F. Bonini, A. P. Turner, S. A. Piletsky, *Biosens Bioelectron* **22**, 1131 (2007).
94. Y. Uludag, S. A. Piletsky, A. P. Turner, M. A. Cooper, *Febs J* **274**, 5471 (2007).
95. A. H. Wu, M. J. Syu, *Biosens Bioelectron* **21**, 2345 (2006).
96. F. L. Dickert *et al.*, *Anal Bioanal Chem* **378**, 1929 (2004).
97. D. F. Tai, C. Y. Lin, T. Z. Wu, J. H. Huang, P. Y. Shu, *Clin Chem* **52**, 1486 (2006).
98. T. A. Morton, D. G. Myszka, *Methods Enzymol* **295**, 268 (1998).
99. T. Stein, G. Gerisch, *Anal Biochem* **237**, 252 (1996).

6 Macromolecular interactions

Francis Markey

THE SCOPE OF MACROMOLECULAR INTERACTIONS	143
PROTEIN–PROTEIN INTERACTION MODES	144
Multicomponent complexes	144
Multivalency and avidity	145
Multimer formation	146
Cooperativity and conformational change	146
Enzyme activity	147
SELECTED EXAMPLES FROM THE LITERATURE	147
Measuring interaction kinetics	147
Determining specificity	148
Combined structural and functional studies	149
Epitope mapping	150
Ligand fishing	153
PRACTICAL ASPECTS	154
Design of the active surface	155
Confirmation of surface activity	155
Choice of regeneration conditions	156
Experimental design and control experiments	157
SUMMARY	157
REFERENCES	158

THE SCOPE OF MACROMOLECULAR INTERACTIONS

Interactions between proteins constitute the best-characterized and perhaps largest family of interactions between biological macromolecules. This is partly because of the long-recognized diversity of the protein constituents of cells. The significance of diversity in nucleic acids, lipids, and carbohydrates is becoming more apparent, but proteins have held pride of place as the carriers of biological specificity for many decades. It is also partly because the diversity of protein structure and function is so easily observed in contrast to the more subtle variations that determine the specificity of other macromolecules. It is not surprising, therefore, that the bulk of biomolecular interaction studies performed with optical biosensors involves protein–protein binding. The technology can in principle be used to monitor any molecular interactions, and studies involving nucleic acids answer for a significant fraction of the published literature.

Proteins lend themselves well to biomolecular interaction work with optical biosensors. The wealth of functional groups on a "standard" protein molecule offers a selection of approaches for attaching interactants to the sensor surface, ranging from amine- or thiol-based chemical coupling to specific high-affinity capturing methods. Membrane-associated proteins can be embedded in lipid layers on the sensor surface to preserve a hydrophobic environment. Commonly, proteins are relatively large molecules with molecular weights from tens up to hundreds of kilodaltons, so the response obtained in mass-concentration detection methods like surface plasmon resonance (SPR) is usually well above the detection limit of the instrumentation. In addition, most protein–protein interactions can be broken by exposure to moderately mild conditions, offering good opportunities for efficient surface regeneration.

The literature describing optical biosensor studies on protein–protein interactions poses experimental questions covering an extensive range of conceivable applications, from simple yes/no binding studies to concentration measurement and detailed kinetic and mechanistic analyses. This chapter will give a brief overview of the application of optical biosensors to the study of protein–protein interactions and describe some examples that highlight particularly valuable aspects of the technology.

PROTEIN–PROTEIN INTERACTION MODES

Many interactions between biological macromolecules behave in a relatively simple fashion, involving two binding partners that form a straightforward binary complex. These are the simplest interactions to study with optical biosensors. Others are more complex, however, and may involve for example multiple or multivalent components, dimerization or other aggregation of one or both binding partners, cooperativity in the interaction, or enzymatic activity. Depending on the experiment's design and purpose, the interaction's complexity may be exploited or studied with biosensor technology or may be a disruptive factor that complicates interpretation and should be eliminated if possible.

Multicomponent complexes

Formation of multicomponent complexes lends itself particularly well to analysis with label-free biosensors, because the technology requires no labels and detects all components. Testing the binding of components added in different orders or with critical components present or absent is a simple and direct approach to determining interdependencies between several interactions in a multicomponent complex. The approach has been applied with advantage for example to

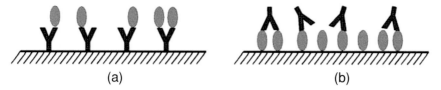

Figure 6–1: A bivalent binding partner attached to the surface **(a)** offers two independent binding sites for the partner in solution. In contrast, a bivalent binding partner in solution **(b)** can interact with surface-attached molecules at either one or two sites, resulting in an avidity effect.

analysis of the large complex that constitutes active deoxyribonucleic acid (DNA) polymerase III from *E. coli*, involving at least ten distinct subunits.[1-3] Simultaneous binding of several molecular species can also be exploited deliberately to map binding sites on the surface of a common partner. This approach is used in epitope mapping studies by pairwise antibody binding, an example of which is described in more detail later.

Multivalency and avidity

Asymmetrical multivalency in an interaction (where one molecule of one partner can bind several molecules of the other) affects design and interpretation of biosensor experiments differently according to the orientation of the system on the sensor surface. Perhaps the most widely encountered interaction family exhibiting this behavior is the interaction of bivalent immunoglobulin G (IgG) antibodies with monovalent antigen. If the antibody is immobilized on the surface, antigen molecules bind independently of each other to the two sites on each antibody molecule, and the result is equivalent to analysis of monovalent interactions (assuming there is no steric interference between the two sites). If, however, the antigen is attached to the surface, binding of one antigen molecule at one site on the antibody holds the latter on the surface and increases the chance of binding a second antigen molecule at the second site (Figure 6–1). The result is an avidity effect so that the overall apparent affinity is higher than the affinity at each site independently.

Avidity can be an advantage if the goal of the experiment is to detect the bivalent binding partner in samples, because the higher apparent affinity lowers the threshold of detection. For other purposes, however, avidity is a complicating factor that should be avoided if possible. In kinetic determinations, for example, avidity effects complicate the interaction model and introduce difficulties in extracting kinetic constants from the binding data (see Chapters 3 and 4).

In systems where the monovalent partner is attached to the surface and the multivalent partner is in solution, working with low levels of material attached to the surface can reduce avidity to some extent. The sparse distribution of material on the surface reduces the chances of molecules from solution binding simultaneously to two surface-attached partners. In some situations, avidity can be eliminated entirely by using monovalent fragments of the multivalent partner.

Fab' fragments of IgG would seem to be good candidates for this purpose in antibody contexts. However, it is worth noting that these fragments often show a marked tendency to dimerize so that true monovalency is often not achieved.

Multimer formation

Dynamic formation of dimers or higher-order multimers of one binding partner generally complicates the interpretation of biosensor experiments because the observed behavior is intermediate between simple monomeric binding and multivalent interaction. If for example the binding partner in solution shows a tendency to dimerize without interfering with the binding site for the surface attached partner (so that the dimer behaves as a bivalent species), the observed binding will show a certain level of avidity depending on the extent of dimer formation. In addition, the response from binding of one dimer will be twice that from one monomer, and the dimer may occupy one or two binding sites on the sensor surface. Quantitative analysis of this kind of behavior can be difficult because the interaction scheme involves binding of both monomer and dimer to one- and two-partner molecules and becomes quite complex.

In some cases, multimer formation can be exploited as a tool in the experimental investigation. Dimerization of the human growth hormone receptor attached to the sensor surface as a result of binding of the hormone has been neatly demonstrated, using the difference in binding stoichiometry that results from the formation of dimers (see below). Here, the potential for dimer formation could be regulated either by mutagenesis of the hormone (which itself dimerizes when bound to the receptor) or by choice of attachment point for the receptor to the surface, and the results were fully in accord with a model of ligand-induced receptor dimerization.

Cooperativity and conformational change

Cooperativity, defined as the ability of one interaction to affect the properties of subsequent interactions, necessarily involves forming a complex between more than two molecules and is often associated with induced conformational change in one or more of the components. As such, cooperativity is open to the same kind of investigation using optical biosensors as other multivalent interactions. In the special case where the interaction involves multiple molecules of one interactant binding to one molecule of the other, there is a mechanistic distinction between cooperativity (where the properties of the second interaction differ from those of the first, even if the interacting components are superficially identical) and avidity (where an increase in measured affinity at the second site results solely from the restricted movement imposed by binding at the first site). In practice, however, it may be difficult or impossible to distinguish cooperativity from avidity in the case where the interactant in solution carries multiple cooperative binding sites. If the multivalent component is attached to the surface, however, cooperativity results in an induced heterogeneity of binding sites on the surface as a result of the interaction. Evidence for cooperativity obtained from optical biosensors is largely indirect.

Conformational changes by themselves cannot as a rule be detected directly by optical biosensors because they are not associated with a change in mass concentration on the surface. In special cases, however, it may be possible to deduce the existence of conformational changes resulting from the interaction from detailed analysis of the binding kinetics. Briefly, if initial binding between the interactants is followed by a slower rearrangement resulting in a more stable complex, kinetic analysis may be able to resolve the binding and stabilization into two distinct processes.

Enzyme activity

Enzyme activity involves necessarily transient interactions with the formation of a normally short-lived transition complex between the enzyme and substrate. Except in the most extreme cases, the time scale of the formation of enzyme–substrate complex and release of the end-products is too fast to allow direct monitoring with today's optical biosensor technology. Enzyme reactions that lead to a change in substrate mass can be followed directly if the substrate is attached to the surface: examples are protease activity[4] and the various enzymes involved in the cleavage, ligation, and elongation of nucleic acids.[5,6] In principle, the resolution of current biosensor instrumentation is sufficient to detect the mass change resulting from phosphorylation or dephosphorylation of the ligand on the sensor surface. It is also possible to follow enzyme activity indirectly if either the substrate or product can be uniquely monitored using reporter molecules or if the activity leads to a change in the interaction properties of the substrate.[7]

SELECTED EXAMPLES FROM THE LITERATURE

An exhaustive review of the diversity of macromolecular interaction studies using optical biosensors is beyond the scope of this book. Current literature reference lists are provided on the Internet sites of the major technology suppliers. The examples described below are each chosen to highlight aspects of research where the use of optical biosensors has provided insights that were difficult or impossible to obtain by other means.

Measuring interaction kinetics

The real-time monitoring capabilities of optical biosensors, particularly those with a controlled continuous flow system for supplying samples to the sensor, permit resolution of separate kinetic parameters for the association and dissociation processes that comprise a molecular interaction. Examples of kinetic measurements are frequent in optical biosensor literature. In one recent study, for example,[8] the authors examined the kinetic properties of the repressor protein involved in copper homeostasis in the bacterium *Enterococcus hirae*. Expression of the copper homeostasis system is under the control of a copper-responsive repressor named CopY that binds to a promoter sequence in the bacterial DNA in a classical repressor–promoter expression control system. Addition of copper to the system increased the dissociation rate of the repressor from the promoter by an order of magnitude sufficient to explain the underlying mechanism of copper-dependent control of expression of the homeostasis proteins.

Resolution of binding affinities into separate association and dissociation rate constants is fairly common in the investigation of interaction mechanisms with optical biosensors. In another example, Katsamba et al.[9] argued for a two-step "lure and lock" mechanism for the binding of spliceosomal protein U1A to ribonucleic acid (RNA) from nuclear ribonucleoprotein on the basis of the kinetic effects of different mutations. Two alterations of lysine to alanine reduced the binding affinity primarily through an effect on the association rates, suggesting ionic components in the recognition or "lure" stage of the interaction. In contrast, replacement of a phenylalanine residue with alanine, which also reduced the affinity, increased the dissociation rate, indicating a role for non-ionic interactions with the phenylalanine residue in stabilizing or "locking" the complex. It is the nonlabel, direct monitoring of interaction processes with optical biosensors that enables this kind of dissection of interaction mechanisms.

Determining specificity

Protein–protein interactions range from promiscuous (proteins that show some degree of binding to a wide range of other molecular species) to highly specific (proteins that bind to each other but to little else). The question of specificity is particularly relevant in the context of protein families, where a set of closely similar but not identical proteins differs in binding profiles and therefore in function. Antibodies provide perhaps the best-known and most impressive example, with highly selective specificity in the framework of structurally similar proteins.

Optical biosensors are well suited to the determination of interaction specificity between proteins because the absence of labeling and the immediate monitoring of the binding process mean that results can be obtained rapidly without necessarily purifying the components first. In addition, separate recording of association and dissociation processes and of equilibrium binding levels allow the somewhat fuzzy term "specificity" to be quantitated in terms of interaction rates and affinities. This is well illustrated in work by Ober et al.,[10] who approached the problem of why therapeutic antibodies based on mouse IgG are rapidly cleared from the human circulation though pharmacokinetic tests in mice indicate that the antibodies should persist in the circulation.

The major histocompatibility complex (MHC) class 1-related receptor FcRn on endothelial cells seems to play an instrumental role in delaying clearance of

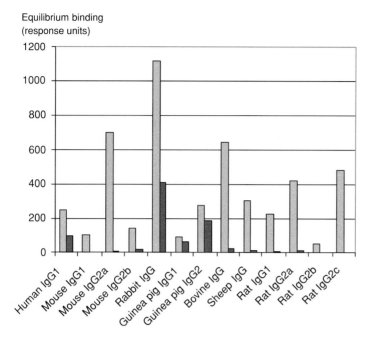

Equilibrium binding
(response units)

Figure 6–2: Equilibrium binding levels of equivalent concentrations of mouse FcRn (light bars) and human FcRn to different IgG species (drawn from data reported by Ober et al.[10]).

IgG from the circulation. Antibodies that bind to the receptor have significantly longer clearance times, and the serum half-life of injected IgG or Fc fragments in mice correlates directly with the affinity for the receptor. Ober et al. extended this knowledge by screening IgG from various sources for binding to mouse or human FcRn. Both association and dissociation processes were fairly rapid, and Ober et al. kept the experiment simple by using the same concentration of different IgGs and by measuring the steady-state binding level as an indicator of the relative affinities. Because all of the IgG species used have essentially the same molecular weight, the response levels can be compared directly with no need to correct for different molar contributions to the response.

The results, summarized in Figure 6–2, clearly show that the mouse receptor is fairly promiscuous, binding IgGs from a range of different species, though the human receptor shows a strong selectivity for certain species but binds only weakly to mouse IgG. In addition to explaining the observed clearance characteristics for mouse IgG, this simple analysis has wide-reaching implications for the design of therapeutic antibodies and for the choice of animal model for preclinical pharmacokinetic work.

Combined structural and functional studies

Investigation of structural and functional epitopes on human growth hormone receptor (hGHr)[11] dates from the early days of optical biosensor applications but still stands as an eloquent testimony to the power of the technology for protein interaction studies.

The major aim of this work was to examine the functional contributions of residues known from crystallographic work to be in the structural binding site for the hormone on the receptor. To this end, selected amino acid residues in the binding site area were replaced by alanine, and an SPR biosensor measured the consequences of the exchange on the kinetics of hormone binding. The approach using biosensor technology had the double advantage of eliminating the need for labeled interactants and providing separate kinetic information for the association and dissociation. The results could be interpreted in terms of the thermodynamic contributions to the interaction of residues in the binding area and showed (perhaps not surprisingly) that only a relatively small number of the apparent contact residues were important for binding.

As part of this work, the authors investigated the binding behavior of wild-type and mutant hormone to the receptor. In common with many cell surface receptors, the action of hGH involves dimerization of the receptor through two distinct binding sites on each hormone molecule. The mutant hormone used was known not to support dimerization of the receptor. To control the orientation of the receptor immobilized on the sensor surface (a potentially important factor in simplifying the interpretation of kinetic measurements), the receptor was attached to the surface with thiol-based chemistry involving a single cysteine deliberately introduced to the molecule at a location distant from the hormone binding site. When the receptor was anchored to the surface by its "tail," mutant hormone bound to roughly twice the level reached for wild-type but dissociated more rapidly (Figure 6–3). We interpret this to reflect the difference in stoichiometry and the avidity effect of the dimerized receptor. However, if the cysteine used to attach the receptor to the surface was introduced in the region of contact for receptor dimerization, both wild-type and mutant hormone bound to the higher level and dissociated at the same elevated rate (Figure 6–4), supporting the conclusion that receptor dimerization does occur on the sensor surface when conditions allow.

This example underlines the flexibility of optical biosensor technology and the way in which creative experimental design can generate detailed information from relatively simple experiments. The technology available today permits considerable refinements in the approach used by Cunningham and Wells,[11] not least in the analysis of interaction kinetics and in the range of sensor surfaces available for exploring the receptor–hormone interaction mechanism. Nevertheless, the work published in 1993 stands as a pioneering effort in the application of biosensors to protein–protein interactions.

Epitope mapping

Antigenic epitopes on proteins can be characterized by a number of biosensor-based approaches, including interaction analysis following site-directed mutagenesis and peptide scanning (see reference 12 for an example). The most direct

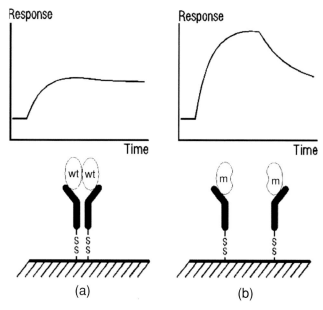

Figure 6–3: Schematic illustration of the design and results of experiments investigating the binding of human growth hormone (hGH) to its receptor.[11] When the hGH receptor is attached to the surface through a thiol group at the "tail" of the receptor, the wild-type hormone can induce receptor dissociation on the surface **(a)**. Dissociation of the hormone from the surface proceeds slowly. A mutant hormone that in itself cannot induce receptor dimerization **(b)** binds to a higher level but dissociates more rapidly because a higher stoichiometry is achieved but avidity effects are absent.

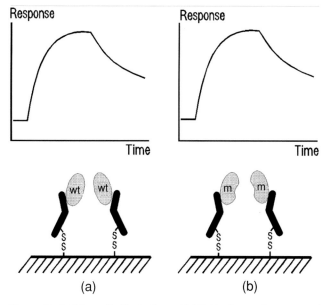

Figure 6–4: Schematic illustration of hGH binding to its receptor (continued). When the thiol group used to attach the receptor to the surface is positioned so that dimerization of the receptor is not possible, both the wild-type **(a)** and mutant **(b)** hormones bind with the same characteristics and do not show avidity effects.

way, which also illustrates the power of optical biosensors in analyzing multicomponent complex formation, is to test the ability of different monoclonal antibodies to bind simultaneously to the antigen: only antibodies recognizing structurally distinct epitopes can bind at the same time. This approach provides information on the epitope specificity of the antibodies (which can be valuable in selecting or designing antibodies as biomolecular reagents) as well as mapping the epitopes in relation to each other on the surface of the antigen.

Robbio et al.[13] used a panel of ten monoclonal antibodies (MAbs) for epitope mapping of human apolipoprotein B-100, a constituent of serum lipoproteins, and a ligand for the low-density lipoprotein (LDL) receptor. The work provides a straightforward illustration of the technique of simultaneous binding analysis for epitope mapping. The basic experimental protocol is simple: (1) one antibody is attached to the sensor surface; (2) antigen is captured on the bound antibody; and (3) the ability of other antibodies to bind to the captured antigen is tested. To allow the same sensor surface to be used for tests of different antibody combinations, the first antibody is captured on immobilized rabbit antimouse IgG-Fc (RAM-Fc) rather than covalently immobilized. Surface regeneration then removes the captured first antibody as well as antigen and subsequent antibodies so that the same or a different first antibody can be tested in the next cycle (the principle is illustrated in Figure 6–5). This refinement introduces an additional step, blocking of unoccupied RAM-Fc with an irrelevant antibody after capture of the first MAb.

Using this approach, we can accomplish pairwise testing of 10 antibodies in 100 cycles. Homologous pairs (where the same antibody is used both first and second) are expected to show no binding of the second antibody and provide a control that RAM-Fc sites on the surface are fully occupied and that the antigen does not carry multiple copies of the same epitope. Reciprocal pairs (where the roles of first and second antibody are reversed) are expected to give equivalent results and provide a control for complications such as cooperativity between binding of different antibodies. Robbio et al. also used an extended variant of the pairwise interaction scheme where a series of antibodies instead of a single second antibody was tested for binding to the captured antigen. From the results of the binding analysis, the panel of ten MAbs could be divided into seven groups related to partially or wholly distinct epitopes. Immunoblotting experiments with defined fragments of the apolipoprotein supported the conclusions, identifying four regions of the molecule where distinct epitopes were located.

This approach to epitope mapping involves the construction of multicomponent complexes. To identify two separate epitopes a complex containing at least three molecules has to be observed (the antigen and two different antibodies). In contrast to other immunoanalytical methods for epitope mapping, optical biosensors offer the powerful advantage that every step in the attempted construction of the complex is observed with the same detection technique and sensitivity. Bound levels of second antibody can be related directly to the amount of antigen captured by the first antibody, greatly reducing uncertainties in interpretation of the results.

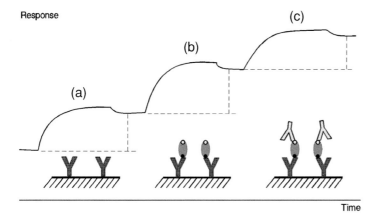

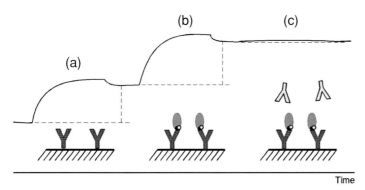

Figure 6–5: Schematic illustration of the principle of pairwise epitope mapping. One antibody is attached to the surface **(a)** and binds the antigen **(b)**. The complex is then challenged with a second antibody **(c)**. A response is obtained from the second antibody if the two antibodies recognize distinct epitopes (*top panel*) but not if the epitopes are identical or interfere sterically with each other (*bottom panel*). Steps that are important for the practical execution of epitope mapping but are not essential to the principle are omitted from this illustration (see text).

Ligand fishing

Optical biosensors can be used as both analytical and preparative tools in combination with other laboratory techniques. Catimel et al.'s work[14] establishing conditions for using biosensors with chromatographic techniques in fishing for orphan ligands may serve as an illustration. Orphan ligands are binding partners for molecules whose receptor function is deduced from structural or functional analogy with known receptors.

Catimel et al. used biosensors in combination with microscale chromatography in what they term "bidirectional synergy," where on the one hand chromatographic techniques prepare material for biosensor experiments and on the other hand biosensors identify and characterize fractions from chromatographic separations. With the supposed receptor attached to the sensor surface, the biosensor can be used to determine the presence of orphan ligands in cell culture media, cell extracts, or other sources and to locate fractions containing the ligand during purification.

A natural extension of this application is using the biosensor itself as a micropreparative affinity surface. Although the flow system in continuous-flow Biacore systems has been refined to allow micropreparative work, the small surface area of the flow cell restricts the amount of material that can be handled in one interaction cycle. Here a well-based system such as that in ForteBio instruments, with a surface area ten times that of the Biacore flow cell, comes to advantage: the improved capacity for micropreparative work outweighs the higher precision and kinetic resolution offered by a flow system.

To establish the usefulness of the approach, Catimel et al. used the A33 epithelial antigen, identified as a possible target for antibody treatment of colon cancer, as the "orphan ligand," and humanized A33 antibodies or F(ab)$'_2$ fragments as the "receptor." Though not a true ligand fishing experiment, this setup establishes proof-of-principle because the components are adequately characterized. Detergent extracts of transfected cells expressing the A33 antigen were fractionated on a Mono Q™ anion exchange column, and biosensor analysis with A33 antibody immobilized on the surface was used to identify fractions containing the antigen. The same surface was then used in repeated cycles of binding and recovery to enrich the antigen in the pooled Mono Q fractions, resulting in recovery of approximately 1 μg of protein from 50 automated cycles. The resulting material was not pure antigen but was sufficiently enriched to allow identification of the antigen by gel electrophoresis and western blotting.

In further experiments, the same antigen was recovered using a similar approach from nontransfected colon carcinoma cells. Using F(ab)$'_2$ fragments instead of IgG on the sensor surface resulted in lower contamination of the antigen with other proteins in the micropreparative biosensor step, and the recovered material could be further characterized and identified using reverse-phase high performance liquid chromatography (HPLC).

Catimel et al. used a known system to establish that micropreparative ligand fishing with optical biosensors is a workable approach. Several other authors have used the technique on an analytical scale to identify true orphan ligands.[15–17] Though the basic principle of ligand fishing with biosensors is no different from nanoscale affinity chromatography, the striking advantage of the technique is that the interaction is monitored directly and quantitatively throughout the experiment.

PRACTICAL ASPECTS

In view of the wide diversity of experimental purposes and molecules involved in protein–protein interaction studies, it is not particularly useful to review worked examples of individual studies. Instead, this chapter reviews the central practical considerations common to most protein–protein investigations (and indeed to investigation of many other types of interaction) using optical biosensors.

Design of the active surface

Preparing the active sensor surface is the first step in any biosensor investigation, and appropriate design of the surface may be critical to the outcome of the experiments. One issue is the choice of which interaction partner should be attached to the surface. This is sometimes dictated by the type and conditions of the experiment and at other times is a parameter in experimental design. Thus, ligand fishing experiments require the orphan receptor to be immobilized because the ligand is unknown; characterization of monoclonal antibodies in cell culture medium is best performed with the antigen attached to the surface, because a single antigen will be used to characterize a range of antibodies. On the other hand, careful kinetic determinations performed with purified components leave the question of which partner to immobilize open to experimental design, and the decision will be made on the basis of criteria such as availability, ease of attachment, ability to regenerate without loss of activity, and interaction mechanism. In some cases it can be valuable to perform reciprocal determinations, switching the roles of attached and free partners, to establish whether attachment of one or the other molecule in itself affects the experiment's outcome.

Suppliers of optical biosensor equipment offer a range of surfaces with different surface properties and a range of approaches for chemical or capture-based protein attachment. This flexibility allows extensive opportunity to optimize the active surface design with respect to mode and orientation of attachment as well as surface environment for the interaction, ranging from indiscriminate attachment through amine coupling to site-directed thiol immobilization (see, for example, reference 11) and maintenance of hydrophobic environments for membrane proteins.

The amount of protein attached to the surface is important in experimental design. Increasing this amount to a certain level increases the binding capacity of the surface and enables detection of lower amounts of binding partner in the sample but also shifts the balance between binding rate and transport of the soluble partner to the surface, limiting the range of kinetic determinations. As a general rule of thumb, the amount of protein attached to the surface should be high for experiments aimed at detecting or quantitating binding partner in solution and low for experiments aimed at measuring interaction rates.

Confirmation of surface activity

Once the chosen binding partner has been attached to the surface, it is important to establish that the biological activity of the attached molecule is preserved as far as possible. In practical terms, this is a matter of testing the binding of a known interaction partner under controlled conditions, and for some simple experiments it may be sufficient just to establish that the surface can bind enough material for confident detection of the interaction partner in solution. For more demanding applications, the surface activity can be expressed quantitatively as the maximum binding capacity for a known analyte in relation to the expected capacity. Because optical biosensors respond to changes in mass concentration on the surface, the expected capacity can be calculated from the response corresponding to the amount of immobilized partner and the size ratio and binding stoichiometries for the partners:

$$\text{expected capacity} = \text{response}_{\text{immobilized}} \times \frac{\text{MW}_{\text{soluble}}}{\text{MW}_{\text{immobilized}}} \times \text{stoichiometry},$$

where $\text{MW}_{\text{soluble}}$ and $\text{MW}_{\text{immobilized}}$ are the molecular weights of the partners in solution and on the surface, respectively.

In practice, the binding capacity of the active surface seldom reaches the level expected from this theoretical calculation for a number of reasons: the original protein sample used to prepare the surface may not be fully active, some activity may be lost as a result of the attachment procedure, and so on. Activity levels of 60% to 70% of the expected or higher are often acceptable: for quantitative determinations such as kinetic measurements it may be more important to determine the actual binding capacity of the surface than to strive for the highest possible level of activity. Surfaces with a binding capacity less than about 50% of the expected should in general be treated with caution because the interaction properties of the apparently inactive fraction of the attached material are unknown.

The maximum binding capacity of a surface may be determined approximately from the response obtained with a sample containing a high concentration of known binder. This is not always practicable: if the known binder is only available at relatively low concentrations, the response will be lower than the maximum capacity (for a 1:1 interaction, the surface will be 50% saturated when the sample contains binding partner at a concentration equal to the equilibrium dissociation constant). In such cases, it may not be possible to quantitate the activity of the surface: it is, however, still important to determine that the surface is sufficiently active to give an adequate response for the experimental purpose.

A similar situation can arise when the binding partner for the surface is not available in purified form or is unknown as in the case of ligand fishing experiments. Some kind of positive control must be devised for these experiments: it is not satisfactory to simply attach a protein to the sensor surface and assume that it retains activity relevant to the experimental purpose.

Choice of regeneration conditions

Practically all optical biosensor experiments involve repeated cycles of analysis on the same surface, and efficient regeneration of the surface between cycles is essential in experimental design. Ideal regeneration removes all sample-derived material from the surface and leaves the attached binding partner intact and fully active. In experiments that attach the binding partner using a high-affinity capturing approach, regeneration commonly removes the binding partner as well, leaving the surface free to capture fresh binding partner in the next cycle. This has the advantage that regeneration conditions do not have to be optimized to preserve the activity of the binding partner.

Appropriate regeneration conditions can vary widely depending on the nature of the binding partners and their interaction. For protein–protein interactions, washing with a low pH solution (e.g., glycine buffer at pH around 2–3) is often effective, although a balance usually has to be established between efficiently removing bound material and preserving the activity of the attached protein. Chaotropic agents, detergents, and organic compounds such as ethylene glycol can also be used as regeneration solutions in some cases. Though it is possible to generalize on the kind of solutions that may be used, optimal conditions have to be empirically established for every individual experimental situation. The essential criterion for determining optimal regeneration conditions is that repeated binding of a test sample gives consistent responses over at least five to ten cycles. The number of cycles tested and the degree of tolerance in response variation should be chosen in relation to the purpose of the experiment.

Experimental design and control experiments

As in all experimental work, control experiments are necessary to establish the validity of conclusions drawn from the results. In some respects, optical biosensors simplify the design of control experiments by their ability to monitor intermediate steps in a procedure as well as to provide a final result. This is made clear by a comparison with the format of enzyme-linked immunosorbent assay (ELISA), which involves a series of binding and washing steps before the final enzyme-based reporting step. All intermediate steps in an ELISA assay are performed blindly, and a lack of response at the end of the assay may be attributed to failure at any one of the previous steps. In contrast, optical biosensors monitor all stages of the procedure with the same technology, giving a continuous readout of the status of the experiment.

The same feature of the technology that offers an advantage in controlling intermediate steps also introduces a need for additional controls that other technologies may not need. Because the detection principle is indiscriminate, binding of irrelevant components is observed with the same characteristics as binding of the molecule of interest, and it is impossible to identify the nature of the component responsible for the signal from the response alone. Controls for nonspecific or irrelevant binding are therefore particularly important in optical biosensor work; they can also in some cases prove especially difficult to perform with conclusive results. Tools available for this kind of control include tests of binding to unmodified surfaces or surfaces prepared with an irrelevant protein, reversal of the roles of attached and soluble binding partners where this is feasible, and the use of independent secondary reagents to confirm the identity of the material bound to the surface.

SUMMARY

In this chapter we have given a brief and selective overview of optical biosensor technology in biomolecular interaction analysis with a focus on protein–protein interactions. The technology was introduced commercially as late as 1990 with the first system from Biacore AB (now a part of GE Healthcare), and the contribution of optical biosensors to research in life sciences and development of biotechnological and pharmaceutical products may be seen from the number of peer-reviewed publications exploiting the technology that have appeared. Many papers report work that could not have been carried out without access to the technology. Many others have found that the use of optical biosensors has streamlined and simplified investigations that otherwise would have been costly and time-consuming. From being a somewhat esoteric and specialist technology in the early 1990s, label-free biosensors have moved into the sphere of established laboratory tools.

REFERENCES

1. B. P. Glover, C. S. McHenry, *J Biol Chem* **273**, 23476 (1998).
2. M. S. Song, H. G. Dallmann, C. S. McHenry, *J Biol Chem* **276**, 40668 (2001).
3. F. P. Leu, M. O'Donnell, *J Biol Chem* **276**, 47185 (2001).
4. P. Steinrucke *et al.*, *Anal Biochem* **286**, 26 (2000).
5. I. K. Pemberton, M. Buckle, *J Mol Recognit* **12**, 322 (1999).
6. P. Nilsson, B. Persson, M. Uhlén, P.-Å. Nygren, *Anal Biochem* **224**, 400 (1995).
7. S. C. Schuster, R. V. Swanson, L. A. Alex, R. B. Bourret, M. I. Simon, *Nature* **365**, 343 (1993).
8. R. Portmann *et al.*, *J Biol Inorg Chem* **9**, 396 (2004).
9. P. S. Katsamba, S. Park, I. A. Laird-Offringa, *Methods* **26**, 95 (2002).
10. R. J. Ober, C. G. Radu, V. Ghetie, E. S. Ward, *Int Immunol* **13**, 1551 (2001).
11. B. C. Cunningham, J. A. Wells, *J Mol Biol* **234**, 554 (1993).
12. C. Uthaipibull *et al.*, *J Mol Biol* **307**, 1381 (2001).
13. L. L. Robbio *et al.*, *Biosens Bioelectron* **16**, 963 (2001).
14. B. Catimel, J. Weinstock, M. Nerrie, T. Domagala, E. C. Nice, *J Chromatogr A* **869**, 261 (2000).
15. S. Davis *et al.*, *Cell* **87**, 1161 (1996).
16. T. D. Bartley *et al.*, *Nature* **368**, 558 (1994).
17. C. Williams, *Curr Opin Biotechnol* **11**, 42 (2000).

7 Interactions with membranes and membrane receptors

Matthew A. Cooper

INTRODUCTION	159
SUPPORTED LIPID MONOLAYERS	160
TETHERED BILAYER MEMBRANES	162
BLACK LIPID MEMBRANES	163
POLYMER-SUPPORTED LIPID LAYERS	163
AFFINITY-CAPTURED LIPOSOMES	165
MICROARRAYED LIPID LAYERS	165
G-COUPLED PROTEIN RECEPTORS	166
GPCR interactions monitored with flow-based biosensors	167
GPCR interactions analysis with cell impedance measurement	168
KINETIC ANALYSIS OF LIPID BINDING	169
RECONSTITUTION OF MEMBRANE RECEPTORS	171
General experimental considerations	171
WORKED EXAMPLES	172
Example 1: Preparation of vesicles and membrane fragments	172
Example 2: Cholera toxin binding a ganglioside–lipid monolayer	174
REFERENCES	177

INTRODUCTION

Many interactions studied in the biological and biomedical sciences occur with receptors at membrane surfaces. Prominent examples are neuroreceptors, cytokine receptors, ligand-gated ion channels, G protein-coupled receptors (GPCRs), and antibody and cytokine receptors. Interactions with these receptors are especially important to academics and the pharmaceutical industry as almost half of the 100 best-selling drugs on the market are targeted to a membrane receptor. To better understand the binding mechanisms of ligands with these receptors, the ligand-receptor interactions must be probed directly *in vivo* or in reconstituted membrane systems.[1,2] Most techniques for detailed kinetic analysis of molecular recognition events are applied in solution phase using a truncated, soluble form of the receptor. Membrane receptors, however, possess significant hydrophobic domains and are likely to have different tertiary structures and binding affinities in solution relative to those occurring in a membrane environment. This approach is limited to receptors containing a single transmembrane domain and does not allow the study of signaling cascades triggered by ligand binding to a receptor or the investigation of complex membrane proteins that often homo- or heterodimerize. However, in the last 10 years there has been significant progress in the development of techniques that allow the analysis of membrane-associated ligand–receptor interactions in a model resembling their native membrane environment.

Biophysical techniques such as patch clamping, magic angle spinning nuclear magnetic resonance (MAS-NMR), fluorescence correlation spectroscopy, fluorescence resonance energy transfer, and analytical ultracentrifugation have been applied to the analysis of binding to whole cells, membrane protoplasts, and proteoliposomes.[3,4] However, interactions with membrane receptors are surface-related processes that are difficult to study with bulk techniques. Furthermore, there are relatively few methods that allow quantitative, noninvasive determination of both the affinity *and* kinetics of such interactions. The successful approaches using biosensors are reviewed briefly in this chapter. More extensive reviews of other systems such as lipid monolayers at the air–water interface and freestanding lipid bilayers or black lipid membranes can be found elsewhere.[5-7]

SUPPORTED LIPID MONOLAYERS

The simplest method for the immobilization of membranes on a sensor surface is simply to adsorb the lipid onto a hydrophobic surface.[8,9] This results in the formation of a supported lipid monolayer in which the hydrophobic acyl chains of the lipids contact the hydrophobic surface and the polar lipid head groups are presented to solution (Figure 7–1). This method is limited to receptors anchored only in the outer leaflet of a native membrane or cases in which the analyte binds to the lipid itself. Lipid monolayers are normally formed from small unilamellar vesicles that spontaneously adhere to a hydrophobic self-assembled monolayer (SAM) with concomitant release of their strain energy. The kinetics of vesicle "unrolling" depend strongly on the composition of the SAM, suggesting that the adsorption of the monolayer, not diffusion of the vesicle to the surface, is the rate-limiting step.[10] More hydrophilic surfaces promote formation of lipid bilayers.[6,10] Gel-like vesicles constituted from the synthetic lipid dimyristoylphosphatidylcholine (DMPC) adsorbed much more rapidly to the surface than more robust, fluid, liquid crystal-like vesicles constituted from natural egg phosphatidylcholine (PC).[10] Somewhat surprisingly, studies using labeled lipids and proteins have shown that supported lipid monolayers possess similar lateral

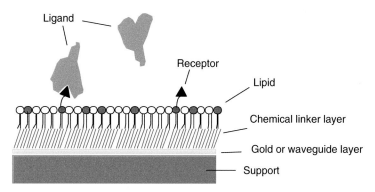

Figure 7–1: A supported lipid monolayer that has been formed on top of a hydrophobic self-assembled monolayer on a gold surface or a waveguide layer. *See color plates.*

diffusion rates to those found in supported lipid bilayers.[11] It has also been shown that membrane proteins incorporated into lipid monolayers can functionally interact with each other.[12] Lipid monolayers are highly homogeneous, possess very few surface defects, and resist nonspecific binding of proteins such as BSA.[8,13–15] The fractional surface coverage by phosphatidylcholine vesicles of a hydrophobic gold–octadecanethiol surface has been shown by cyclic voltammetry (CV) and electrochemical impedance spectroscopy (EIS) to be greater than 98%.[16]

Vogel and coworkers first exploited surface plasmon resonance (SPR) with supported lipid monolayers to study the interaction of the cholera toxin B subunit with the cell surface ganglioside GM_1,[8] a receptor–ligand pair widely employed to validate model membrane systems.[17–20] In a typical application, a gold-coated glass slide was immersed in an ethanolic solution of octadecanethiol or hexadecanethiol to render the surface hydrophobic for the adsorption of vesicles composed of palmitoyloleoylphosphatidylcholine (POPC) doped with GM_1.

In the Biacore biosensor, the correlation between RU and adsorbed lipid has been determined[15] using radiolabeled dipalmitoylphosphatidylcholine (DPPC) to be 0.92 ng.mm^2.RU^{-1}; an identical value to that reported[21] for absorption of proteins onto the dextran hydrogel-derivatized CM5 sensor chip. A flowcell covered with a confluent lipid monolayer corresponds to a level of 2200 RU or ca. 2 ng.mm^{-2}. This corresponds to a surface lipid density of 2 ng.mm^{-2} or 2.6 pmol.mm^{-2} and an area per lipid molecule of 64 Å^2. The calculated surface density is approximately half that reported[22] for a supported lipid bilayer (5.5 pmol.mm^{-2}) and the calculated lipid head group area agrees well with the value determined[23] for hydrated DPPC by continuous X-ray scattering (66 Å^2). Lipid loading levels of ca. 1500–2000 RU are routinely observed that resist BSA binding.[15] Once formed, the lipid monolayers resist many reagents used for the dissociation of protein–protein and protein–nucleotide interactions and can be used with a wide variety of detergent-free buffers.[15] In general, small ligands such as lipopeptides and gangliosides can be introduced directly into preformed lipid layers by injection of dilute solutions over the surface. Transfer of lipids and receptors to the surface may also be achieved by dipping the surface in a lipid monolayer at the air–water interface of a Langmuir trough. This also allows control of the packing density of the lipids,[6] but this method is not widely used in commercial biosensors.

A hydrophobic sensor surface was commercialized by Biacore and released as the HPA (hydrophobic association analysis) sensor chip in 1996. Since then there have been numerous applications of this very simple, yet robust model membrane chip. It has been used to study interactions between proteins and lipids, for example, factor VIII and phosphatidylserine,[24] protein kinases with phosphoinositol lipids,[25] the antibacterial peptide cecropin with vesicles of different compositions,[26] and cholera toxin with a range of gangliosides and other cell surface receptors.[18,27] The hydrophobic association (HPA) chip has also been used to study interactions of drugs with membrane-bound lipopeptides, for example, vancomycin binding to bacterial peptidoglycan analogues.[28,29] GPI-anchored proteins such as amino peptidase N have been immobilized in lipid monolayers on the HPA chip and shown by SPR to bind to the *Bacillus thurengiensis* toxins.[30]

Supported lipid monolayers also provide an ideal surface on which to study more complex, multivalent interactions with cell surface receptors. Multivalent binding is an important, inherent feature of many biological systems. Bivalent molecules such as antibodies behave differently in solution than at a surface. Studies of antibody binding using radiolabeled antibodies have shown that measured affinity constants at surfaces are sensitive to numerous experimental variables, such as the volume of incubation and the antigen surface density. Receptors in a lipid layer should be mobile in the plane of the lipid and therefore able, in principle, to adopt the required geometry for cooperative binding at the surface. This approach has been used to study bivalent binding of antibodies to lipoylated receptors[31] and polyvalent bacterial toxins with glycolipids.[32] Finally, it has been demonstrated that the cytochrome C oxidase complex formed between horse heart cytochrome C and bovine cytochrome C oxidase can be reconstituted into a supported planar egg phosphatidylcholine monolayer containing varying amounts of cardiolipin.[33]

TETHERED BILAYER MEMBRANES

A lipid layer directly adsorbed onto a sensor surface has the significant drawback that it cannot accommodate transmembrane proteins with sizeable cytosolic or extracellular domains. To overcome this limitation several methodologies have been developed to distance a lipid bilayer at some distance away from the surface. These "tethered bilayer membranes" (tBLMs) are attached in a variety of ways to a solid support (Figure 7–2). They are readily formed by self-assembly, are very stable, and can be probed by SPR, surface plasmon fluorescence spectroscopy (SPFS), and also via electrical measurement if the surface is conducting (e.g., metals, indium-tin oxide, and conducting polymers). Much of the pioneering work

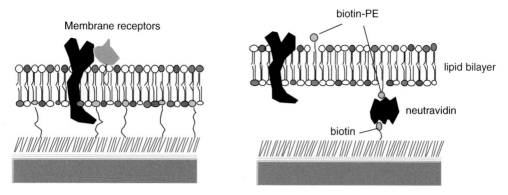

Figure 7–2: Two examples of tethered lipid bilayers that contain an integral (transmembrane) receptor. The bilayer is either captured on the surface using synthetic phospholipids that are tethered to the support by flexible, hydrophilic linkers (*left*) or via immobilized neutravidin in conjunction with biotinylated lipids or a biotinylated receptor. *See color plates.*

in this area has been carried out by Vogel and coworkers, who employed a thio-phospholipid that possesses a triethyleneglycol spacer unit to capture membranes and membrane proteins.[34] This approach has enabled the functional reconstitution of GPCRs such as rhodposin[1,35] and the nicotinic acetylcholine receptor,[36] and ion channels such as OmpF.[37] Cornell and coworkers used a similar strategy with the addition of a membrane-spanning thiophospholipid that greatly improved the stability of the tBLM.[38] Finally, mixed self-assembled monolayers of hydroxyl- and cholesterol-terminating thiols were employed to capture lipid bilayers,[39] which could also be microarrayed using microcontact printing techniques.[40]

BLACK LIPID MEMBRANES

Freestanding planar lipid bilayers or "black lipid membranes" are formed by spreading a small amount of a lipid solution over millimeter-sized holes in a solid support. They have long been used to study membrane receptors, particularly ion channels. However, there are few examples in which this approach has been combined with SPR. Tollin and coworkers have used a Teflon® (DuPont) spacer to support bilayers over a silver film to enable SPR detection of binding of transducin to the GPCR rhodopsin.[41] Unfortunately, these structures require a high degree of technical skill to prevent multilayer formation and to maintain the physical integrity of the membrane for subsequent analysis. As such, they have not yet been developed in a "ready to use" commercially available system.

POLYMER-SUPPORTED LIPID LAYERS

To overcome problems associated with surface roughness, the groups of Sackmann, Ringsdorf, and Knoll have investigated the formation of lipid bilayers

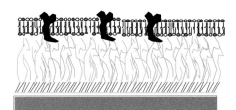
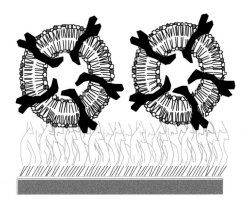

Figure 7–3: Flexible, amphipathic polymer cushions support membranes as either supported lipid bilayers (*left*) or captured proteoliposome layers (*right*). *See color plates.*

bound to, but structurally decoupled from, the solid support by a flexible polymer (Figure 7–3). These soft polymer cushions provide a lubricating layer between the surface and the membrane and enable the "self-healing" of surface defects that increase the degree of nonspecific binding to the surface. Three different basic strategies have been employed: (a) the chemical grafting to the solid surface of an ultrathin film of a water-soluble natural polymer such as dextran or hyaluronic acid, which has been derivatized with lipophilic groups that insert into and anchor membranes[42–44]; (b) coupling to the surface lipopolymers, which possess functionalized head groups[45]; and (c) the deposition of soft hydrophilic multilayers of rodlike molecules with lipidic moieties chains that insert into and anchor membranes.[46–48] Of these alternate approaches, the former has been most widely adopted due to the development by Biacore of an amphipathic carboxymethyldextran surface in which dextran is derivatized with alkyl chains, released as the L1 sensor chip in 2000.[43]

There has been a dichotomy in recent literature regarding the morphology of the model membranes formed on surfaces such as the L1 chip. This is important as the behavior of lipids and membrane receptors in a "fluid mosaic" integral plane, rather than in discretely immobilized liposomes, may affect analyte binding behavior. Confocal fluorescence microscopy (CFM) used together with sulphorhodamine-containing egg PC vesicles were used to show that upwards of 8000 RU of lipid could be immobilized on the chip in the form of intact vesicles.[43] In contrast to this report, it was later shown that synthetic POPC vesicles encapsulating tetramethylrhodamine isothiocyanate (TRTIC) labeled dextran burst open upon loading on the L1 chip to form a supported lipid bilayer.[49] Similar behavior was also observed when using a quartz surface and vesicles composed of synthetic POPC or of natural *E. coli* lipids: the synthetic lipid vesicles spread on the surface to form a supported lipid bilayer, whereas natural lipids were deposited as intact vesicles.[22] When vesicles composed of synthetic DMPC containing 25 mM tetramethylsulforhodamine B were deposited on the L1 chip to a level of 6500 RU, CFM of the surface showed no fluorescence, indicating that the vesicles had "burst" open on the surface. DMPC vesicles containing 1% w/w of the fluorescent lipid 1-oleoyl-2-[6-[(7-nitro-2-1,3-benzoxadiazol-4-yl)amino]hexanoyl]-sn-glycero-3-phosphoethanolamine (NBD-PE) were deposited on the chip to a level of 6900 RU. CFM of this surface showed homogeneous fluorescence suggesting a supported lipid bilayer. To explain these results it is important to note that lipids from natural sources are a complex mixture of compounds with saturated and unsaturated acyl chains of varying length. They possess much lower transition temperatures and form much more liquid crystal-like layers and more robust vesicles than synthetic lipids.[3,50] Using a lipid block copolymer support Setitz et al.[46] have recently shown that hypotonic osmotic stress of lipid vesicles stimulates the transition from captured intact vesicles to a homogeneous lipid bilayer on the polymer cushion. Hence, the apparently conflicting results in the literature regarding morphology of the model membrane formed on the L1 chip may simply reflect the type of lipids used and the conditions employed during vesicle deposition.

There is significant additional experimental evidence[2,42,46,51] that supported lipid bilayers and captured vesicles are located on the periphery of the supporting polymer cushion. Note that if there is insufficient coverage of the supporting polymer layer with lipids, it is possible that resultant defects in the lipid bilayer/captured vesicle layer could allow proteins and small molecules to penetrate the hydrogel and bind nonspecifically to the polymer interior. Despite this potential limitation, polymer-supported lipid "chips" have been used extensively in biosensor analysis of molecular recognition at membrane surfaces. Of note are studies of the interaction of kinases and regulators of signal transduction with phosphoinositol lipids,[52–54] low molecular weight drugs and brush border membrane vesicles,[55] lytic peptides with neutral and cationic lipids.[56] Applications of the L1 chip for use together with SPR have been reviewed in detail.[57]

AFFINITY-CAPTURED LIPOSOMES

Methodologies have also been developed to indirectly capture intact vesicles, initially via chemically coupling antibiotin antibodies to a polymer hydrogel, which in turn captures vesicles containing biotinylated-phosphatidylethanolamine.[58] Others have chemically coupled antibacterial-lipopolysaccharide antibodies to the hydrogel, which then captured vesicles containing lipopolysaccharide[32] or employed a high-density streptavidin monolayer formed above a mixed monolayer of biotin-terminated and hydroxyl-terminated poly(ethylene oxide) alkylthiolates tethered to a gold surface to capture biotin-phospholipid vesicles to form a high-density, planar layer of intact vesicles.[59]

MICROARRAYED LIPID LAYERS

With almost half of the drugs on the market today targeting membrane receptors,[60] there is considerable interest in methods to create spatially addressable

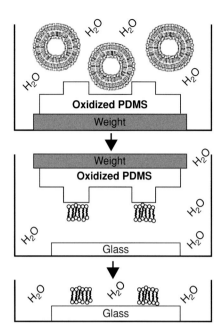

Figure 7–4: Schematic illustration of the micro-contact printing process. Vesicles are brought into contact with an oxidized PDMS stamp and fuse to form supported bilayers. The pattern is then transferred to a glass surface where the shape is retained. Typically, the raised PDMS features are on the order of 1 μm and the bilayer is on the order of 5 nm, so the drawing is not to scale.

arrays of lipid suitable for screening applications. Much of the pioneering work in this area was carried out by the groups of Boxer[61,62] and Cremer.[63] Both groups used a process called microcontact printing (μCP) in which a patterned stamp made from poly(dimethylsiloxane), or PDMS, is brought into contact with a planar-supported lipid bilayer formed by physisorbtion or Langmuir-Blodgett transfer on a glass slide or silica wafer. This resulted in the displacement of adsorbed lipid at the areas of contact between the stamp and the slide, which could then be filled with a blocking molecule such as bovine serum albumin (BSA). Alternatively, the PDMS stamp could be used to pattern the surface with a filler molecule such as BSA and the remaining areas backfilled with lipid. More recently Boxer's group has shown that supported lipid layers can be formed directly onto hydrophilic oxidized PDMS and the lipid transferred onto a planar surface with no loss of structure (Figure 7–4).

Arrays of membrane receptors have also been fabricated by directly printing membrane preparations of neurotensin (NTR1), adrenergic, and dopamine receptors onto surfaces coated with aminopropylsilane.[64]

G-COUPLED PROTEIN RECEPTORS

GPCRs constitute a large family of proteins that controls many physiological processes and is the target of many effective drugs. GPCRs are said to be highly "druggable" targets, (recent review data indicates that over a quarter of present drugs have a GPCR as a target[65]), and are hence of considerable pharmacological importance. Unfortunately, GPCRs are generally unstable when isolated, and despite considerable efforts as of today it has been possible to crystallize only very limited number of GPCRs. These include bovine rhodopsin, which is exceptionally stable, and the beta-2 adrenergic receptor, which could only be crystallized as a fusion protein, as a thermostable mutant, or in complex with an antibody fragment. In the face of limited structural data related to drug–GPCR interactions, there is a significant unmet demand for label-free methods that can characterize in detail the interactions of drugs and drug candidates with GPCRs.

GPCR interactions monitored with flow-based biosensors

Following pioneering work that demonstrated analysis of binding of analytes to the nicotinic acetylcholine receptor[36] and rhodopsin,[1,35] the concept of "on-chip" dialysis for reconstitution of functional GPCRs from detergent-solubilized preparations was developed.[66] Here rapid immobilization and reconstitution of GPCRs on carboxylated dextran surfaces modified with lipophilic groups was achieved after direct amine coupling of a detergent-solubilized receptor to a carboxymethyldextran surface. Lipid/detergent-mixed micelles are adhered as they are injected over the immobilized surface, taking advantage of integrated flow cells present in many biosensor systems. The detergent can then be eluted in the subsequent buffer flow, leaving functional, intact mutant GPCRs for subsequent screening and analysis. Solubilized receptor preparations are typically made from a suitable inducible cell line (e.g., Sf9, XL10), which can be harvested by centrifugation, aliquoted, and then stored frozen. The cells are then broken by multiple freeze–thaw cycles and resuspended in physiological buffer containing a protease inhibitor, lysosyme, and DNase. They are then solubilized with nonionic detergents such as 0.5% to 2% w/v decyl maltoside, dodecyl maltoside, trilauryldimethylamine oxide, nonyl glucoside, or octyl glucoside. These preparations can be centrifuged to remove insoluble material and then transferred to new tubes and either used immediately or (depending on the receptor) kept frozen at –80°C until analysis. Resultant solubilized receptors are then captured onto a suitable amine-reactive support such as a Biacore CM4 or CM5 sensor chip. The dextran matrix of the sensor chip is first activated by 35 uL of 50 mM N-hydroxysuccinimide and 200 mM N-ethyl-N-[(dimethylamino)propyl]carbodiimide at a flow rate of 5 uL/min, followed by a 7-min injection of 0.1 mg/mL detergent-solubilized GPCR receptor. Any remaining reactive carboxyl groups are deactivated using a 7-min pulse of 1 M ethanolamine hydrochloride, pH 8.5. After the injection, the biosensor chip is washed at high flow rate with the SPR running solution until a stable baseline is restored (ca. 30 min). This washing step works like a flow dialysis procedure and ensures the removal of the detergent from the sensor chip surface; however, hydrophobic parts of the transmembrane receptor rhodopsin will still be attached to some lipid or detergent molecules to maintain functional integrity.

Such an approach is ideally suited for use in other flow-based biosensors, such as quartz crystal microbalance, planar waveguide, SPR, surface Raman, or grating-coupled biosensors. For example, it has recently been reported that protease activated receptors (PARs), which consist of a family of four G protein-coupled receptors, can be analyzed using a resonant waveguide grating biosensor (the Corning Epic system)[67] in conjunction with human epidermoid carcinoma A431 cells. In this case the biosensor measures dynamic mass redistribution resulting from drug-induced receptor activation in adherent cells. In A431 cells, both PAR1 and PAR2 agonists, but neither PAR3 nor PAR4 agonists, triggered dose-dependent Ca^{2+} mobilization as well as G_q-type dynamic mass redistribution (DMR) signals (see Chapter 9 in this volume for more details on this effect). The same biosensor system has been used in endogenous receptor panning by determining the efficacies of a set of family-specific agonists.[68] In this case three major types of optical signatures were identified that correlated with the activation of a class of GPCRs, depending on the G protein with which the receptor is coupled (i.e., G_q, G_s, and G_i). This approach provides an alternative readout for examining receptor activation under physiologically relevant conditions.

GPCR interactions analysis with cell impedance measurement

Cell-based assay technologies that provide high-content information (high-content screening [HCS]) have recently generated significant interest in the hit-to-lead discovery process. Most of the cell-based systems on the market today are based on fluorescence detection; however, three groups have recently developed innovative cellular assay technologies based on radio frequency spectrometry and bioimpedance measurements that may offer a complementary approach to traditional HCS. Impedance measurements have long been exploited in commercial biosensors,[69] and the basic principles for application to cell analysis were first reported by Giaever and Keese, then at the General Electric Corporate Research and Development Centre.[70] In contrast to other methods of monitoring cellular signal transduction, impedance measurement of cellular responses can provide high-information content in a simplified, label-free, and noninvasive fashion. Major application areas include cancer biology, cell adhesion and spreading, receptor–ligand binding and signal transduction analysis, cell proliferation, cytotoxicity, cellular differentiation, and environmental toxicology. General applications in the drug discovery field including cell proliferation assays and the elucidation of compound toxicity are covered in another chapter in this volume; a brief overview of applications involving agonists, partial agonists, and antagonists of GPCRs is summarized here.

Systems from Applied Biophysics, ACEA Biosciences, and MDS Sciex that use cellular dielectric spectroscopy (CDS) can measure, quantitatively and kinetically, endogenous GPCR responses to analytes in live cells. Using this technology, a series of receptor-specific, frequency-dependent impedance patterns, sometimes termed CDS response profiles, which result from changes in cellular bioimpedance across a spectrum of frequencies (typically 1 kHz to 10 MHz), are collected. The characteristics of the CDS response profiles determine the identity of the signaling pathway being activated by a receptor–analyte interaction, which in turn provides facile access to information on compound selectivity. In addition, these profiles allow quantitative pharmacological analyses such as potency

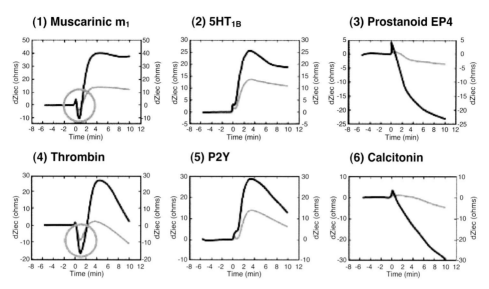

Figure 7–5: Graphs illustrating characteristic GPCR-mediated response profiles in CHO$_{m1}$ cells. Stimulation of different classes of GPCRs leads to unique kinetic response profiles such as that shown for transfected muscarinic m1 receptor (1), and the endogenous serotonergic 5HT$_{1B}$ (2), prostanoid EP4 (3), thrombin (4), P2Y (5), and calcitonin (6) receptors. The CDS response profiles representing G$_q$ GPCRs typically demonstrate an initial decrease in Z$_{iec}$ and Z$_{itc}$ (circled) followed by an increase in Z$_{iec}$ and Z$_{itc}$, whereas over the entire time period Z$_{iec}$ and Z$_{itc}$ decrease in the context of G$_s$ GPCR-mediated responses. In contrast, Z$_{iec}$ and Z$_{itc}$ increase over the entire time period in the context of G$_i$ GPCR-mediated responses. Modified with permission from Figure 7–3, reference 71.

and Schild analyses. Recently published work[71] (Figure 7–5) demonstrates the effectiveness of the system in profiling many endogenous ligand-induced cellular responses mediated by the three major classes of G-protein-coupled receptors, G$_s$, G$_i$, and G$_q$, as well as a number of protein tyrosine kinase receptors in many different cell types including primary cells.[71] In addition, the data gained from the CDS response profiles provide information about the selectivity of compounds for the target of interest. In the near future it may be possible to use this approach to assist in receptor panning, identification of G-protein coupling mechanisms, deconvolution of signal transduction pathways, and the pharmacological analysis of partial, full, and inverse agonist and antagonist action.

KINETIC ANALYSIS OF LIPID BINDING

This calculation of affinities from a kinetic analysis of analyte binding to lipid is based on the simplifying and inaccurate assumption that there is a one-to-one association between analyte and lipid.[26] However, it is not valid to use models normally applied for binding of an analyte to multiple, independent ligands or binding sites at a surface in the case of binding of proteins to immobilized membranes. Protein binding to membranes is a complicated process that often involves cooperative interaction of many lipid molecules with the protein and structural rearrangements such as compression or invagination of the lipids. Qualitatively biosensor data can often fit very well to a simple 1:1 Langmuir isotherm, in principle allowing curve fitting to provide k_{ass} and k_{diss}. However, for interactions in which the analyte interacts partially or entirely with lipids (rather than a discrete receptor), the number of lipid binding sites for each molecule of analyte is either unknown or variable (and often dependent on electrostatic as well as hydrophobic forces as well as the analyte footprint on the membrane). Hence a direct comparison of partitioning coefficient (K_p) and the affinity (K_d) derived from a biosensor assay cannot be made directly. Nevertheless, one can measure the observed association rate at a certain concentration; k_{obs} and the half-life or dissociation rate constant k_{diss}, values that are independent of the number of binding sites.

Before embarking on a complicated kinetic analysis of binding data, a simple steady-state Scatchard analysis of binding levels at equilibrium should be done to determine affinity and to check there is no steric limitation of binding levels at the lipid surface (which would produce a nonlinear Scatchard plot). In practice this means using a low concentration of analyte. The relationship between the partitioning coefficient (K_p), concentration (C), and association constant affinity (K_a) can be expressed in terms of the number of sites bound (N_b) and the total number of binding sites (N_t):

$$N_b = \frac{K_a \cdot N_t \cdot C}{1 + K_a \cdot C} + K_p \cdot C. \qquad (7\text{--}1)$$

Further, it is possible to define K_p (M^{-1}) as the proportionality factor between the mole fraction of protein bound to the membrane and the molar concentration (χ) of an analyte such as a protein, P, free in the bulk aqueous phase ([P_f]):

$$\chi = \frac{[P_b]}{([L] + [P_b])} = K_p[P_f], \qquad (7\text{--}2)$$

where [P_b] is the molar concentration of protein bound to the membrane and [L] is the concentration of lipid accessible to the protein. When [P_b] << [L], Equation 7–1 may be written as:

$$K_p = \frac{[P_b]}{[P_f][L]}. \qquad (7\text{--}3)$$

The total molar concentration of protein in the solution, [P_o], is the sum of bound and free protein concentrations:

$$[P_o] = [P_b] + [Pf]. \qquad (7\text{--}4)$$

Combining[72] Equations 7–3 and 7–4, we obtain an expression for K as a function of known [L] and measured membrane-bound and membrane-free quantities [P_b]/[P_o]:

$$\frac{[P_b]}{[P_o]} \cdot 100\% = \frac{K_p[L]}{(1 + K[L])}. \qquad (7\text{--}5)$$

RECONSTITUTION OF MEMBRANE RECEPTORS

Successful use of model membrane systems described previously requires diligent preparation and sometimes purification of the membrane-associated receptor. Membrane proteins are best deposited on the surface together with lipid in the form of proteoliposomes. The most commonly employed methods for the formation of proteoliposomes have been reviewed[22], and synopses appear in the worked examples section. Starting from either prokaryote or eukaryote cells, membrane proteins can be isolated either in a mixed micelle with detergent, dissolved, or aggregated as membrane fragments by extrusion or sonication in buffer. Once isolated and purified, the membrane proteins can be reconstituted into vesicles by (a) reconstitutions (reverse-phase evaporation, rehydration of lipid–protein films), (b) mechanical means (sonication, French press, freeze–thaw), or (c) detergent-mediated means (dialysis, dilution, or direct incorporation into preformed vesicles or into bicelles).

General experimental considerations

- Always try to use lipids above their gel–liquid crystal phase transition temperature. At 25°C all dimyristoyl lipids (e.g., DMPC), most unsaturated lipids (e.g., POPC), and most lipids from natural sources (e.g., egg, bacteria, bovine) will be in a fluid, liquid crystal state.
- When using a serial multiflow cell instrument, always load the lipids in reverse order to prevent carryover between flow cells.
- Always clean surfaces with detergent before loading with lipid and before undocking the chip at the end of an experiment. For planar surfaces (e.g., sensor chip HPA) use 40 mM octyl-D-glucoside or 40 mM octyl-D-maltoside. For hydrogel-supported surfaces (e.g., sensor chip L1) use 20 mM 3-[(3-cholamidopropyl)dimethylammonio]-1-propanesulfonate (CHAPS).
- Hydrophobic chips are best stored dry under nitrogen or argon at 4°C. Hydrogel chips are best stored under humid nitrogen or argon at 4°C (just add some water to the bottom of a falcon tube, flush with nitrogen, then seal).
- Small ligands (e.g., gangliosides, lipopeptides, phosphoinositol lipids) can be deposited by direct injection of very dilute solutions (below the critical micelle concentration) over preformed lipid layers. The amount of ligand deposited can be observed directly in real time by a biosensor.
- Larger ligands (e.g., GPI-anchored proteins, integral membrane proteins) should be deposited together with lipid in the form of detergent-solubilized receptors, proteoliposomes formed by detergent dialysis, or in the form of membrane fragments formed by sonication. There are extensive reviews of methodologies for the reconstitution of membrane proteins.[3,22,73]
- When using mass-based detection biosensors, high concentrations and purity of membrane receptor are generally required, particularly in the case of GPCRs. For pathway-based detection with whole cells (e.g., resonant waveguide grating sensing or cell impedance sensing), better results are obtained

with transfected rather than parental cell lines; however, in some cases responses involving endogenous receptors can be studied.

- Always check the surface coverage of lipid by checking for nonspecific binding of a control protein such as BSA or casein (5 min exposure at 0.1 mg/ml).

WORKED EXAMPLES

Example 1: Preparation of vesicles and membrane fragments

When working with membranes and membrane receptors preparations must be handled carefully. Many lipids and membrane receptors are sensitive to heat, oxidation, and hydrolysis and should, if possible, be kept cool in deoxygenated solvents during sample preparation. However, vesicles should always be *formed* above the phase transition temperature of the constituent lipids.[3] Unsaturated and most natural-source lipids can be processed at low temperatures to minimize degradation (e.g., egg PC has a T_c of $-15°C$), but saturated lipids such as DMPC (T_c of $24°C$) must be processed at higher temperatures. Once formed, vesicle suspensions are generally stable for weeks at $4°C$ in an inert atmosphere. Freeze–thawing induces rupture and re-fusion of vesicles, which increases their size. For a more detailed description of the methods described below and other methods of vesicle formation please refer to the comprehensive review: "Liposomes: a practical approach," Ed. R.C. New, Oxford University Press, Oxford, 1990. There is also a comprehensive review of the reconstitution of membrane proteins into vesicles.[73]

Vesicles by extrusion. Egg yolk L-α-phosphatidylcholine (16 mg, 0.1 mmol) together with 0.1% w/w ligand was dissolved in chloroform (10 ml) in a 100–ml round-bottom flask. (Note that more polar lipids and ligands may be dissolved in ethanol/chloroform mixtures.) The mixture was deposited as a thin film in the flask by removing the solvent under reduced pressure on a rotary evaporator (if this is not available then use a drying stream of nitrogen while rotating the flask). The resultant thin lipid film is then further dried under high vacuum (e.g., on a lyopholizer) for 2 hr. Two milliliters of a detergent-free buffer at close to neutral pH and physiological ionic strength such as 120mM PBS, pH 7.4, was then added to give a suspension of multilamellar vesicles at 1.25 mM lipid concentration. This milky white suspension was shaken for 30 min then passed 17 times through a 50-nm polycarbonate filter in an extrusion apparatus (available from Avanti Lipids or Avestin). It is important to pass the vesicles through the filter an odd number of times so that the final sample is not contaminated with the initial aliquot containing multilamellar vesicles.

Vesicles by sonication. A suspension of multilamellar vesicles was formed as described previously, then subjected to probe sonication in an MSE Soniprep 150 (10 pulses of 20 s duration at 6 μ or 200 W amplitude using a 3-mm microtip probe). The sample should be kept cool with a beaker of ice water, and the probe should be ca. 0.5 cm below the surface of the lipid suspension. It is important to wait 20 s between pulses to allow the sample to cool. The sample was then centrifuged at 100 000 g for 20 min to sediment multilamellar vesicles and contaminant titanium particles. The resultant supernatant contains small unilamellar vesicles with a mean diameter of 25 nm. Note this procedure is not suited for sensitive membrane receptors such as GPCRs.

Vesicles by detergent dialysis. A suspension of multilamellar vesicles was formed as a suspension in phosphate buffer solution (PBS) as described above, then a detergent such as Triton X-100 or octyl-D-glucoside was added. After equilibration for 15 min, a solution of solubilized membrane protein was added and the mixture stirred gently at room temperature for 5 min. Note that the optimal lipid/detergent/protein ratio for protein activity varies widely according to the particular protein, hence a range of such mixtures is usually prepared and assayed. The detergent was then removed by dialysis in 2 liters of PBS at 4°C overnight (with a low-volume dialysis cartridge such as a Slide-a-Lyzer available from Pierce).

Bacterial membrane fragments by sonication

- Spin a bacterial culture at midlogarithmic phase ($OD_{600} = 0.5$–0.6) at 1000 g for 10 min, then resuspend pellet in ca. 2–3 times the pellet volume of a suitable ice-cold buffer (e.g., 50 mM Tris.HCl, 2 mM $MgCl_2$, 1 mM EGTA, pH 7.4). Include 10 µg/ml DNAse and RNAse in the buffer to hydrolyze released oligonucleotides.
- Sonicate the sample in an MSE Soniprep 150 (10 pulses of 20 s duration at 6 µ or 150–200 W amplitude using a 3-mm microtip probe). The sample should be kept cool with a beaker of ice water and the probe should be ca. 0.5 cm below the surface of the lipid suspension. It is important to wait 20 s between pulses to allow the sample to cool.

Bacterial membrane spheroplasts by sonication

- Spin a bacterial culture at midlogarithmic phase ($OD_{600} = 0.5$–0.6) at 1000 g for 10 min, then resuspend pellet in ca. 2–3 times the pellet volume of a suitable ice-cold buffer (e.g., 50 mM Tris.HCl, 2 mM $MgCl_2$, 20% w/v sucrose and 50 µg/ml chloramphenicol, pH 7.4).
- Lyse the bacterial peptidoglycan with 0.5 mg/ml lysozyme and 10 µg/ml DNAse and RNAse in the buffer to hydrolyze released oligonucleotides.
- Sonicate the sample in a sonicator such as an MSE Soniprep 150 (10 pulses of 20 s duration at 6 µ or 150–200 W amplitude using a 3-mm microtip probe). The sample should be kept cool with a beaker of ice water and the probe should be ca. 0.5 cm below the surface of the lipid suspension. It is important to wait 20 s between pulses to allow the sample to cool.

Cell membrane fragments by homogenization. This method is adapted from earlier work.[4]

Solution A: 20 mM tris-HCl pH 8, 1 mM EDTA, 1 mM EGTA, 0.1 mM PMSF, 2 µg/ml aprotinin, and 10 µg/ml leupeptin

Solution B: 20 mM tris-HCl pH 8, 3 mM MgCl, 10 ug/ml DNAse I plus PMSF, 2 μg/ml aprotinin and 10 μg/ml leupeptin

- Spin 500 ml of Sf9 an overnight bacterial culture 1000 g for 10 min, remove the supernatant and then resuspend the pellet in ca. twice the pellet volume of ice-cold solution A.
- Homogenize with Dounce homogenizer (type A) for 10 strokes, then centrifuge at 30 000 g for 20 min.
- Resuspend pellet with 50 ml of solution B and rehomogenize.
- Centrifuge at 30 000 g for 20 min and resuspend pellet in 20 ml of solution B and homogenize again.
- Store at 4°C for immediate use. Store after snap freezing in liquid nitrogen at –80°C.

From the above preparation solubilized receptors can be captured as described earlier on an amino-reactive polymer-coated surface such as the Biacore CM4 or CM5 sensor chip. In this latter case, the dextran matrix of the sensor chip is activated by 35 uL of 50 mM *N*-hydroxysuccinimide and 200 mM *N*-ethyl-*N*-([dimethylamino]propyl)carbodiimide at a flow rate of 5 uL/min, followed by a 7-min injection of 0.1 mg/mL detergent-solubilized membrane receptor. Any remaining reactive carboxy groups are deactivated using a 7-min pulse of 1 M ethanolamine hydrochloride, pH 8.5. After the injection, the biosensor chip is washed at high flow rate with the SPR running solution until a stable baseline is restored (ca. 30 min). This washing step works like a flow dialysis procedure and ensures removal of the detergent from the sensor chip surface; however, hydrophobic parts of the transmembrane receptor rhodopsin will still be attached to some lipid or detergent molecules to maintain functional integrity.

Detergent-solubilized GPCRs. The GPCR must first be expressed in a suitable cell line (e.g., XL10 cells, Stratagene), which is then cultured in 50 ml of 2 × TY medium containing ampicillin (100 μg/ml) at 37°C with shaking until $OD_{600} = 3$ and then induced with 0.4 mM isopropyl-beta-D-thiogalactopyranoside (IPTG). Induced cultures were incubated at 25°C for 4 h, and then the cells were harvested by centrifugation at 13 000 g for 1 min (aliquots of 2 ml) and stored at –20°C. Cells were then broken by freeze–thaw (five cycles) and resuspended in 500 μl of buffer (20 mM Tris, pH 8, 0.4 M NaCl, 1 mM EDTA, protease inhibitor [Roche]). After incubation for 1 h at 4°C with 100 μg/ml lysozyme and DNase I (Sigma), samples were solubilized with 2% dodecylmaltoside on ice for 30 min. Insoluble material was removed by centrifugation (15 000 for 2 min at 4°C), and the supernatant was used directly for assays or stored in 1-ml aliquots at –80°C after flash-freezing in liquid nitrogen.

Example 2: Cholera toxin binding a ganglioside–lipid monolayer

Equipment and reagents needed

Biacore or similar flow-based biosensor
Hydrophobic sensor chip (e.g., Biacore HPA chip)

PBS (120 mM NaCl, 10 mM NaH_2PO_4/Na_2HPO_4, pH 7.4, filtered and degassed)

Dimyristoylphosphatidylcholine (DMPC)

40 mM octyl D-glucoside

10 mM sodium hydroxide

10 mM glycine at pH 2.0

0.1 mg/ml bovine serum albumin in PBS

Cholera toxin B subunit

Ganglioside GM_1

Sample preparation

- Small unilamellar DMPC vesicles (SUV) containing 1% w/w ganglioside GM_1 in PBS were prepared by extrusion as described above. A control batch of vesicles was also prepared lacking the GM_1.

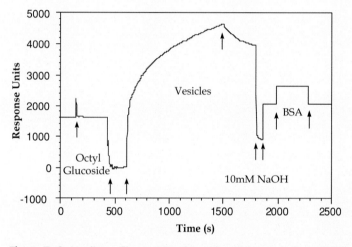

Figure 7–6: Loading a Biacore HPA chip with vesicles made from DMPC. Note the removal of loosely associated lipid by the basic injection and the subsequent lack of binding of BSA.

Surface preparation. Proteins adsorb strongly to the hydrophobic surfaces of the HPA chip. The system must be cleaned with detergent using the appropriate instrument protocols and then thoroughly rinsed with water before beginning the experiment. Detergents must obviously be excluded from all running buffers. Layer the GM_1-containing vesicles on flow cell 2 and control lipid-only vesicles on flow cell 1 of a Biacore HPA chip (Figure 7–6) as follows:

- Clean the surface with a 5-min injection of 40 mM octyl D-glucoside at 20 μl/min.
- Inject the small unilamellar vesicles across the cleaned surface for 30-min at 2 μl/min.
- Wash the surface with a 30-s pulse of 10 mM sodium hydroxide at 40 μl/min.
- Check that surface is completely covered with lipid by a 5-min injection of 0.1 mg/ml of BSA in PBS (<100 RU should bind).

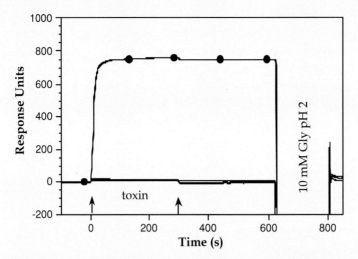

Figure 7–7: Duplicated injections of cholera toxin at 172 nM binding to a lipid monolayer-only control (*bottom*) and a 1% GM$_1$/DMPC monolayer (*top*) on an HPA chip.

- This surface is stable for up to 48 hr while the sensor chip is docked in the instrument. The chip should be thoroughly cleaned with two 5-min injections of 40 mM octyl D-glucoside before undocking.

Binding assay
- Flow 20 µl/min multichannel detection in Fc 1 and 2.
- Inject cholera toxin B subunit across Fc 1 and 2 for 3 min at 172 nM.
- Regenerate the surface with a 30-s injection of 10 mM glycine at pH 2.0.
- Repeat to make sure the binding is reproducible (Figure 7–7).

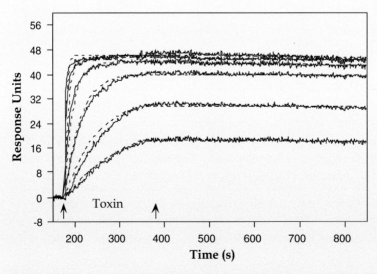

Figure 7–8: Injections of cholera toxin at 172–5.4 nM binding to a 0.1% GM$_1$/DMPC monolayer corrected for bulk refractive index changes and globally fitted to a 1:1 binding model.

Once we have shown that the binding is specific and that we can completely regenerate the surface, we can carry out a kinetic analysis. Toxin is diluted twofold from 172 to 5.4 nM, then passed in series across the lipid control surface, then the 0.1% w/w GM_1/DMPC surface (note the low receptor levels needed for robust kinetic analysis). The data are then corrected for bulk refractive index changes by subtraction of the control lipid-only data (Figure 7–8).

REFERENCES

1. S. Heyse, T. Stora, E. Schmid, J. H. Lakey, H. Vogel, *Biophysica Acta* **1376**, 3, 319 (1998).
2. E. Sackmann, M. Tanaka, *Trends Biotech* **18**, 58 (2000).
3. R. C. New, Liposomes: A Practical Approach. D. Rickwood, Ed., *Practical Approach Series* (Oxford University Press, Oxford, 1990).
4. J. M. Graham, J. A. Higgins, *Membrane Analysis*, (Springer-Verlag, New York, 1997).
5. R. B. Gennis, *Biomembranes*, (Springer-Verlag, Heidelberg, 1989).
6. D. Marsh, *Handbook of Lipid Bilayers*, (CRC Press, Boca Raton, FL, 1990).
7. W. Knoll *et al., J Biotechnol* **74**, 137 (2000).
8. S. Terrettaz, T. Stora, C. Duschl, H. Vogel, *Langmuir* **9**, 1361 (1993).
9. A. L. Plant, *Langmuir* **9**, 2764 (1993).
10. L. M. Williams *et al., Langmuir* **13**, 751 (1997).
11. F. M. Linseisen, M. Hetzer, T. Brumm, T. M. Bayerl, *Biophys J* **72**, 1659 (1997).
12. E. Torchut, C. Bourdillon, J. Laval, *Biosens Bioelectron* **9**, 719 (1994).
13. A. L. Plant, *Langmuir* **15**, 5128 (1999).
14. A. L. Plant, M. Brigham-Burke, E. C. Petrella, D. J. O'Shannessy, *Anal Biochem* **226**, 342 (1995).
15. M. A. Cooper, A. C. Try, J. Carroll, D. J. Ellar, D. H. Williams, *Biochimia et Biophysica Acta* **1373**, 101 (1998).
16. P. Diao *et al., Bioelectrochem Bioenerg* **48**, 469 (1999).
17. R. A. Reed, J. Mattai, G. G. Shipley, *Biochemistry* **26**, 824 (1987).
18. G. M. Kuziemko, M. Stroh, R. C. Stevens, *Biochemistry* **35**, 6375 (1996).
19. N. Athanassopoulou, R. J. Davies, P. R. Edwards, D. Yeung, C. H. Maule, *Biochem Soc Trans* **27**, 340 (1999).
20. G. Puu, *Anal Chem* **73**, 72 (2001).
21. E. Stenberg, B. Persson, H. Roos, C. Urbaniczky, *J Colloid Interface Sci* **143**, 513 (1991).
22. P. Nollert, H. Kiefer, F. Jahnig, *Biophysical Journal* **69**, 1447 (1995).
23. B. A. Lewis, D. M. Engelman, *J Mol Biol* **166**, 211 (1983).
24. E. Saenko *et al., J Chromatogr A* **852**, 59 (1999).
25. R. A. Currie *et al., Biochem J* **337**, 575 (1999).
26. W. Wang, D. K. Smith, K. Moulding, H. M. Chen, *J Biol Chem* **273**, 27438 (1998).
27. G. Puu, I. Gustafson, *Biochim Biophys Acta* **1327**, 149 (1997).
28. M. A. Cooper, D. H. Williams, *Chem Biol* **6**, 891 (1999).
29. M. A. Cooper, D. H. Williams, Y. R. Cho, *J Chem Soc Chem Commun* **17**, 1625 (1997).
30. M. A. Cooper, J. Carroll, E. R. Travis, D. H. Williams, D. J. Ellar, *Biochem J* **333**, 677 (1998).
31. M. A. Cooper, D. H. Williams, *Anal Biochem* **276**, 36 (1999).
32. C. R. MacKenzie, T. Hirama, K. K. Lee, E. Altman, N. M. Young, *J Biol Chem* **272**, 5533 (1997).
33. Z. Salamon, G. Tollin, *Biophys J* **71**, 858 (1996).
34. H. Lang, C. Dushcl, H. Vogel, *Langmuir* **10**, 197 (1994).
35. C. Bieri, O. P. Ernst, S. Heyse, K. P. Hofmann, H. Vogel, *Nature Biotech* **17**, 1105 (1999).
36. E. K. Schmidt *et al., Biosens Bioelectron* **13**, 585 (1998).
37. T. Stora, J. H. Lakey, H. Vogel, *Angew Chem Int Ed Engl* **38**, 389 (1999).
38. B. A. Cornell *et al., Nature* **387**, 580 (1997).

39. N. Boden *et al.*, *Tetrahedron* **53**, 10939 (1997).
40. A. T. A. Jenkins *et al.*, *J Am Chem Soc* **121**, 5274 (1999).
41. Z. Salamon, Y. Wang, J. L. Souagales, G. Tollin, *Biophys J* **71**, 283 (1996).
42. J. Spinke *et al.*, *Biophys J* **63**, 1667 (1992).
43. M. A. Cooper, A. Hansson, S. Löfås, D. H. Williams, *Anal Biochem* **277**, 196 (2000).
44. C. Erdelen *et al.*, *Langmuir* **10**, 1246 (1994).
45. E. Sackmann, *Science* **271**, 43 (1996).
46. M. Seitz *et al.*, *Langmuir* **16**, 6067 (2000).
47. D. Beyer *et al.*, *Angewante Chemie* **35**, 1682 (1996).
48. A. Kloboucek, A. Behrisch, J. Faix, E. Sackmann, *Biochem J* **77**, 2311 (1999).
49. E.-M. Erb *et al.*, *Anal Biochem* **280**, 29 (2000).
50. M. E. Jones, B. Lentz, *Biochem* **25**, 567 (1986).
51. J. Majewski *et al.*, *Biophys J* **75**, 2363 (1998).
52. C. D. Ellson *et al.*, *Nat Cell Biol* **3**, 679 (2001).
53. S. Krugmann *et al.*, *Mol Cell* **9**, 95 (2002).
54. S. H. Ridley *et al.*, *J Cell Sci* **114**, 3991 (2001).
55. K. Kim *et al.*, *Pharm Res* **21**, 1233 (2004).
56. N. Papo, Y. Shai, *Biochem* **42**, 458 (2003).
57. M. Besenicar, P. Macek, J. H. Lakey, G. Anderluh, *Chem Phys Lipids* **141**, 169 (2006).
58. L. Masson, A. Mazza, R. Brousseau, *Anal Biochem* **218**, 405 (1994).
59. L. S. Jung, J. S. Shumaker-Parry, C. T. Campbell, S. S. Yee, M. H. Gelb, *J Am Chem Soc* **122**, 4177 (2000).
60. G. C. Terstappen, R. Angelo, *Trends Pharmacol Sci* **22**, 23 (2001).
61. J. S. Hovis, S. G. Boxer, *Langmuir* **17**, 3400 (2001).
62. S. G. Boxer, *Curr Op Chem Biol* **4**, 704 (2000).
63. T. Yang, E. E. Simanek, P. Cremer, *Anal Chem* **121**, 8130 (2000).
64. Y. Fang, A. G. Frutos, J. Lahiri, *J Am Chem Soc* **124**, 2394 (2002).
65. J. P. Overington, B. Al-Lazikani, A. L. Hopkins, *Nat Rev Drug Discov* **5**, 993 (2006).
66. O. P. Karlsson, S. Löfås, *Anal Biochem* **300**, 132 (2002).
67. Y. Fang, A. M. Ferrie, *BMC Cell Biol* **8**, 24 (2007).
68. Y. Fang, G. Li, A. M. Ferrie, *J Pharmacol Toxicol Methods* **55**, 314 (2007).
69. E. Katz, I. Willner, *Electroanalysis* **15**, 913 (2003).
70. I. Giaever, C. R. Keese, *Proc Natl Acad Sci U S A* **81**, 3761 (1984).
71. G. Leung *et al.*, *J Assoc Lab Automat* **10**, 258 (2005).
72. D. Murray *et al.*, *Biophys J* **77**, 3176 (1999).
73. J. L. Rigaud, B. Pitard, D. Levy, *Biochimica et Biophysica Acta* **1231**, 223 (1995).

8 Application of SPR technology to pharmaceutical relevant drug–receptor interactions

Walter Huber

INTRODUCTION	179
ANALYSIS OF BIOPHARMACEUTICALS	180
Experimental considerations	180
Overview on applications	181
WORKED EXAMPLES	182
Monoclonal anti-β-amyloid antibodies; screening, characterization, anti-immunogenicity testing	182
Kinetic and thermodynamic analysis of the binding of pegylated interferon 2α (IFNα) to interferon receptor (IFNAR2)	185
SMALL SYNTHETIC DRUG MOLECULES	187
Experimental considerations	187
Overview on applications	188
Examples from hit and lead finding	192
Virtual and SPR screening for HPPK inhibitor	192
Screening for allosteric site binders of a heterotrimeric protein	193
Validation of DPP-IV lead series using high-resolution kinetic analysis and modified binding site controls	197
Mechanistic aspects of inhibition of DHNA	199
Consequence of inhibitor binding on the interaction of tissue factor and FVIIa	202
SUMMARY	202
REFERENCES	203

INTRODUCTION

Selection of promising, well-characterized hits and leads is essential for success in the drug discovery process. To this end, new technologies that enable screening for new hits and profiling them with respect to their interaction with the targeted biological system are highly demanded in the pharmaceutical industry.[1–3] The biological system can be a single biomolecule[4,5] or a cascade of biomolecules including the target or a whole cellular system.[6] Independent from the targeted biological system, detailed information on the interaction of potential drug

179

candidates with the targeted biomolecule form the basis for understanding more complex schemes. Binding assays provide such information on affinity, kinetics, and thermodynamics.[7,8]

Until recently, obtaining reliable data was often difficult and time-consuming because most of the methods available used labeled compounds that had to be specially synthesized. Biophysical binding assays generate label-free, high-quality data on the interaction between a target and a potential drug candidate. Label-free screening methods include isothermal titration calorimetry (ITC),[9] analytical ultracentrifugation (AUC),[10] nuclear magnetic resonance spectroscopy,[11,12] mass spectroscopy,[13] and biosensor.[14-21]

Biosensors, as described in this book, are the tools most often used. They offer rapid access to these relevant binding data without the need for labeling interacting molecules. These biosensors measure in real time the quantity of the complex formed between a molecule immobilized on the sensor surface and a molecule in solution.

The different underlying detection principles, the different surface chemistries for the immobilization of the targets, the experimental details to perform binding studies, and the ways to extract meaningful data have all been discussed in different articles of this monograph. The goal of this chapter is to present to the reader different opportunities such biosensors offer in the pharmaceutical industry along the drug discovery and development process. It is clearly outside the scope of this chapter to give a complete literature citation. Such surveys appear yearly in the *Journal of Molecular Recognition*.[9,21,22]

ANALYSIS OF BIOPHARMACEUTICALS

Experimental considerations

Sensitivity is generally not a major issue when working with biomolecules as analyte molecule in solution because the surface plasmon resonance (SPR) signal directly scales with the molecular weight of the molecule that binds to the surface. Immobilizing sufficient amounts of the active target biomolecule on the surface can be achieved in most cases. It has even been demonstrated that when using purified, solubilized protein, sufficient amounts of reconstituted GPCRs can be immobilized to monitor ligand binding.[23] One main point to be considered is steric hindrance that occurs from overcrowding of immobilized ligands on the surface. Capturing methods that support an exposition of the recognition site into the solution are recommended. They increase accessibility of the binding sites and reduce artifacts arising from interaction of ligand and analyte with the surface. An additional source of erroneous results is the tendency of many isolated biomolecules to form aggregates. When such aggregates contribute to the monitored binding event, it is impossible to extract meaningful kinetic and equilibrium data by applying simple models. A careful characterization by analytical ultracentrifugation or light scattering of the biomolecule preparation is a must to exclude such erroneous results.

Overview on applications

Some of the earliest applications of SPR biosensors were as immunosensors to characterize antibody activity and to map binding epitopes. Though the use of biosensors has expanded into other application domains, antibody characterization remains a major area. It is now well accepted that the technology has a number of advantages over traditional immunoassays such as Enzyme labeled immuno sandwich assay (ELISA) or radiolabeled imuuno assay (RIA). The possibility to monitor the binding in real time provides more detailed information about reaction kinetics, affinity, and epitope maps. Kinetic information can help select from a panel of antibodies with similar affinity those that possess the ideal kinetic behavior for a given application.

Monoclonal antibodies continue to be the most important biomolecular agents for therapeutics. Advanced antibody production technology[24,25] makes it possible to create large batches of antibodies, thereby creating a bottleneck in characterizing antibody-binding activity. SPR can improve the pace of antibody characterization.[26–29]

A recent publication by Canziani and colleagues[29] reports on the development of experimental and data analysis protocols to screen antibodies directly from hybridoma culture supernatants. The Flexchip system from Biacore uses externally spotted protein chips[30] and grating-coupled SPR detection to profiling up to 400 protein–protein interactions simultaneously. This enables large numbers of antibodies to be ranked based on their kinetic behavior in a single experiment. Monitoring the immune response quickly and directly makes this technology valuable for following passive and active vaccination.

The steadily increasing role of biopharmaceuticals in medicine has placed growing importance on the ability to characterize immune response during preclinical trials. Protein interaction analysis using SPR biosensors has increasingly demonstrated its value in this area. Several groups have reported advantages in using this technology.[31,32] These advantages concern the quality of data with lower inter- and intra-assay variation, earlier detection of the immune response because antibodies of lower kinetic stability are detected by this method, and higher information content because of kinetic profiling of antibodies.

Antibodies have carbohydrate chains (glycosylation) attached covalently to their polypeptide backbone. This posttranslational modification takes place in the endoplasmic reticulum and Golgi apparatus and is sensitive to many environmental parameters. Because changes in the glycosylation pattern can affect biologically important parameters such as activity, stability, solubility, and biological half-life, there is increasing interest in determining the glycosylation pattern. Lectins are useful for studying glycosylation. In combination with SPR this method has proved to be superior to conventional ELISA technology.[31]

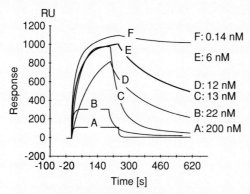

A: original clone
B: clone after 1.maturation cycle
C+D: clones after 2.maturation cycle
E: clones produced by cross-cloning heavy and light chains from C&D
F: IgG antibody from clone E

Figure 8–1: Sensograms observed for Fab fragments isolated during the selection and subsequent maturation of a Fab-producing clone. The K_D values of the different constructs A–F are shown on the left side.

WORKED EXAMPLES

Monoclonal anti-β-amyloid antibodies; screening, characterization, anti-immunogenicity testing

The project discussed in this chapter aimed at finding and improving antibodies against β-amyloid plaques that could be used for analytical and/or therapeutic purposes in the field of Alzheimer's desease. In this example, SPR technology is used in the many phases of the development of a therapeutic antibody: selection of suitable starting points from a human combinatorial antibody library,[24,33,34] characterization of the antibodies during their maturation, and immunogenicity testing during *in vivo* experiments. A special focus of the example is on handling complex antigens such as β-amyloid fibers that are heterogeneous in size and have a high density of repetitive epitopes.

Synthetically produced antibody fibers were used as an antigen mimic. These fibers could be produced from the monomeric precursors and immobilized on CM5 sensors from Biacore using amine coupling chemistry. Fibers corresponding to a sensor response of about 300 RU were immobilized. Different antibodies were produced in large numbers as Fab-fragments using the phage display technology of Morphosys.[24,34]

Fab fragments preselected at Morphosys using a setup similar to that described by Canzani[29] were further characterized in-house. Figure 8–1 depicts sensorgrams of Fab fragments of a clone selected and matured in two different maturation cycles and a so-called cross-cloning experiment during which heavy and light chains of a Fab fragment were exchanged. The clone first identified by a random screen originally produced antibodies of weak affinity. It is not unusual to find such weak antibodies in a first run. It would, however, have been impossible to find this Fab fragment using conventional ELISA technology. Because of the low affinity and especially because of the fast off-rate this Fab fragment would have been lost during washing procedures. The continuing affinity improvement of this clone during the first two maturation cycles is obvious from both the amplitude of the signal and the dissociation profile. The two Fab fragments selected from the second maturation cycle had similar affinities but different kinetic behaviors. One of them showed fast on but also fast off kinetics, while the other one showed slow on and slow off kinetics. An ideal higher-affinity Fab should have combined fast on kinetics with slow off kinetics. It is shown in Figure 8–1 that the two desired properties could be obtained by a so-called cross-cloning experiment during which heavy and light chains of the two clones were exchanged. The resulting Fab possessed a nanomolar affinity, and the complete antibody derived from this Fab fragment showed subnanomolar affinity.

A special problem within this project was the extraction of meaningful equilibrium-binding data from the binding curves observed for the human immunoglobulin G^{h-IgG} antibodies prepared from the selected Fab fragments. The problem arises because of the occurrence of a mixture of monovalent and bivalent binding of the antibody to the immobilized antigen. Conventional suggested experimental setups to avoid such bivalent binding failed in this project. Immobilization of the antibody and the antigen as an analyte in solution could not be considered because it is nearly impossible to reproducibly prepare well-defined solutions of β-amyloid fibers that are generated in different lengths and that have a tendency for sedimentation. Moreover, because of the length of these fibers and the exposition of numereous epitopes, multivalent binding could not have been avoided by using this setup. The other alternative – diluting the antigen and hence the epitopes on the surface – does not offer a real alternative. Because of the structural feature of fibers mentioned previously, bivalent binding events and heterogeneity with respect to affinity on a surface due to steric hindrance cannot be avoided even at infinite dilution of fibers on the sensor surface. It is important to emphasize that this difficulty is not specific to SPR; this would arise with any other technique including ELISA.

The observed binding curves (see Figure 8–2) can be satisfactorily fitted with a kinetic model considering mono and bivalent binding. K_D values, however, cannot be calculated, as in the case of monovalent binding from the kinetic rate constants. To estimate such values for the bivalent high-affinity complex, equilibrium binding levels have to be determined that can be subjected to a Scatchard analysis.

In determining such equilibrium value a third difficulty arises because of the slow association process at subnanomolar concentration. It is obvious from the simulation of the curve using the determined kinetic constants (Figure 8–2B) that it takes more than 3 hr to reach equilibrium at a concentration of 1 nM. Myszka's group has suggested a modification to Biacore instruments that includes the use of peristaltic pumps to measure long association phases.[35] This approach, however, eliminates only the problems arising from limitations in the maximum injectable volume of the flow system and not the problem of low throughput

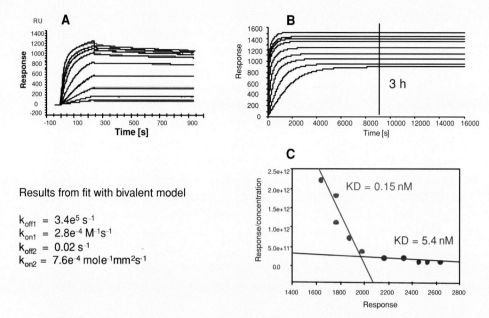

Results from fit with bivalent model

k_{off1} = $3.4e^5$ s^{-1}
k_{on1} = $2.8e^{-4}$ M^{-1}s^{-1}
k_{off2} = 0.02 s^{-1}
k_{on2} = $7.6e^{-4}$ mole^{-1}mm^{2}s^{-1}

Figure 8–2: Determination of K_D values for antibodies with subnanomolar affinity binding in mono- and bivalent binding mode to antigen on the surface. **A**: Concentration-dependent binding of the antibody to immobilized fibers. The experimental curves are overlaid with the curves emerging from a fitting procedure based on a bivalent model. **B**: Extrapolation of the curves for long contact intervals using the kinetic rate constants determined by the fitting procedure. **C**: Evaluating high- and low-affinity component by Scatchard analysis. *See color plates.*

due to long contact intervals. We worked with relatively short contact intervals and determined the equilibrium binding levels by extrapolating experimental binding curves (Figure 8–2A) to infinite time intervals (Figure 8–2B) using the kinetic rate constants determined by a fitting of the experimental curves with a bivalent model. Using this protocol, we could speed up the process of K_D determination for the high-affinity antibodies. It could be demonstrated that such a fast protocol leads to K_D values for the high-affinity, bivalent complex that differ only within the limits of error from K_D values determined with monitored equilibrium binding responses.

In this project SPR measurements were also used to perform immunogenicity testing. Because the Morphosys approach leads to full human antibodies, *in vivo* experiments in animals could be influenced by the immune response of the animals against the foreign protein. In this project *in vivo* testing was performed in mice and SPR was used to analyze mice plasma with respect to the presence of antibodies against the human antiamyloid antibody. The assay includes immobilization of the antiamyloid antibody on the surface. In this assay the antibody was covalently immobilized and regeneration could be performed with 10 mM HCl solution. This immobilized antibody is then contacted with the plasma from the test animals and subsequently with an antimouse-Fc-antibody. Figure 8–3 shows typical curves monitored for positive and negative plasma.

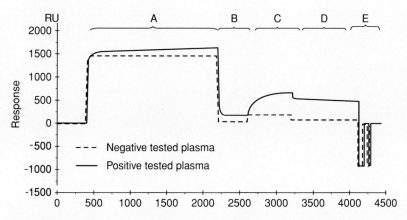

Figure 8–3: Immunogenicity testing. At time point zero the antiamyloid antibody is already covalently immobilized. During the time intervals (A–E) the sensor is in contact with plasma containing 50 mM carboxymethyldextran (A), with buffer (B), with a solution containing antimouse-FC antibody (C), and again with buffer (D). Regeneration of the sensor is performed during the interval E.

Kinetic and thermodynamic analysis of the binding of pegylated interferon 2α (IFNα) to interferon receptor (IFNAR2)

Pegylation is one of the most prominent artificial postmodifications of biomolecular drug molecules. It can improve solubility, temperature stability, stability against enzymatic degradations, serum half-life, immunogenicity, and thus biological effectiveness when compared with the unmodified protein. Because the poly(ethyleneglycol) polymers (PEG polymers) used for this modification have molecular weights and sizes comparable to the therapeutic protein, they have a strong influence on drug/target interaction and therefore a considerable influence on efficacy. A full kinetic and thermodynamic characterization of such an interaction using SPR and an interpretation of the data are described in the following example.

Superior efficacy of a pegylated over nonpegylated drug is demonstrated for pegylated interferon α (PEGASYS), approved in 2002 by the FDA for the treatment of HCV. The mode of action of the therapeutics is based on the interaction of INFα with its receptor. PEGASYS contains IFNα randomly pegylated with a 45-KDa branched PEG-polymer. The PEG-polymer is attached to the protein surface-exposed ε-amino groups of lysine residues. The final product is a mixture of monopegylated IFNa with 11 potential lysine positions of modification. Using ion-exchange chromatography, the four major and two minor isomers were purified.

For the kinetic and thermodynamic analysis of the binding of pegylated IFNα to the extracellular domain of the receptor IFNAR2, the receptor IFNAR2 was immobilized on a CM5 sensor using amine coupling chemistry. Pegylation does not abolish the binding to the receptor but has a significant influence on the kinetics of the interaction. This becomes most obvious when presenting the rate

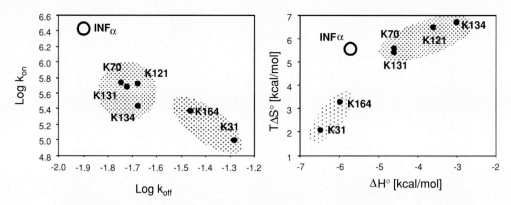

Figure 8–4: Influence of pegylation at different lysines on kinetic (*right*) and thermodynamic parameters (*left*). O nonpegylated IFNα; • INFα pegylated at the indicated lysine.

constants determined via concentration-dependent measurement in a k_{on}/k_{off} diagram (Figure 8–4). The different positional isomers cluster in two different areas of such a plot. For any lysine position pegylation leads to a lowering of the association velocity and a destabilization of the complex between INFa and IFNAR2. However, this effect is much more pronounced for the isomers K31 and K164 than for the K70, K121, K131, and K134. This clustering already reflects the structural characteristics of the interaction. For the isomers K31 and K164 the pegylation site is located in close proximity to the IFNAR2 binding site, whereas for the isomers K70, K131, K121, and K134 this pegylation site is placed remotely from this binding interface. The reduction in k_{on} values could be explained using a theoretical model based on self-avoiding walk. This model indicates that the general observed drop in k_{on} and hence K_D as observed for the cluster consisting of K70, K131, K121, and K134 is explained by a reduction in the ratio of the overall surface of the molecule and the interacting surface due to pegylation. The additional reduction in k_{on} observed for the cluster consisting of K31 and K164 is due to a lowering in the accessibility of the binding area due to the close proximity of the PEG-chains.

SPR technology was also used in this case for measuring the temperature dependence of K_D values. The analysis of these temperature-dependent K_Ds using the integrated van't Hoff equation led to ΔH, ΔS, and ΔCp values. ΔH and TΔS are graphically presented in Figure 8–4. Similar clusters of positional isomers are observed in such a ΔH/TΔS plot as in the k_{on}/k_{off} plot. For the cluster K70, K131, K121, and K134 pegylation has nearly no influence on the entropy gain upon binding. However, this cluster displays a mean enthalpy loss of about 30%. An interpretation of this enthalpy loss is difficult. Whether this difference in ΔH is due to alterations in remotely located surface charges and/or charges involved in the binding cannot be decided with the data available. The difference could also be attributed to differences in the hydration of the protein and the PEG polymer involved.[36] Interestingly, K164 and K31 show a mean enthalpy gain of about 10% and a significant mean entropy loss of about 50%. An obvious possibility to explain this significant loss of entropy is the reduced number of conformations accessible to the PEG-polymer due to the steric hindrance of mobility arising from IFNa and IFNAR2, and vice versa.

In summary, any pegylation of a biomolecule lowers the binding strength to its relevant binding partner due to two effects: reduction of the number of successful collisions between the molecules and reduced number of accessible conformations. The former could be minimized by using smaller PEG-polymers and the latter by site-specific pegylation that avoids pegylation at positions in close proximity to the interaction area.

SMALL SYNTHETIC DRUG MOLECULES

Experimental considerations

When working with small synthetic drug molecules (200 Da < MW < 600 Da) one has to consider that the SPR response observed is related to the refractive index change caused by an adsorbing molecule. The refractive index change per molecule is related linearly to its molecular size, a property that limited the application of this method for a long time to the binding of larger molecules such as proteins or oligonucleotides. The study of direct binding of low molecular weight compounds to immobilized biomolecules was published first by Karlsson from Biacore AB in 1994/1995.[37,38] In the meantime, many publications from this field appeared, and monitoring the interaction of low molecular weight analytes with immobilized ligands has become an accepted application for SPR. Although the sensitivity of the method has increased during the last decades mainly by improvements in the optical detection system, the flow system, and data evaluation procedures, special care still needs to be taken when setting up SPR-based binding assays with small molecules. Some of the most important ones will be discussed in the following section. For a more comprehensive discussion the reader is referred to other publications.[17,39–41]

Experimental factors that critically influence the results of binding studies with small molecules are the density of active protein on the surface of an SPR device, the quality of buffers and reagents used, nonspecific binding, carryover effects, and regeneration procedures. From the data evaluation side it is mainly the precise separation of signals due to binding from those due to refractive index mismatch between running buffer and analyte solution. How to cope with these factors has been already discussed in many chapters of this book. The present introduction to work with small molecules will only briefly emphasize the importance of using several filters for selecting true positive hits. In general the experimentalists only apply an affinity filter to their results. This means that they select their positive hits only according to the strength of binding, which is generally proportional to the observed signal height. With the graphs in Figure 8–5 one can demonstrate that this often leads to erroneous selection of the true positives.

Results presented in the graphs are taken from a screen performed with 600 compounds on a Biacore A100 instrument. The target of the screen is a serine

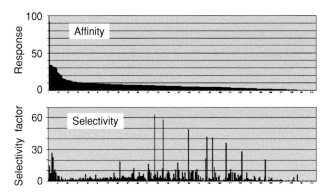

Figure 8–5: The top graph represents the responses monitored for 350 compounds during a screen against the active form of a serine protease. The compounds are ranked according to the signal intensity. In the bottom graph the compounds are ranked from left to right identically. The height of the bars is proportional to a so-called selectivity factor calculated as the ratio of the sensor responses monitored on the active protein and the zymogen.

protease that is a well-established pharmaceutical target in the cardiovascular disease area. The protein is expressed and isolated in an inactive form, the zymogen, in which the active site is blocked by a prodomain attached to the C-terminal end of the peptide chain. This propeptide is cleaved off upon activation of the protein at acidic pH. Zymogen and active protein are immobilized on different spots of the sensor surface and contacted in parallel with the compounds at a concentration of 100 µM. For reference compounds that bind selectively to the active site of the protease one observes a selectivity factor (here described as [response active]/[response zymogen]) of 8–10. We summarize results for the test compounds qualitatively in Figure 8–5.

When applying only an affinity filter (Figure 8–5, top), compounds are selected according to the signal height. These compounds are located at the left side of the top graph in Figure 8–5. With such a pure affinity filter one would have selected primarily compounds that show a poor selectivity. This means most of the signal intensity for these compounds is not due to binding to the active site but mainly to unspecific binding to the sensor or the protein surface. Only when considering the selectivity filter will one select from this set of compounds the real positive ones, namely those that bind with a high selectivity to the active site of the protein. Because this selectivity criterion is of such high importance, discussion of the following examples will always focus on the way selectivity is addressed in the experimental setup.

Overview on applications

Screening for hits: In principle, a direct binding assay as offered by the SPR method would be the simplest method by which to perform screening for new hit structures. The advantage of such an approach would mainly be the fast development time for such assays and the fact that results clearly show whether the compounds interact with the target (provided that the assay is designed with proper referencing or if additional competition assays are performed). Despite obvious advantages, not many applications of the SPR technology in hit finding have been reported to date.

This is due to moderate throughput. Only recently, several companies have started to develop instruments that allow higher throughput.[42] Most of them use SPR as the monitoring principle. A few are based on waveguide and/or grating coupler technology. Biacore offers an instrument with four parallel flow cells in which up to four different proteins can be immobilized per cell. This raises the throughput up to about 1500 binding events per day. Corning Inc. offers on its homepage a system that uses single waveguide grating sensors as the detection technology to directly monitor in the wells of disposable 384-well plates. Corning claims that they can monitor up to 40 000 binding events per day. This will be in a range already comparable to a high throughput screening (HTS) assay. So far, data have only been presented at conferences.

Despite the limitation in throughput, there are a few applications published in the literature or presented at meetings that demonstrate the use of SPR in hit finding. For example, SPR screening has been performed with focused libraries that contain only a small fraction of the compounds available from the depositories of pharmaceutical companies. Specially designed and assembled target-specific libraries and/or fragment libraries are also frequently used. Böhm et al.[43] were among the first to publish the successful use of physical detection methods including SPR to screen for compounds of novel structures that bind to the ATP binding site of bacterial gyrase. Several new structure classes for inhibitors were identified via this route that could be chemically modified to give nanomolar inhibitors. An analogous approach will be discussed to find inhibitors for bacterial hydroxymethyl-pterin pyrophosphokinase (HPPK).[19]

Secondary screening: SPR-based assays are more often used in secondary screening where one has to deal with fewer compounds. Hit lists from HTS often contain false positives that show up during the screen because the compounds interact not with the target protein but with any other auxiliary reagent in the assay solution. Especially in fluorescence-based assays, interference of the fluorescence of the tested compounds is often a cause for false positives. It is therefore important to show whether the active compounds really interact with the target protein before starting a chemical program around a compound class. SPR-based assays are highly suitable for such secondary screening with many different target protein classes. Only a few of the numerous applications published in this field can be discussed here.

One of the targets for which such secondary screening was performed most extensively was HIV-1 protease.[19,44–51] The work on this target is pioneering because it demonstrates all the different facets and opportunities of this kind of screening. The work on HIV-1 protease is one of the few examples where not only binding activity but also enzymatic activity was tested to confirm that the immobilization strategy did not modify the function of the enzyme.

SPR technology was also used to characterize binding of small molecules to the ligand binding domain of human estrogen receptor.[52,53] The assay discerned specific binders from nonbinders. The time-dependent responses during association and dissociation could be fitted to a simple 1:1 kinetic model to extract kinetic rate constants. Association rate constants for antagonists were found to be 500-fold slower than those determined for agonists. These findings indicate that the antagonists bind to a different conformation of the receptor. By using the biosensor assay, subtle differences were found in how the same compound binds to the different isoforms of the receptor.

An SPR-based binding assay evaluated binding of compounds to CD80, which is found on the surface of antigen-presenting cells and is believed to be a target for a therapy of autoimmune disease such as rheumatoid arthritis.[54,55] Huxley et al.[55] examined binding properties of hit compounds to identify lead candidates. Two hundred fifty-nine compounds from five different chemical compound classes were examined.[54] The CD-80 binding properties of each compound were characterized by the determination of association and dissociation constants. The data were globally interpreted by presenting them in a k_{on}/k_{off} map. The general picture again indicates that high affinity is predominantly the result of slower dissociation constants.

Protein kinases are emerging as one of the most intensively studied classes of enzymes in pharmacological research because of their central role in physiologically and clinically important cellular signaling. Despite this fact, there are only a few examples in the literature where small-molecule interactions with kinases have been analyzed using SPR technology. Those studies mainly concern the mitogen-activated protein kinase p38.[56] It is believed that the immobilization of kinases is quite tricky due to the low stability of this type of protein. Recently a methodology package was published[57] that can be taken as a guideline for immobilization conditions that preserve the binding capacity and activity of the immobilized kinase. The authors report the immobilization of nine different kinases. They obtained binding capacities in the range of 25% to 90%. Besides a normal binding assay to screen for inhibitors and for kinetic characterization, an ATP competition assay was used.

Plasma protein binding: It is widely accepted that the effect of a drug is related to the exposure of a patient to the unbound concentration of the drug in plasma rather than to total concentration. Binding of drugs to plasma proteins such as human serum albumin (HSA) and α-acid glycoprotein (α-AGP) influences the free concentration. HSA is the most abundant protein in plasma and serves as a major transport protein for drugs. Traditional binding to HSA has been examined using equilibrium dialysis or chromatographic techniques, with results reported as either percent bound and/or binding constants. The two data sets are well correlated for low affinities. At higher affinities ($K_D < 32$ μM) the percent bound data (>95%) fails to rank the binders.[58,59] Because it is mostly strong binding that might have an influence on the therapeutic effect of drugs, the recognition of high binders is essential.

Several publications deal with the possibility of using SPR experiments to examine HSA binding of drug molecules.[60–65] It was demonstrated by all publications that the SPR method is suitable for ranking drug molecules with respect to HSA binding. Weak, medium, and strong binders can be distinguished when immo-bilizing HSA on a sensor surface, contacting it with 80-μM drug solutions and taking the equilibrium response at this concentration as a relative measure for the affinity to HSA. Such a ranking correlates well with rankings based on data from equilibrium dialysis or chromatography. However, a differentiation between different strong binders (>95% protein binding) is again not possible using such a crude ranking. Such differentiation must be based on K_D values determined via concentration-dependent measurements.

The problems of determining such values using SPR methods have been critically discussed. Saturation behavior and clear 1/1 stoichiometry is seldom observed. Biphasic, multiphasic, and linear isotherms were observed for many drug molecules, which made the estimation of saturation levels and determination of equilibrium binding constants difficult or even impossible. Recently a published work introduced a high-resolution and high-throughput protocol for the determination of K_D values of HSA–drug interactions.[63] Using warfarin as a model compound it was demonstrated that highly accurate and reproducible data (K_D = 3.7 μM) can be obtained by using this protocol. The protocol that can be scaled up for high throughput was used to determine binding data for ten test compounds.

Only a few articles report binding of drug molecules to other plasma proteins.[65] Results published so far for α-AGP indicate that binding studies enable a similar ranking of compounds into low-, medium-, and strong-affinity binders as with HSA. Quantitative kinetic studies and K_D determinations that would enable more mechanistic interpretations have also failed for α-AGP.[65]

Interactions of drugs with liposomes: An important aspect of drug discovery and development is identifying drugs that are sufficiently soluble in the aqueous environment and yet pass through the hydrophobic intestinal mucosa into the bloodstream. Several *in vitro* techniques assess the lipophilicity of a drug to predict how well it will be adsorbed in the body. The most widely used predictor is the partition coefficient determined with aqueous buffer and n-octanol. Many researchers also work with liposomes: lipid vesicles that mimic the bilayer geometry and the ionic characteristics of the biological membrane. Several researchers have demonstrated the successful application of SPR techniques to study liposome/drug interaction.[66–68] The authors illustrated how liposomes could be captured on chemically modified sensor chips. One of the publications could clearly classify the compounds with respect to their kinetic behavior.[66] Type I drugs are primarily negatively charged at physiological pH and rapidly dissociate from the lipid surface containing lipids with choline head groups. It is proposed that such drug molecules do not form a stable complex with the phosphatidylcholine molecules and pass freely into and out of the lipid bilayer. Type II molecules are primarily positively charged at physiological pH and display two distinct kinetic phases during association and dissociation. This finding suggests that these drugs both diffuse across and intercalate into the phospholipid bilayer. With its unique kinetic abilities, SPR is the only method that could reveal this mechanism. Recently a publication appeared that reports the outcome from experiments with liposomes of different composition.[67]

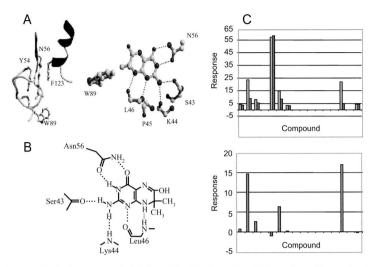

Figure 8–6: A: Structural details of the binding site of HPPK from X-ray structure of a protein substrate analogue complex. **B**: The relevant polar interactions in the binding pocket. **C**: Responses of 16 compounds monitored in contact with the wild-type protein and the mutated protein (*top*). Difference in signal between wild-type and mutated protein (*bottom*).

Examples from hit and lead finding

Virtual and SPR screening for HPPK inhibitor[19]

Because no biological assay was available for HPPK, a focused library using *in silico* screening was generated, and a direct binding assay had to be developed to screen these compounds for HPPK binding. The X-ray structure of a protein substrate analogue complex was determined in-house and the identification of the binding site and generation of a pharmacophore hypothesis was straightforward. The most prominent features of the binding site included polar interactions between the pterin ring of the substrate and asparagine 56, serine 43, lysine 44, and leucine 46, and nonpolar interactions with phenylalanine 123 and tyrosine 54 (Figure 8–6). Two hundred fifty compounds were selected from an in-house library. These compounds fully or partly matched the binding pattern identified through the pharmacophore hypothesis developed from structural knowledge.

A mutant form of the protein designed to have impaired binding capability was generated by exchanging asparagine 56 with alanine. Though the binding site had been successfully altered, and binding affinity for the substrate analogue was fully destroyed by this single point mutation, protein conformation remained virtually unchanged as confirmed by circular dichroism spectroscopy.

The binding activity profiles of the wild-type and mutant proteins were elucidated using SPR. Proteins were immobilized on a Biacore sensor chip using standard amine coupling chemistry. The saturation signal for the binding of small molecules (MW ~400 Da) was approximately 25 resonance units (RUs). Comparing this saturation response with the amount of protein immobilized indicated that 30% of the immobilized wild-type protein remained active. No binding was observed for the substrate analogue to the immobilized mutant protein.

To facilitate the rapid data processing required for the daily screening of hundreds of compounds, we evaluated real-time binding responses using a limited number of discrete "report points." These are software-predetermined time points that identify specific signs of binding activity. Figure 8–6C compares the responses at such report points monitored for the wild-type and mutant protein for 16 out of the 250 compounds. Of these, 8 bind to wild-type HPPK, but only 4 bind with significantly different affinity to the mutant (Figure 8–6C, bottom). These four representatives can now be classified as selective binders, that is they interact with the target protein at the desired active site. This is an essential advantage of the assay method. Again, selectivity is the more valuable parameter for compound selection and development than absolute target binding affinity.

The direct binding assays separated a total of 15 compounds that showed selective binding to the HPPK binding site. These hits belong to four different classes of compounds, and their affinities, determined by SPR, were between 400 nM and 200 μM. Although these hits are still far from being drug candidates, they are established selective target binders and represent ideal starting points for the synthesis of derivatives with improved binding activity, improved toxicity, bioavailability, and metabolic profiles.

Screening for allosteric site binders of a heterotrimeric protein

The protein in the following example is an accepted target for new medicines in the metabolic disease area. This example is key to learning the application of this kind of technique when working with a complex target with several drugable sites but no functional biological assay available. It is demonstrated that data produced by testing compounnds against many reference proteins enable the application of stringent selectivity filters necessary for the selection of hits that bind with high selectivity to the desired binding site.

The protein, called heterotrimeric protein (HTP), consists of three different peptide chains (A, B, X) that form the three domains of the trimeric complex (see schematic representation in Figure 8–7). For each peptide chain different isoforms (1, 2, etc.) exist. The active site of the protein is located on domain A. Reference compound 2 (R2) is a natural ligand that binds to the active site and inhibits the protein activity. An allosteric site is located on domain X. Reference compound 1 (R1) binds to this allosteric site and enhances the protein activity. The function of domain B is not known. One of its functions is probably stabilization of the protein complex. It is unclear how much B is involved in the regulation of protein activity. The drug discovery program was looking for allosteric site binders that enhance protein activity. Biological screening failed probably due to the complicated activity regulation system in which upstream proteins are involved. As a consequence SPR-based binding assay was considered as an alternative to screen a limited number of compounds selected by *in silico* screening.

Figure 8–7 lists the different constructs of HTP used in this work. In addition, CA (carbonic anhydrase) and GST (glutathione transferase) were used as indicators for a general stickiness of the tested compounds. All the constructs and dummy proteins were immobilized in parallel on the sensor chip of a Biacore

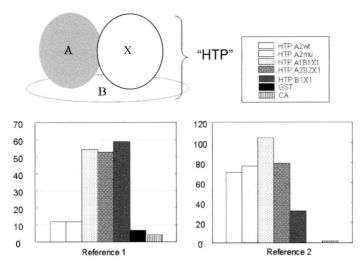

Figure 8–7: Schematic representation of the protein HTP. Grey-scale coded responses monitored for the different constructs or fragments of the HTP in contact with the reference compound 1 (R1) and 2 (R2) at a concentration of 10 mM. The names given to the constructs of HTP (see list) contain the different domains (A, B, X) that form a given construct. GST (gluthathiontransferase) and CA (carbonic anhydrase) are reference proteins. Domain A was used as wild-type (HTP A2wt) protein and an active site mutated protein (HTP A2mu).

A100 instrument and contacted in parallel with R1 and R2. Different response patterns were monitored for the two reference compounds in contact with the different constructs. It is obvious that R1 binds preferentially to constructs containing the allosteric site while R2 binds preferentially to constructs containing the active site. The selectivity of the two compounds, however, is not 100%. R1 shows a measurable affinity for constructs that contain no allosteric site (for instance HTP A2), and R2 shows a measureable affinity for constructs with no active site (HTP B2X2). This finding is due to the similar structure of the two natural ligands R1 and R2 and indicates that similar binding modes are realized in the active and the allosteric site. Selectivity is thus a critical issue in this project, and selection of allosteric site binders must be carefully based on experimental data that enable selectivity considerations.

The focused library contained 1200 compounds. They were selected from the internal library by a similarity search using R1 as the reference structure. These compounds were first screened in a fast screen on a Biacore A100 instrument that allows injection of four different test compounds in parallel. In each of the four channels of the instrument, HTP A1B1X1 and HTP A2wt were immobilized in parallel on two separated spots. Amine coupling chemistry was used. The compounds were injected at a concentration of 100 μM. Hits were selected based on their affinity for the construct HTP A1B1X1 (with response larger than three times the standard deviation of a negative control) and according to a selectivity filter (response on HTP A1B1X1 two times the response on HTP A2wt).

Eighty compounds were then selected from this primary screen and subjected to a more detailed study of their selectivity for the allosteric site. The five HTP constructs, GST, and CA were immobilized pairwise on different spots in the four

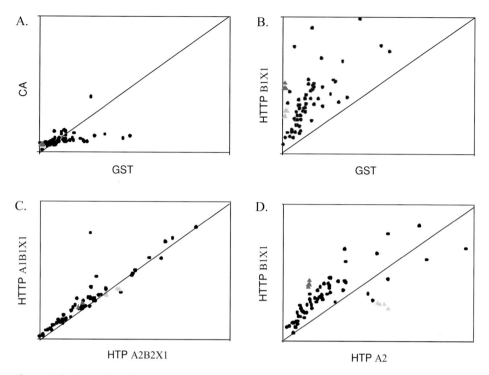

Figure 8–8: Correlation of sensor responses observed for compounds in contact with different constructs and dummy proteins. Black spots represent the responses observed for the test compounds, the red and green spots represent those monitored for the references R1 and R2, respectively. *See color plates.*

different channels of the A100 instrument, and the protein was contacted in parallel with the solutions of the 80 primary hits.

A correlation of the responses for the different constructs and dummy proteins gives a clear indication about the selectivity of the binding of the primary hits. In Figure 8–8A, such a plot is presented for the responses monitored on GST and CA. Most of the data points are located at the origin. The hits therefore do not exhibit a general stickiness to proteins. Only a few show a certain affinity for GST. Figure 8–8B shows an analogous correlation for HTP B1X1 and GST. It is clearly obvious that the compounds have much higher affinity for the HTP construct than for GST. The affinity of the reference compounds for GST is negligible, but R1 binds as expected stronger to the construct then R2. In Figure 8–8C responses of the different isoforms are correlated with each other. Nearly all the data points are located on the diagonal. This clearly indicates that the selected compounds have no preference for one of the isoforms.

The correlation in Figure 8–8D is the most important in the context of the project. Most of the compounds are located above the diagonal and therefore possess a higher affinity for the construct HTP B1X1 that incorporates the allosteric site. The experimental setup was therefore suitable to find compounds that bind to the targeted domain of the protein.

The conclusion, however, does not implicate that these compounds are allosteric site binders. Specific binding to the allosteric site could only be tested by performing competition experiments using R1 as the competitor compound.

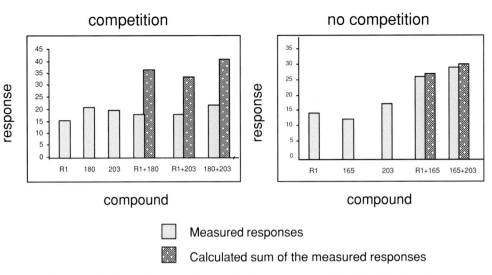

competition   no competition

Measured responses

Calculated sum of the measured responses

Figure 8–9: Competition experiments with test compounds 180, 203, 165, the reference compound R1 and HTP B1C1 as the immobilized protein construct. The light grey bars show the monitored responses for either the single compounds or the mixture. The patterned bars are responses calculated for the mixture if each of the compounds would occupy independent sites.

These experiments included injection of the reference compound, injection of the test compound (hits from secondary screen), and injection of the mixture of reference and hit. To fully block the allosteric site, R1 was injected at a concentration of 10 mM that corresponds to 100 times its K_D. The selected hits were injected at a concentration of 10 μM. The mixtures contained reference and hit compound at the same concentrations. With this setup, the allosteric site is always fully blocked by R1, and the observed signal for the mixture in case of competition should always correspond to the saturation signal measured for R1. Figure 8–9 shows a few examples of competitive and an example of a noncompetitive compound.

When injecting a mixture of R1 and compound 180 we observe a signal that is nearly identical to the saturation signal of R1. This clearly indicates that compound 180 occupies the same allosteric binding site as R1 and is displaced from this site by R1. The same behavior was also observed and the same conclusion drawn for compound 203. When injecting a mixture of 180 and 203 we monitored a signal that is much smaller than the calculated sum of the individual ones. The signal is comparable to the saturation signal. This unambiguously proves that R1, 180, and 203 occupy the identical binding site.

Compound 165 behaves totally different. When injecting a mixture of R1 and 165 or a mixture of 165 and 203 we monitor a response nearly identical to the calculated sum of the individual responses monitored for the compounds. This clearly indicates that compound 165 does not bind to the same site as R1 and 203 and therefore not to the allosteric site. This compound binds selectively to HTP B1X1 but not to the allosteric site on this construct.

By this stepwise approach that included a fast primary screen, a more detailed secondary screen using more reference proteins and a characterization with a

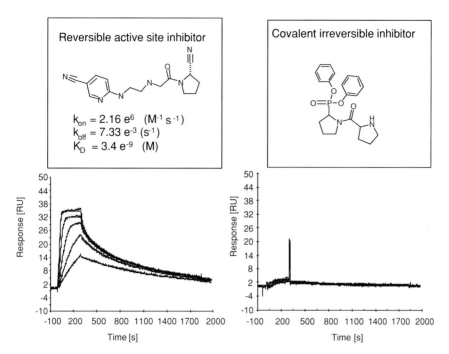

Figure 8–10: Concentration-dependent sensorgrams monitored for a reversible active site inhibitor in contact with DPP-IV (*right side*) and in contact with DPP-IV treated in situ with the covalent irreversible inhibitor. The relevant kinetic data determined by fitting the experimental curves with a 1/1 kinetic model are given in the figure.

competition assay, five classes of compounds could be identified. Cellular assays indicated that some of these compounds also show the expected enhancement of the enzyme activity. Because this was achieved without performing HTS, this is another example that demonstrates the power of the combination of modeling (for generating focused libraries) and methods enabling direct binding assays.

Validation of DPP-IV lead series using high-resolution kinetic analysis and modified binding site controls[19,20,69]

Dipeptidyl peptidase-IV (DPP-IV) is an accepted target for the treatment of diabetes type II. Drug discovery programs are looking for inhibitors that bind to the active site of the enzyme. In this project, an SPR method was used to validate three different lead classes with respect to selective binding to the active site, reversibility of binding, binding kinetics, and binding affinity. Kinetic parameters were especially helpful for sorting out a "promiscuous" compound class[70] from the set of possible lead candidates.

Within this project, the strategy chosen for generating a suitable reference protein to test for selectivity of binding is based on the use of covalent, irreversible inhibitors. A phosphinate ester, the structure of which is shown in Figure 8–10, was selected to irreversibly block the active site of DPP-IV. The flow system design of the Biacore instrument allows this blocking reaction to be carried out on proteins already immobilized to a specific region of the sensor surface, enabling parallel analysis of blocked and unmodified targets in the same assay.

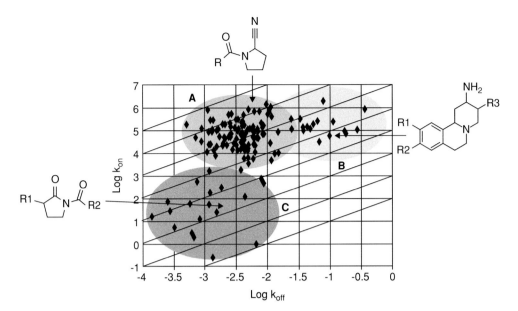

Figure 8–11: k_{off}/k_{on} plot including the kinetic rate constants determined for representatives of the three structure classes shown at the top of the figure.

The outcome of such an experiment is shown in Figure 8–10. The unmodified protein yields information on reversibility of binding along with kinetic data, with all interactions monitored in real time. From the time course of these reactions it is possible to extract the relevant kinetic data (k_{on}, k_{off}) and to calculate the equilibrium binding constant, K_D ($K_D = k_{off}/k_{on}$). The relevance of this high-resolution binding characterization was clearly validated by the lack of response seen with the modified protein, showing that the kinetic data obtained for the nonblocked DPP-IV represent compound binding to the active site (Figure 8–10).

This setup is routinely used at Roche as the method to further characterize active inhibitors synthesized during the hit expansion steps. The compounds could be assigned to three different structure classes represented in Figure 8–11. In general, K_D values correlate well with IC_{50} values and confirm the biological results.

The significant additional value of characterizing lead compounds with an alternative technique is readily demonstrated by looking at a so-called $k_{on}/k_{off}/K_D$-plot,[46,47] in which the two rate constants are plotted against each other on a logarithmic scale and the diagonal iso-affinity lines correspond to the equilibrium dissociation constant value, K_D (Figure 8–11). Inspection of such a plot points to several important conclusions. The plot clearly indicates clustering of the compounds in three major areas (A, B, C). These clusters contain compounds from three different structure classes, cyanopyrrolidines (A), benzoquinolizines (B), and pyrrolidinones (C). The cyanopyrrolidines that cluster in the upper left corner exhibit fast association and slow dissociation rate constants. The compounds with highest affinities (K_D ~1 nM) have on-rates around 10^6 M^{-1}s^{-1} and off-rates of around 10^{-3} s^{-1}. This is probably the class from which a potential drug might emerge if all other prerequisites for developing a successful medicine are fulfilled. The benzoquinolizines in the upper right corner comprise compounds with fast on rates, but in comparison to the first class, also fast off rates. These compounds can be optimized mainly through structural modifications that slow down the dissociation process because it appears from these two compound classes that an increase in K_D is mainly obtained by a decrease in k_{off}. The third cluster (C) is somewhat diffuse and is mainly located in the lower left corner, indicating slow on and off rates. Previously, this behavior with extremely slow on rates had not been observed for low molecular weight ligands, and it was suspected that there could be something wrong with the compound class.

Analytical ultracentrifugation experiments were used for this compound class to study protein, inhibitor, and protein/inhibitor solutions. The pure solutions of these compounds showed interesting behavior, with all demonstrating sedimentation at low rotational speed: a process that increased with higher speed. The sedimentation losses indicate that these compounds exist preferentially as very large aggregates in solution and not as monomers. These aggregates must have molecular weights larger than 10^6 and explain why slow association is seen in the SPR experiments. The slow association is due to the low amount of monomer in solution and to the fact that there is no real driving force for the molecules to leave these aggregates and to bind to the immobilized protein as a monomer. The kinetics of this process must therefore be very slow and must be coupled to a high-activation barrier.

The behavior is totally different in protein solution, a condition comparable to that of the biological assay, where the compounds seem to have "normal" activity. These are prepared by mixing dimethyl sulfoxide (DMSO) solutions of the compounds directly with the protein solution. In this case, no precipitation of large compound aggregates or loss of material was seen in the AUC experiment. However, independently from the concentration ratio of protein and ligand, all of the lead compound was found to be bound to the protein. Obviously the presence of protein in solution reduces the concentration of free ligand such that no aggregates can be formed. It is not surprising that enzymatic activity is no longer observed for a protein fully covered with a compound that binds nonspecifically. The compounds appear as inhibitors in the enzymatic assay, but the inhibitory effect may not be due to the binding of the molecules to the active site. Consequently, such compounds are not druglike, and the substance class C was rejected as a potential lead class. Compounds exhibiting such behavior have been classified recently in the literature as "promiscuous inhibitors" or "frequent hitters".[70] These types of compounds are one of the main sources of false positives found in high-throughput screens. Because the behavior of this compound class is independent of the protein (it is, for instance, similarly solubilized by HSA), it will probably be active in every protein assay with a similar format.

Mechanistic aspects of inhibition of DHNA

In the present example the focus is not on monitoring direct binding of small molecules to a target protein. Binding of true, positive inhibitors has a strong

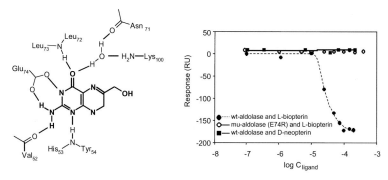

Figure 8–12: *Left*: Details of the binding of substrate analogues into the active site of aldolase. The substrate analogues are shown in bold. Substitution of Glu 74 by arginine leads to a total loss of binding affinity for such substrate analogues. *Right*: Dose–response curves of biologically active L-biopterin injected over wt DHNA (solid circle). The dose-dependent decrease in signal indicates cleavage of the octameric wt DHNA. Such cleavage is not observed when in contact with the biologically inactive D-neopterin or when the mutated DHNA is in contact with L-biopterin.

influence on the quaternary structure of the protein that can be monitored using SPR. The results from such monitoring help to understand the real mechanism of inhibition. This knowledge is critical for an interpretation of the structure–activity relationship.

Dihydroneopterin aldolase (DHNA) is an essential bacterial protein of the folate pathway. SPR was used to perform binding studies and to investigate the influence of active site binders on the quaternary structure of the protein. Only these structural consequences and their influence on inhibition are discussed. An active site mutant was designed to have a reference protein to address the problem of the selectivity of binding. The design of this active site mutant is derived from the 3D crystal structure of DHNA that was solved in-house.[71] It shows that this aldolase exists as an octamer; four monomers form a disk shape tetramer that dimerizes face-to-face. This quaternary structure was confirmed in solution by AUC. The protein is only active in its octameric form. A crystal structure of a substrate/protein complex enabled the design of mutants with impaired binding sites. Two mutants were constructed, with the glutamic acid 74 replaced by arginine or alanine (Figure 8–12). These mutations eliminated binding activity but did not change the folding or physico-chemical properties of the protein.

In the course of lead selection, structure–activity analysis showed different activities for a series of compounds with similar structures (especially those with similar pharmacophores). L-biopterine, for example, is a highly active compound, though D-neopterine is not.

Biopterin

Neopterin

The parts of the molecules responsible for the polar interactions with the amino acids in the active site are identical in both compounds. Both bind to the wild-type protein but not to the mutant, indicating that the compounds really bind to the active site.

In the process of optimizing such compounds for high activity, it is essential to know why such minor structural changes influence the activity of a lead compound to such an extent. This problem was tackled using SPR assays, with the aldolase biotinylated and immobilized on a streptavidin-coated sensor chip surface. This enabled full regeneration of the surface and immobilization of a new batch of aldolase before each binding experiment. The experimental protocol included capture of the biotinylated aldolase, injection of the analyte solution, and regeneration of the surface with ethanolamine solution and short pulses of 50 mM sodium hydroxide and 3 M sodium chloride. This regeneration step removed all the aldolase from the surface without destroying the binding capacity of the streptavidin. After regeneration, fresh aldolase could be captured on the sensor chip. The octameric form is only stable at salt concentration above 1 M and at basic pH (>8.0), conditions often found in bacteria. This experimental setup using captured target allows the evaluation of varying salt or pH on the cleavage of the octamer. The amount of protein loss due to the cleavage of the octamer could directly be measured.

The reason why one compound is active and the other inactive becomes obvious when comparing the sensorgrams. For the active compound, a significant amount of aldolase is lost from the surface when contacting the captured aldolase with the compound solution. It is straightforward to conclude that this compound cleaves the octamer into tetramers, data that were confirmed by AUC experiments. Such protein loss is not observed for the other compound. The efficiency of octamer cleavage is concentration-dependent. Using a titration curve it is possible to extract information that correlates with the inhibition constant. Moreover, using the mutant protein with an impaired binding site as a reference, it can also be demonstrated that the compound must bind to this site to cleave octameric aldolase. Figure 8–12 shows the concentration-dependent cleavage of the wild-type protein by L-biopterin, the active compound. No cleavage is observed when the binding site is impaired, as in the mutant protein E74R-aldolase, or when wt-aldolase is contacted with D-neopterin, the inactive compound in inhibition experiments.

Investigating the influence of inhibitor binding on DHNA quaternary structure shows that structure–activity interpretation has to consider areas not directly involved in forming polar interactions to binding site amino acids. It has to consider inhibitor moieties that protrude from the binding site to influence intramolecular interactions responsible for the formation of the octamer. This conclusion is confirmed by another series of compounds, again with identical moieties protruding from the binding site but with different fragments in the binding site. All bind to aldolase, but only those that cleave the octamer in a concentration-dependent manner show biological activity.

Consequence of inhibitor binding on the interaction of tissue factor and FVIIa

The term *induced fit* is used when a protein changes its conformation to optimize the interaction with another molecule. Such structural adaptation can be significant and often also affects parts of the protein not directly involved in the binding of small molecules but in an interaction with a receptor protein. In the following example SPR technology addresses the consequences of inhibitor binding on a protein/protein interaction.

The Factor VIIa–tissue factor (FVIIa-TF) complex is an excellent target for anticoagulants because FVIIa inhibition would act at the initiation of the amplification cascade for coagulation and would minimize interference with the extrinsic coagulation pathway. High FVIIa activity is only observed if the protein forms a complex with the tissue factor. It is expected that the complex formation leads to conformational changes in the active site. It was therefore of interest to check how FVIIa/TF interaction is influenced by the binding of typical FVIIa inhibitors such as benzamidine derivatives. The Factor FVIIa interaction could be investigated using SPR technology. Soluble tissue factor (sTF) was immobilized on a CM5 sensor chip and the surface contacted with solutions of FVIIa. Figure 8–13A shows typical concentration-dependent sensograms from which k_{on}, k_{off}, and K_D values could be determined. The influence of a benzamidine-type inhibitor on the interaction of sTF and FVIIa was investigated by injecting inhibitor at various concentrations during the dissociation phase. The amount of FVIIa captured by the sTF (~150 RU) is thereby so small that monitoring of direct binding of the inhibitor can clearly be excluded (the ratio of the molecular weights is ~300). It is, however, obvious from Figure 8–13B that the binding of the inhibitor causes a significant alteration of the dissociation process. Binding of such benzamidine inhibitors leads to a significant stabilization of the sTF–FVIIa complex. This stabilization is dependent from the inhibitor concentration as shown in Figure 8–13B. A plot of the slopes of the dissociation curve during inhibitor injection leads to a sigmoid curve from which IC50 values can be extracted. Such IC50 values correlate well with the IC50 values determined by activity assays.

The stabilization of the complex induced by the binding of the small molecule is significant. The dissociation rate constant is reduced in the presence of the inhibitors by two orders of magnitude. Determination of K_D values in the presence of the inhibitors at saturating concentration shows that the affinity is also raised by two orders of magnitude. The significant influence of a small molecule on the energy of a protein/protein interaction is even more surprising when considering the large distance in FVIIa between the active site and the area that interacts with the soluble tissue factor.

SUMMARY

Biosensors have become a routine tool to support drug discovery projects. The restriction in this chapter to SPR devices does not indicate that such a conclusion holds only for this type of biosensors. Many other read-out technologies

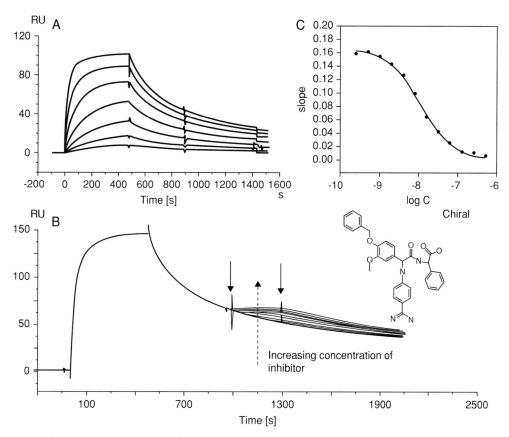

Figure 8–13: Investigation of inhibitor binding on the interaction of the sTF and FVIIa. **A**: Concentration-dependent sensorgrams monitored during a contact of FVIIa with immobilized sTF. K_D value was determined to be 10 nM. **B**: Injection of a bezamidine type of inhibitor during the dissociation of FVIIa from sTF. The injection interval for the inhibitor is marked by arrows. Different curves were monitored during the injection of different inhibitor concentrations. Sensor surface was regenerated after each run. **C**: Plot of the slope determined during the inhibitor injection versus inhibitor concentration. The IC50 value determined from the sigmoid curve is 2.5 nM.

have now been developed to a matured stage that includes high sensitivity, high reproducibility, and routine and automated handling in an industrial environment. The basic experimental strategies presented in this chapter are, however, applicable to all these different types of biosensors to extract useful information. Each of these surface-sensing devices is only fully functional when adequately immobilizing the target biomolecules and when properly referencing to exclude bulk effects. All the technologies have to be pushed to their limits when performing screenings with small synthetic molecules. In this case selectivity filters are important to exclude nonspecific binders in hit selection. Several approaches to integrate such filters have been discussed.

REFERENCES

1. A. M. Davis, D. J. Keeling, J. Steele, N. P. Tomkinson, A. C. Tinker, *Curr Top Med Chem* **5**, 421 (2005).

2. A. Alanine, M. Nettekoven, E. Roberts, A. W. Thomas, *Comb Chem High Throughput Screen* **6**, 51 (2003).
3. S. Wang, T. B. Sim, Y.-S. Kim, Y.-T. Chang, *Curr Opin Chem Biol* **8**, 371 (2004).
4. D. Hayes, G. Mellor, *Enzyme Assays* (2nd Ed.), 235 (2002).
5. J.-P. Goddard, J.-L. Reymond, *Curr Opin Biotechnol* **15**, 314 (2004).
6. V. C. Abraham, D. L. Taylor, J. R. Haskins, *Trends Biotechnol* **22**, 15 (2004).
7. J. S. Marks, D. S. Burdette, D. A. Giegel, *Methods Mol Biol* **190**, 51 (2002).
8. X. Wu, J. F. Glickman, B. R. Bowen, M. A. Sills, *J Biomol Screen* **8**, 381 (2003).
9. M. J. Cliff, A. Gutierrez, J. E. Ladbury, *J Mol Recognit* **17**, 513 (2004).
10. C. Ebel, *Prog Colloid Polym Sci* **127**, 73 (2004).
11. T. Carlomagno, *Annu Rev Biophys Biomol Struct* **34**, 245 (2005).
12. C. Fernandez, W. Jahnke, *Drug Discov Today* **1**, 277 (2004).
13. W. M. A. Niessen, *J Chromatogr A* **1000**, 413 (2003).
14. R. L. Rich, D. G. Myszka, *Drug Discov Today* **1**, 301 (2004).
15. M. A. Cooper, *Curr Opin Pharmacol* **3**, 557 (2003).
16. M. J. Cannon, D. G. Myszka, *Recent Res Develop Biophys Biochem Recent Research Developments in Biophysics & Biochemistry, Coden* **3**, 333 (2003).
17. M. J. Cannon *et al.*, *Anal Biochem* **330**, 98 (2004).
18. S. Lofas, *Modern Drug Discovery* **6**, 47 (2003).
19. W. Huber, *J Mol Recognit* **18**, 273 (2005).
20. W. Huber, F. Mueller, *Curr Pharm Des* **12**, 3999 (2006).
21. M. Cooper, V. T. Singleton, *J Mol Recognit* **20**, 154 (2007).
22. R. L. Rich, D. G. Myszka, *J Mol Recognit* **19**, 478 (2006).
23. I. Navratilova, M. Dioszegi, D. G. Myszka, *Anal Biochem* **355**, 132 (2006).
24. A. Knappik *et al.*, *J Mol Biol* **296**, 57 (2000).
25. J. E. Stacy *et al.*, *J Immunol Methods* **283**, 247 (2003).
26. U. Schlattner, C. Reinhart, T. Hornemann, M. Tokarska-Schlattner, T. Wallimann, *Biochim Biophys Acta* **1579**, 124 (2002).
27. G. L. Scheffer *et al.*, *Br J Cancer* **86**, 954 (2002).
28. K. Kramer, *Environ Sci Technol* **36**, 4892 (2002).
29. G. A. Canziani, S. Klakamp, D. G. Myszka, *Anal Biochem* **325**, 301 (2004).
30. K. Usui-Aoki, K. Shimada, M. Nagano, M. Kawai, H. Koga, *Proteomics* **5**, 2396 (2005).
31. A. Fotinopoulou, T. Meyers, P. Varley, G. Turner, *Biotechnol Appl Biochem* **37**, 1 (2003).
32. M. Vernersson, A. Ledin, J. Johansson, L. Hellman, *FASEB J* **16**, 875 (2002).
33. S. Cesaro-Tadic *et al.*, *Nat Biotechnol* **21**, 679 (2003).
34. J. Hanes, C. Schaffitzel, A. Knappik, A. Pluckthun, *Nat Biotechnol* **18**, 1287 (2000).
35. I. Navratilova, E. Eisenstien, D. G. Myszka, *Anal Biochem* **344**, 295 (2005).
36. W. Ito, H. Yasui, Y. Kurosawa, *J Mol Biol* **248**, 729 (1995).
37. R. Karlsson, *Anal Biochem* **221**, 142 (1994).
38. R. Karlsson, R. Stahlberg, *Anal Biochem* **228**, 274 (1995).
39. D. G. Myszka, *J Mol Recognit* **12**, 279 (1999).
40. D. Myszka, *Abstracts of Papers, 227th ACS National Meeting, Anaheim, CA, United States, March 28–April 1*, 2004, ANYL (2004).
41. D. G. Myszka, *Anal Biochem* **329**, 316 (2004).
42. R. L. Rich, D. G. Myszka, *Anal Biochem* **361**, 1 (2007).
43. H.-J. Boehm *et al.*, *J Med Chem* **43**, 2664 (2000).
44. P.-O. Markgren, M. Hamalainen, U. H. Danielson, *Anal Biochem* **265**, 340 (1998).
45. P.-O. Markgren, M. Hamalainen, U. H. Danielson, *Anal Biochem* **279**, 71 (2000).
46. P.-O. Markgren *et al.*, *Anal Biochem* **291**, 207 (2001).
47. P.-O. Markgren *et al.*, *J Med Chem* **45**, 5430 (2002).
48. C. F. Shuman, P.-O. Markgren, M. Hamalainen, U. H. Danielson, *Antiviral Res* **58**, 235 (2003).
49. C. F. Shuman, L. Vrang, U. H. Danielson, *J Med Chem* **47**, 5953 (2004).
50. C. F. Shuman, M. D. Hamalainen, U. H. Danielson, *J Mol Recognit* **17**, 106 (2004).
51. M. D. Hamalainen *et al.*, *J Biomol Screen* **5**, 353 (2000).

52. E. Jisa *et al.*, *Biochem Pharmacol* **62**, 953 (2001).

53. R. L. Rich *et al.*, *Proc Natl Acad Sci U S A* **99**, 8562 (2002).

54. S. Lofas, *Assay Drug Dev Technol* **2**, 407 (2004).

55. P. Huxley *et al.*, *Chem Biol* **11**, 1651 (2004).

56. R. L. Thurmond, S. A. Wadsworth, P. H. Schafer, R. A. Zivin, J. J. Siekierka, *Eur J Biochem* **268**, 5747 (2001).

57. H. Nordin, M. Jungnelius, R. Karlsson, O. P. Karlsson, *Anal Biochem* **340**, 359 (2005).

58. N. A. Kratochwil, W. Huber, F. Muller, M. Kansy, P. R. Gerber, *Biochem Pharmacol* **64**, 1355 (2002).

59. N. A. Kratochwil, W. Huber, F. Mueller, M. Kansy, P. R. Gerber, *Curr Opin Drug Discov Devel* **7**, 507 (2004).

60. A. Frostell-Karlsson *et al.*, *J Med Chem* **43**, 1986 (2000).

61. D. G. Myszka, R. L. Rich, *Pharm Sci Technol Today* **3**, 310 (2000).

62. R. L. Rich, D. G. Myszka, *Curr Opin Biotechnol* **11**, 54 (2000).

63. R. L. Rich, Y. S. N. Day, T. A. Morton, D. G. Myszka, *Anal Biochem* **296**, 197 (2001).

64. Y. S. N. Day, D. G. Myszka, *J Pharm Sci* **92**, 333 (2003).

65. S. Cimitan, M. T. Lindgren, C. Bertucci, U. H. Danielson, *J Med Chem* **48**, 3536 (2005).

66. E. Danelian *et al.*, *J Med Chem* **43**, 2083 (2000).

67. A. Frostell-Karlsson *et al.*, *J Pharm Sci* **94**, 25 (2005).

68. Y. N. Abdiche, D. G. Myszka, *Anal Biochem* **328**, 233 (2004).

69. R. Thoma *et al.*, *Structure* **11**, 947 (2003).

70. A. J. Ryan, N. M. Gray, P. N. Lowe, C.-W. Chung, *J Med Chem* **46**, 3448 (2003).

71. M. Hennig *et al.*, *Nat Struct Biol* **5**, 357 (1998).

9 High-throughput analysis of biomolecular interactions and cellular responses with resonant waveguide grating biosensors

Ye Fang, Jack Fang, Elizabeth Tran, Xinying Xie, Michael Hallstrom, and Anthony G. Frutos

INTRODUCTION	206
USING RWG BIOSENSORS TO MONITOR BIOMOLECULAR INTERACTIONS	207
Corning Epic System: Theory of operation	207
Measuring the binding affinities of small-molecule/protein interactions	209
Screening small-molecule/protein interactions	209
Functional enzymatic assays: Detecting the proteolytic cleavage of biomolecules	210
USING RWG BIOSENSORS TO MONITOR CELLULAR RESPONSES	213
Theory of detection (dynamic mass redistribution)	213
Panning endogenous receptors in a cell system	214
Analyzing systems cell biology of receptor signaling	215
Evaluating systems cell pharmacology of ligands	216
Screening compounds with living cells	216
CONCLUSION	216
WORKED EXAMPLES	218
Worked Example 1: Determining binding affinity of small-molecule inhibitors for carbonic anhydrase II	218
Worked Example 2: Determining potencies of small-molecule ligands for endogenous receptors in human embryonic kidney 293 cells	219
REFERENCES	222

INTRODUCTION

Assays for high-throughput screening (HTS) and hit validation play a key role in drug discovery and development. Currently, most HTS assays rely on fluorescent or radioactive labels to measure biochemical reactions or to monitor cellular responses. However, labels can have adverse effects on the binding interactions being investigated, leading to false or misleading conclusions about the binding properties of an analyte. For example, fluorescent labels or cell engineering can, in some cases, interfere with the detection, molecular interaction, or cell biology

a

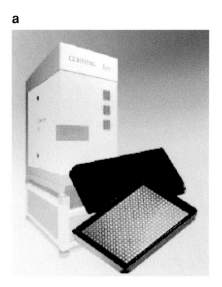

b

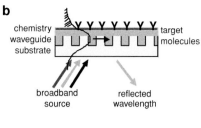

c

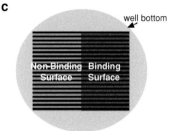

Figure 9–1: The Corning Epic System and principle of detection for biochemical assays. **(a)** The Epic System consists of an SBS standard 384-well microplate with optical biosensors integrated in each well and an HTS-compatible optical reader. **(b)** The Epic System detects a resonant wavelength shift caused by a biochemical binding event. **(c)** Self-referencing technology is utilized in each well for highly sensitive biochemical assays.

of target receptors, thus resulting in spurious results.[1] Furthermore, most cell-based assays based on a single pathway or activity could lead to false negatives, given that many ligands often exhibit collateral efficacies in activating different signaling pathways through a single receptor.[2] In all situations, labels introduce additional complexity and assay development time to the drug discovery process.

Label-free technologies based on a variety of operating principles (heat, refractive index, electrical impedance) have been developed to address the limitations of labels. Although these technologies have helped researchers overcome some of the primary limitations with labels, most label-free biosensor systems suffer from two limitations: (1) relatively low throughput, and (2) limited applicability to analyze both biomolecular interactions and cellular activities.[3–6] Corning has developed and commercialized the Epic System as the first high-throughput label-free biosensor system suitable for both biochemical and cell-based assays. This chapter will illustrate how the Epic System, based on resonant waveguide grating (RWG) sensors, can be used to monitor both biomolecular interactions and cellular responses.

USING RWG BIOSENSORS TO MONITOR BIOMOLECULAR INTERACTIONS

Corning Epic System: Theory of operation

The Epic System (Figure 9–1a) consists of two basic components: (1) a disposable Society for Biomolecular Screening (SBS) standard 384-well microplate with optical biosensors integrated in each well, and (2) an optical reader capable of measuring 40 000 wells in an 8-hr period. The optical sensor in each well of the microplate is an RWG sensor and consists of a substrate with an optical grating and a high index of refraction waveguide coating (Figure 9–1b). When illuminated with broadband light at a fixed angle of incidence, these sensors reflect only a narrow band of wavelengths that is a sensitive function of the index of refraction of the waveguide and is governed by the following equation:

$$\sin\Theta = n_{eff} - \lambda/\Lambda, \tag{9–1}$$

where Θ is the angle of incidence, n_{eff} is the effective index of refraction of the waveguide, λ is the resonant wavelength, and Λ is the grating period. The light that is coupled into the waveguide film propagates parallel to the surface in the plane of the waveguide film and creates an electromagnetic field (i.e., an evanescent wave) in the liquid adjacent to the interface. The amplitude of the evanescent wave decays exponentially with increasing distance from the interface. The distance from the sensor surface at which the electric field strength has decreased to 1/e of its initial value is the penetration depth; for the RWG sensors described here, the penetration depth is ∼150 nm.

For high-throughput biomolecular interaction analysis and screening, RWG biosensors are advantageous because they allow light at nominally normal incident angle to illuminate the biosensor.[6] This is an important design parameter for illuminating large numbers of biosensors simultaneously – a prerequisite for HTS assays in microplates.

For biochemical assays, the microplate surface is coated with a preactivated, amine coupling chemistry for direct immobilization of target molecules (proteins, peptides, DNA, small molecules) via primary amine groups. RWG sensors are sensitive to small changes in temperature and changes in the bulk index of refraction of the solvent. For high-sensitivity measurements, it is critical that these effects be controlled or referenced out. The Epic System employs a self-referencing technology to address this issue in which each well in the microplate has a reference region used to reference out the aforementioned factors (see Figure 9–1c). This is enabled by a patent-pending process that provides protein binding chemistry on only half of the sensor surface so that when a solution of target is added to the well, it only binds to this half of the biosensor, leaving the other half as an in-well reference. The detector uses dual scans to detect shifts in resonant wavelength in both regions. The binding of an analyte to the immobilized target induces a change in the effective index of refraction of the waveguide, which is manifested as a shift in the wavelength of light reflected from the sensor. The unwanted effects, largely due to temperature fluctuation and mismatch in bulk index of refraction between solutions, can be referenced out by subtracting the wavelength shift in the reference region from that in the sample region. The magnitude of the resultant wavelength shift is proportional to the amount of analyte binding to the immobilized target. With this self-referencing approach, wavelength shifts in the picometer range can be measured with noise levels on the order of 0.1 pm (∼7×10^{-7} refractive index units).

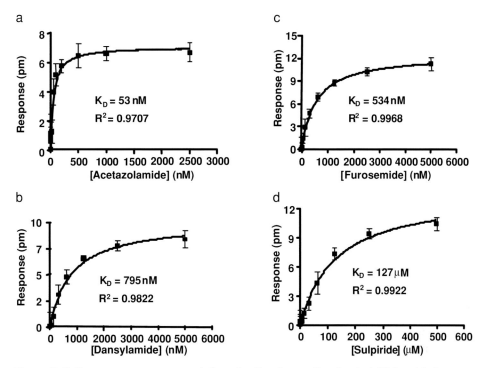

Figure 9–2: Dose–response curves and K_D estimation for small-molecule inhibitors binding to immobilized carbonic anhydrase II: **(a)** acetazolamide; **(b)** dansylamide; **(c)** furosemide; **(d)** sulpiride. Each concentration was run in quadruplicate.

Measuring the binding affinities of small-molecule/protein interactions

Because of its high throughput and high sensitivity, the Epic System can determine binding affinities of large numbers of small molecules under equilibrium binding conditions. The microplate-based Epic System does not utilize continuous flow cells/channels; as a consequence, measurement of the kinetics of binding is mass-transport or diffusion limited.

Figure 9–2 shows an example of dose–response curves for four different compounds to immobilized carbonic anhydrase isosyme II (CAII). Various concentrations of four small-molecule inhibitors in binding buffer (phosphate-buffered saline [PBS]/0.1% dimethyl sulfoxide [DMSO]), acetazolamide, dansylamide, furosemide, and sulpiride, were added to wells containing immobilized CAII. Final readings were performed after 30-min incubation at 25°C. Saturation binding experiments were analyzed by nonlinear regression software (GraphPad Software, San Diego, CA, USA). All binding curves were dose-dependent and saturable. As shown in Table 9–1, the four binding constants obtained were similar to published data.[7–10]

Screening small-molecule/protein interactions

The Epic System is designed for high-throughput screening applications; this is enabled by the 384-well microplate format and the ability of the instrument to

Table 9–1. Comparison of K_D values for CAII binders

Compound	MW (Dalton)	Epic K_D (nM)	Literature K_D (nM)
Acetazolamide	222.2	53	26[10]
Dansylamide	250.3	795	760[8]
Furosemide	330.7	534	513[8]
Sulpiride	341.4	127000	186000[8]

be integrated with robotics. To demonstrate the robustness of a high-throughput small-molecule binding assay, the Epic System determined the yes/no binding of furosemide (330.7 Da) to immobilized bovine carbonic anhydrase II. The Z′ factor is a standard measure of the quality of any high-throughput screening assay and takes into account both the variability of the assay and the signal window.[11] Z′ values greater than 0.5 are generally required for an assay to be considered for HTS. As indicated in Figure 9–4a, Z′ was 0.54 for a 384-well plate (192 wells had 5 μM furosemide and 192 wells had buffer-only control). This level of performance indicated that the assay was robust and suitable for high-throughput screening.

Having demonstrated the robustness of the assay, we used the Epic System to screen 80 compounds with molecular weights from 59 to 677 Da from the Sigma LOPAC (Library of Pharmaceutically Active Compounds) library. Figure 9–3b shows the results of the screen. As expected, most compounds showed no binding to the immobilized carbonic anhydrase. In addition to the positive control (furosemide) we identified one hit in the screen, acetazolamide, a known carbonic anhydrase inhibitor.

Functional enzymatic assays: Detecting the proteolytic cleavage of biomolecules

Proteases are a large group of enzymes involved in a multitude of physiological reactions from simple digestion of food proteins to highly regulated cascades. These enzymes are known for digesting long protein chains into short

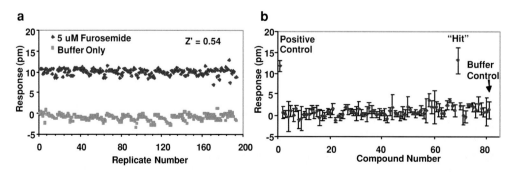

Figure 9–3: Screen of small molecules against carbonic anhydrase II: **(a)** evaluation of assay robustness (Z′); **(b)** screen of 80 small molecules (10 uM in 3% DMSO). Furosemide and buffer were used as the positive and negative controls, respectively. The hit is acetazolamide, a known carbonic anhydrase inhibitor.

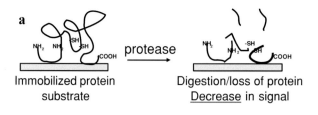

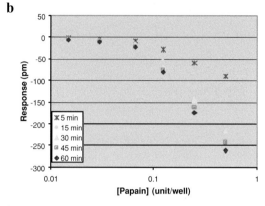

Figure 9–4: Optimization of functional enzymatic assays: **(a)** principle of the assay; **(b)** results of assay optimization for papain concentration and reaction time.

fragments or splitting the peptide bonds that link amino acid residues.[12–14] Conventional protease inhibitor screens involve labeling a substrate with fluorescent probes, which in some cases can alter or inhibit the native functional interactions between enzyme and substrate.[15]

Functional lytic assays are based on the enzymatic cleavage of immobilized biomolecules on the sensor as depicted in Figure 9–4a. The lytic process is observed as a decrease in resonant wavelength signal as a result of loss of mass from the sensor. In these assays, the measured digestion rates are diffusion-limited because the microplate experiments are not carried out under continuous flow conditions.

Figure 9–4b shows assay optimization results for a functional protease assay in which the concentration of enzyme (papain) and its reaction time were varied in wells containing the immobilized substrate human serum albumin. As expected, longer reaction times and higher concentrations of enzyme resulted in larger decrease in signal due to more digestion of the immobilized substrate. Reliable signals were observed in as little as 5 min.

Functional enzymatic assays can be performed in a quantitative manner for the determination of compound potency. Figure 9–5a shows the results of a functional trypsin assay in which the IC_{50} of two different trypsin inhibitors was measured. Table 9–2 summarizes the results for different proteases and inhibitors. The experimentally measured IC_{50} values compared closely with literature IC_{50} values. This agreement provided sufficient confidence that the surface-based digestion profiles could reliably detect inhibitor effects. Because the assays were performed in a microplate format, multiple different assays could be performed in parallel. To demonstrate this throughput capability, experiments were performed

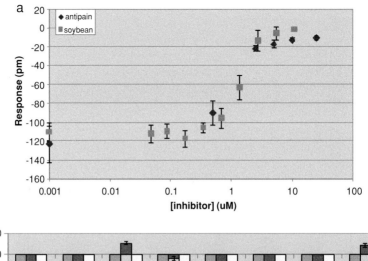

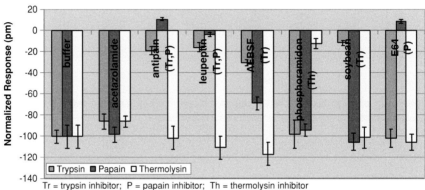

Tr = trypsin inhibitor; P = papain inhibitor; Th = thermolysin inhibitor

Figure 9–5: HTS functional enzymatic assays. **(a)** IC_{50} determination of trypsin inhibitors using human serum albumin as the substrate; **(b)** cross-screening of inhibitors against three different proteases.

to screen seven different compounds against three different enzymes simultaneously. Responses of the three different enzymes were normalized and set at –100 pm. As shown in Figure 9–5b, one can easily identify which compounds inhibit which enzymes and which ones show cross-reactivity. For example, leupeptin shows functional inhibition of both trypsin and papain but not of thermolysin, consistent with the fact that leupeptin is a known inhibitor of trypsin

Table 9–2. Assay performance for functional lytic assays

Enzyme	Inhibitor	Epic IC_{50} (μM)	Reported IC_{50} (μM)	Z'
trypsin	antipain	1.5	0.43[a]	0.64
trypsin	leupeptin	2.5	2.6[b]	0.62
papain	leupeptin	1.9	NA	0.67
thermolysin	phosphoramidon	9	8[c]	0.54

[a] Sigma Aldrich product information sheet.
[b] Chemicon International product information sheet.
[c] FEBS Lett 356, 238–243 (1994).

and papain. All measured Z' values ($N = 40$) were above 0.5, indicating that the assays are robust and suitable for HTS.

USING RWG BIOSENSORS TO MONITOR CELLULAR RESPONSES

The ability of an optical biosensor such as the Epic System to examine living cells is increasingly important in both basic research and drug discovery and development. Although more complex and less specific than biomolecular interaction analysis, biosensor cell-based assays show a superior ability to facilitate the measurements of mode of action, pathway activation, toxicity, and phenotypic responses of cells upon stimulation. Recently we demonstrated the use of these assays for probing cell signaling in both high-throughput and high-information content modes.[5,6, 16–23] Here we highlight several examples in which label-free Epic cellular assays impact basic research and drug discovery.

Theory of detection (dynamic mass redistribution)

In Epic cell assays, whole cells are cultured directly on the sensor surface and exposed to various compounds (agonist, antagonists, etc.). The detection principles for performing whole cell assays are similar to those for biochemical assays in that changes in local index of refraction are manifested as a shift in response of the sensor. The surface sensitivity of RWGs means that only the bottom portion of whole cells cultured on the sensor are monitored during an assay. Because the amplitude of the evanescent wave decays exponentially from the sensor surface, a target or complex contributes more to the overall response when the target or complex is closer to the sensor surface as compared to when it is farther from the sensor surface. Recently, we developed theoretical and mathematical models that describe the parameters and nature of the optical signals of living cells upon stimulation.[16] These models, based on a three-layer waveguide system in combination with known cellular biophysics, provide a link from stimulus-induced optical responses to several cellular processes mediated through a receptor. We found[16] that a ligand-induced change in effective refractive index (i.e., ΔN, the detected signal) is governed by:

$$\Delta N = S(C)\alpha d \sum_i \Delta C_i(t) \left[e^{\frac{-z_i}{\Delta Z_C}} - e^{\frac{-z_{+1i}}{\Delta Z_C}} \right], \qquad (9\text{--}2)$$

where $S(C)$ is sensitivity to the cell layer, ΔZ_c the penetration depth into the cell layer, α the specific refraction increment (about 0.18/mL/g for proteins), $\Delta C_i(t)$ the change in local concentration of biomolecules at the given location and a specific time, z_i the distance where the mass redistribution occurs, and d an arbitrary thickness of a slice within the cell layer. This theory predicts, as confirmed by experimental studies,[16–18,23] that the optical biosensor can monitor cell signaling in real time; a stimulus-induced optical response is dominated by

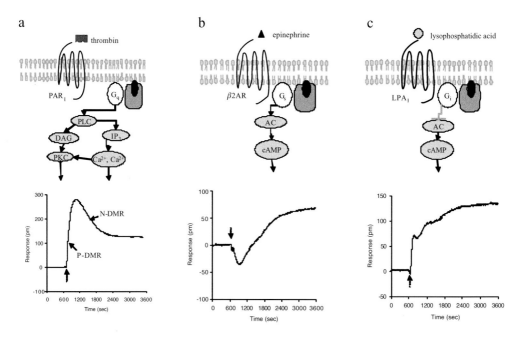

Figure 9–6: GPCR signaling and its characteristic DMR signals in a human cancer cell line A431. **(a)** Gq signaling and its DMR signal (protease-activated receptor subtype 1, 10 U/mL thrombin). **(b)** G_s signaling and its DMR signal (β_2-adrenergic receptors in A431 cells, 2 nM epinephrine). **(c)** Gi signaling and its DMR signal (lysophophatidic acid [LPA] receptor subtype 1 in A431, 100 nM LPA). The broken arrows indicate the time when an agonist solution is introduced. Figure was reproduced with permission from reference 17. (Y. Fang, G. Li, A. M. Ferrie, *J Pharmacol Toxicol Methods* 55, 314 (2007)).

dynamic mass redistribution (DMR), and the resultant DMR signal is an integrated cellular response containing contributions from many cellular events mediated through a receptor.

Panning endogenous receptors in a cell system

The sensitivity of the Epic System enables assays using cells with endogenously expressed levels of receptors, which eliminates the need for cell manipulation/engineering steps such as transfection. Furthermore, these assays are not pathway-biased and apply to many classes of targets, including G protein-coupled receptors (GPCRs), receptor tyrosine kinases (RTKs), ion channels, and others. Importantly, ligand-induced DMR signals are pathway-sensitive and offer a global measure of the entire cell signaling mediated through a target receptor. Thus, receptor panning using a focused ligand library known to activate many families of receptors would allow one to reliably map endogenous receptors in a cell system. Using the Epic System, we have found that the human cancer cell line A431 endogenously expresses multiple families of GPCRs, whose activation mediates signaling through at least three distinct pathways: G_q-, G_s-, and G_i-signaling[16] (Figure 9–6).

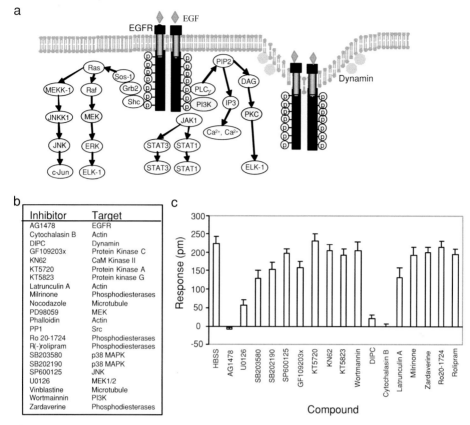

Figure 9–7: EGFR signaling and the modulation profiles of the EGF-induced DMR signal in A431 by panels of inhibitors: **(a)** the EGFR-signaling pathways; **(b)** panels of inhibitors and their corresponding targets; **(c)** the amplitudes of the N-DMR event in the EGF-induced DMR signal as a function of compound. The "HBSS" in the graph indicates that the cells are pretreated with $1 \times HBSS$ buffer (Hank's balanced salt buffer) prior to EGF stimulation and used as a positive control. Figure was reproduced with permission from reference 19. (Y. Fang, A. M. Ferrie, N. H. Fontaine, P. K. Yuen, *Anal Chem* 77, 5720 (2005)).

Analyzing systems cell biology of receptor signaling

Cells rely on highly dynamic network interactions to activate specific responses to exogenous stimuli. Epic's DMR signal can be used to analyze systems cell biology of receptor signaling because of its sensitivity to pathways being activated, its integrative nature, and its physiological relevance. Combined with biochemical, cell biology, and biophysical approaches, we have studied the systems cell biology of epidermal growth factor receptor,[19] bradykinin B_2 receptor,[18] and β_2 adrenergic receptor,[21] all endogenously expressed in A431 human epidermoid carcinoma cells. Analysis of epidermal growth factor (EGF)-induced DMR signals of quiescent A431 cells allows the mapping of signaling pathways and network interaction of the EGF receptor (Figure 9–7). The EGF-induced DMR signal was found to be saturable to EGF concentration and sensitive to the activity of EGFR tyrosine kinase, MEK, and dynamin. This study not only revealed the importance of receptor endocytosis in EGF receptor signaling but also linked the EGF-induced

Epic response of quiescent A431 cells to the Ras/mitogen-activated protein (MAP) kinase pathway, which proceeds primarily through MEK and leads to cell detachment.

Evaluating systems cell pharmacology of ligands

A compound such as a ligand or drug candidate could initiate a variety of responses in a cell system. First, a compound could have off-target effects, such as cross-talk with distinct cellular targets. Second, different ligands could display distinct functional selectivity by inducing an operative bias to activate specific portions of the cell machinery through the same receptor.[2] Unlike most pathway-biased cell assays, Epic's cellular assays do not require prior knowledge of cell signaling and are applicable to wide classes of targets and cell signaling. Thus, a single biosensor assay can be used to study the systems cell pharmacology of a drug candidate acting on cells.

An example is highlighted in Figure 9–8. Protease-activated receptors (PARs) consist of a family of four G protein-coupled receptors. Using the Epic System, we have shown that there are functional interactions between endogenous PAR_1 and PAR_2 in A431 cells, and a PAR_1-activating peptide SFFLR-amide exhibits cross talk with PAR_2.[19] Further analysis suggests that the RWG biosensor-based cellular assays differentiate the signaling of PAR_1 and PAR_2.

Screening compounds with living cells

Assaying large numbers of samples is essential for the most widely adopted screening paradigm in drug discovery, which is centered on the ability to "fish" out hits from a library of millions of compounds. The Epic System is the first biosensor system that has been applied for HTS using both immobilized proteins and living cells.[5,6] Recently we have demonstrated that the Epic System is amenable to high-throughput screening against endogenously expressed GPCRs. Examples include the bradykinin B_2 receptors and the β_2ARs in A431 cells and the protease-activated receptor subtype 1 (PAR_1) in Chinese hamster ovary (CHO) cells.[20] Based on the kinetics of agonist-mediated optical signals, two time points, one before and one after stimulation, were chosen to develop HTS assays for both B_2 receptors in A431 and PAR1 in CHO cells. The results suggest that such end point measurements enable high-throughput screening of compounds using endogenous GPCRs (Figure 9–9).

CONCLUSION

Label-free optical biosensors have migrated from a tool solely for biomolecular interaction analysis to a universal platform for both biochemical and cell-based assays. Together with recent advancement in instruments for high-throughput screening, the newly discovered ability of optical biosensors to assay living cells

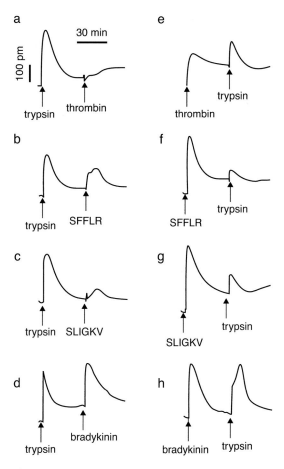

Figure 9–8: Desensitization of A431 cells to repeated agonist stimulation. The cells were subjected to repeated stimulation, separated by ~1-hr intervals with various combinations (a–h) of agonists. The agonist concentration was 40 units/ml, 200 nM, 20 μM, 20 μM, and 100 nM for thrombin, trypsin, SFLLR-amide, SLIGKV-amide, and bradykinin, respectively. Figure was reproduced with permission from reference 23. (Y. Fang, A. M. Ferrie, *BMC Cell Biol* 8, 24 (2007)).

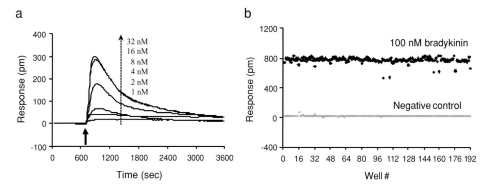

Figure 9–9: (a) The dose-dependent DMR responses of quiescent A431 cells induced by bradykinin. **(b)** The wavelength shifts, measured using end point assays, as a function of wells having quiescent A431 cells. The negative control is obtained by treating the cells with the buffer only. Figure was reproduced with permission from reference 20. (G. Li, A. M. Ferrie, Y. Fang, *J Assoc Lab Automat* 11, 181 (2006)).

will accelerate wide adoption of label-free biosensors in both basic research and drug discovery processes.

The universal nature of this technology makes it amenable to a wide range of additional formats and applications. These formats include 96-well and 1536-well plates, and the incorporation of continuous flow cells will enable high-throughput kinetic analysis of biomolecular interactions. We anticipate future developments in applications ranging from virology and viral detection to absorption, distribution, metabolism, and excretion (ADME) Tox, antibody profiling, and single cell analysis.

WORKED EXAMPLES

We describe in detail two examples that illustrate how the Epic System can be used for monitoring biomolecular interactions and cellular responses. The first example is related to the binding of small molecules to carbonic anhydrase II immobilized on the biosensor surfaces, with an emphasis on how to reach robust detection in a high-throughput platform. The second example is related to the determination of the potencies of small-molecule ligands for two endogenous G protein-coupled receptors, the muscarinic receptor, and protease-activated receptors in human embryonic kidney 293 cells, as well as the evaluation of assay robustness. These examples show distinct assay formats, the end point measurements for biomolecular interaction analysis, and the kinetic measurements for ligand-induced cellular responses. In both instances, assay development is minimal and the assays are robust.

Worked Example 1: Determining binding affinity of small-molecule inhibitors for carbonic anhydrase II

Background. The interaction between bovine carbonic anhydrase II (CAII) and its small-molecule inhibitors has been studied extensively and used as a model system for label-free biosensors.[8-11] Carbonic anhydrase is a zinc metallo-enzyme that catalyzes the reversible hydration of carbon dioxide. The enzyme is implicated in glaucoma, and the drugs acetazolamide and dichlorphenamide that treat this condition target carbonic anhydrase.

Materials needed
- 384-well Epic biochemical assay microplates (Corning)
- Sodium acetate buffer (20 mM sodium acetate, pH 5.5) (Sigma)
- PBS buffer (20 mM NaH_2PO_4-Na_2HPO_4, 150 mM NaCl, pH 7.4)
- Benzenesulfonamide, 4-carboxybenzenesulfonamide, acetazolamide, dansylamide, furosemide, and sulpiride (Sigma)
- Carbonic anhydrase II from bovine erthyrocytes (Sigma)

Immobilization of carbonic anhydrase II. The amine coupling chemistry in the Epic biochemical assay plates is a preactivated chemistry based on polymeric

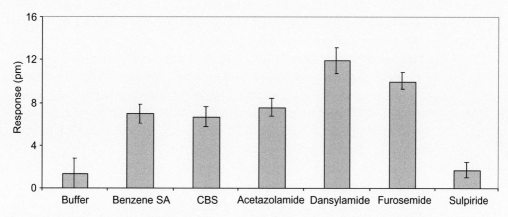

Figure 9–10: Detection of inhibitor binding to the immobilized bovine carbonic anhydrase II. Quadruplicates of 5 μM each small-molecule inhibitors, benzenesulfonamide (benzene SA, 157.2 Da), 4-carboxybenzenesulfonamide (CBS, 201.1 Da), acetazolamide (222.2 Da), dansylamide (250.3 Da), furosemide (330.7 Da), and sulpiride (341.4 Da), were added to the immobilized CAII.

maleic anhydride groups. In general, highest levels of protein immobilization are obtained when using an immobilization buffer ∼0.5 – 1 unit below the isoelectric point of the protein. (For immobilization of peptides and small molecules, an immobilization pH of 7–9 is recommended.)

Briefly, CA II was diluted to 100 μg/ml in 20 mM sodium acetate pH 5.5 and was coupled to the surface overnight at 4°C. The plate was washed several times with PBS. Finally, the microplate having CAII immobilized was maintained using PBS buffer containing 0.1% DMSO.

Analyte preparation. Each inhibitor was dissolved in 100% DMSO to prepare a stock solution of 10 mM, and aliquots of these solutions were stored at –20°C. Before use, all inhibitor stock solutions were diluted using PBS buffer, and all solutions contained 0.1% DMSO.

Binding assays. Figure 9–10 shows the direct binding of six known small-molecule inhibitors, benzenesulfonamide (157.2 Da), 4-carboxybenzenesulfona-mide (201.1 Da), acetazolamide (222.2 Da), dansylamide (250.3 Da), furosemide (330.7 Da), and sulpiride (341.4 Da), to the immobilized CAII. In this example, quadruplicates of 5 μM each inhibitor were incubated with the immobilized CAII at room temperature for 30 min before the final reading. We observed clear binding signals for five of six inhibitors except sulpiride, which could be explained by its high K_D value of 186 μM.[9] The smallest compound, benzenesulfonamide (157.2 Da), had a more than fivefold signal-to-background ratio.

Worked Example 2: Determining potencies of small-molecule ligands for endogenous receptors in human embryonic kidney 293 cells

Assay design. Human embryonic kidney 293 (HEK293) cells were directly cultured in the wells of a 384-well Epic fibronectin-coated cell assay microplate. After a baseline of 2 min was established, compound solutions were then introduced and the cellular responses were continuously monitored for about 1 hr. The solutions included: the negative control (e.g., the assay buffer; columns 1 and 24), 1 μM carbachol (columns 2 to 5), 5 μM carbachol (columns 6 to 9), a twofold dilution series of carbachol (columns 10 to 12), 1 μM PAR$_1$ agonist SFLLR-amide (columns 13 to 16), 5 μM SFLLR-amide (columns 17 to 20), and a twofold dilution series of SFLLR-amide (column 21 to 24). This assay was designed to determine the ligand potencies as well as the assay robustness.

Materials needed

- 384-well Epic cell assay microplate, fibronectin coated (Corning)
- SFLLR-amide (Bachem)
- Carbachol (Sigma)
- Human embryonic kidney 293 cells (American Type Cell Culture)
- Dulbecco's modified Eagle's medium (DMEM) (Invitrogen)
- Fetal bovine serum (FBS) (BD Biosciences)
- 1 × regular Hank's balanced salt solution (HBSS) (Invitrogen)
- T175 tissue-culture treated polystyrene flask (Corning)
- 1 M 4-(2-hydroxyethyl)-1-piperazineethanesulfonic acid (HEPES) buffer, pH 7.1 (Invitrogen)
- BioTek plate washer
- Corning 384-well microplates (Corning)

Cell culture. HEK293 cells were grown in T175 tissue-culture treated polystyrene flasks until a confluency of ∼80% was reached. After harvesting, cells were suspended in a culture medium and seeded into the Epic plate at 12000 cells per 50 μL per well. The culture medium contains DMEM medium, 10% FBS, 4.5 g/liter glucose, 2 mM glutamine, and antibiotics. The cells were then cultured at 37°C under air/5% CO$_2$ for ∼20 hr.

Cell handling. After culture, the cells were subject to a gentle wash using a BioTek plate washer and then maintained in 30 μl assay buffer. The assay buffer consists of 1×HBSS and 20 mM HEPES buffer, pH 7.1.

Compound preparation. Both carbachol and SFLLR-amide were dissolved in dimethyl sulfoxide to prepare stock solutions (each at 10 mM). Before use, the stock solutions were thawed and diluted into 4x desired concentrations using the assay buffer. The resultant compound solutions were dispensed into a 384-well compound source plate.

Epic cell assays. Both the cell plate and the compound source plates were incubated inside the Epic System for about 1 hr to minimize the effect of any temperature mismatch between the cell solution and compound solutions. Afterward, a baseline scan of 2 min ensured that the cells reach quasi-equilibrium state, as shown by the steady net-zero response (Figure 9–11). Compound solution of

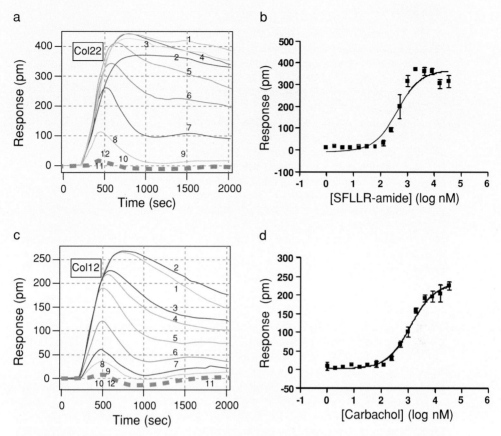

Figure 9–11: The dose-dependent responses of HEK293 cells upon stimulation with SFLLR-amide (**a** and **b**) and carbachol (**c** and **d**). The numbers in (**a**) and (**c**) indicate the concentrations of ligand – 32768, 16384, 8192, 4096, 2048, 1024, 512, 256, 128, 64, 32, 16 nM for the numbers from 1 to 12, respectively.

10 μl was then introduced, and the cell responses were continuously monitored for about 1 hr.

Data process and analysis. All data were processed using Microsoft Excel and Prism software (Graph Pad). Figure 9–11a shows the dose-dependent response of HEK293 cells induced by SFLLR-amide. SFLLR-amide is an agonist for protease-activated receptor subtype 1 (PAR$_1$), which is endogenously expressed in HEK293 cells. SFLLR-amide led to a dose-dependent optical response whose maximal amplitude is saturable, with an apparent potency of 325 ± 20 nM ($n = 3$) (Figure 9–11b). SFLLR-amide at 1 μM led to a maximal response of 282 ± 40 pm ($n = 64$), whereas at 5 μM it resulted in a maximal response of 335 ± 34 pm ($n = 64$).

Figure 9–11c shows the dose-dependent response of HEK293 cells induced by carbachol. Carbachol is a natural agonist for endogenous muscarinic receptors in the cells. Its maximal response was also dose-dependent and saturable, leading to an apparent EC$_{50}$ of 1248 ± 150 nM ($n = 3$) (Figure 9–11d). The response induced by carbachol at 5 μM was 202 ± 24 pm ($n = 64$), higher than that at 1 μM (144 ± 24 pm).

REFERENCES

1. M. A. Cooper, *Drug Discov Today* **11**, 1061 (2006).
2. T. Kenakin, *Nat Rev Drug Discov* **4**, 919 (2005).
3. R. L. Rich, D. G. Myszka, *Anal Biochem* **361**, 1 (2007).
4. M. A. Cooper, *Nat Rev Drug Discov* **1**, 515 (2002).
5. Y. Fang, *Assay Drug Dev Technol* **4**, 583 (2006).
6. Y. Fang, *Sensors* **7**, 2316 (2007).
7. K. Hakansson, A. Liljas, *FEBS Lett* **350**, 319 (1994).
8. D. G. Myszka, *Anal Biochem* **329**, 316 (2004).
9. G. A. Papalia *et al.*, *Anal Biochem* **359**, 94 (2006).
10. M. J. Cannon *et al.*, *Anal Biochem* **330**, 98 (2004).
11. J. Zhang, T. Chung, K. Oldenburg, *J Biomol Screen* **4**, 67 (1999).
12. S. O'Malley, X. Xie, A. Frutos, *J Biomol Screen* **12**, 117 (2007).
13. B. Lin, J. Gerstenmeier, P. Li, H. Pien, J. Pepper, B. A. Cunningham, *Biosens Bioelectron* **17**, 827 (2002).
14. C. Sumner, S. Krause, A. Sabot, K. Turner, C. J. McNeil, *Biosens Bioelectron* **16**, 709 (2001).
15. N. Ramachandran, D. N. Larson, P. R. H. Stark, E. Hainsworth, J. LaBear, *FEBS J* **272**, 5412 (2005).
16. Y. Fang, A. M. Ferrie, N. H. Fontaine, J. Mauro, J. Balakrishnan, *Biophys J* **91**, 1925 (2006).
17. Y. Fang, G. Li, A. M. Ferrie, *J Pharmacol Toxicol Methods* **55**, 314 (2007).
18. Y. Fang, G. Li, J. Peng, *FEBS Lett* **579**, 6365 (2005).
19. Y. Fang, A. M. Ferrie, N. H. Fontaine, P. K. Yuen, *Anal Chem* **77**, 5720 (2005).
20. G. Li, A. M. Ferrie, Y. Fang, *J Assoc Lab Automat* **11**, 181 (2006).
21. Y. Fang, A. M. Ferrie, J. Lahiri, *Trends in Signal Transduction Research*. J. N. Meyers, Ed., Nova Science Publishers Inc. (New York, 2007), p. 145.
22. Y. Fang, A. M. Ferrie, G. Li, N. H. Fontaine, *Biochim Biophys Acta* **1763**, 254 (2006).
23. Y. Fang, A. M. Ferrie, *BMC Cell Biol* **8**, 24 (2007).
24. M. Wu *et al.*, *FEBS Lett* **580**, 5681 (2006).

10 ITC-derived binding constants: Using microgram quantities of protein

Richard K. Brown, J. Michael Brandts, Ronan O'Brien, and William B. Peters

INTRODUCTION	223
BINDING CONSTANTS AND THERMODYNAMICS FROM A SINGLE LABEL-FREE MEASUREMENT	224
Introduction to the ITC experiment	224
Application example – binding thermodynamics of statins to HMG-CoA reductase	228
ITC INSTRUMENT DESIGN	231
Latest developments in ITC miniaturization	234
Reduction in sample quantity requirements	234
Increases in sample throughput	236
Low-protein quantity data comparison	238
ITC EXPERIMENT DESIGN CONSIDERATIONS	241
Experiment design alternatives to minimize protein usage	241
Example: Binding study of carbonic anhydrase II to multiple ligands	243
The ITC experiment design window	244
Example 1: A c value-limited situation	248
Example 2: A heat-limited situation	248
CONCLUSIONS	248
REFERENCES	249

INTRODUCTION

Isothermal titration calorimetry (ITC) has gained wide acceptance in drug discovery and development laboratories throughout the world. Every major pharmaceutical company and most biopharmaceutical companies and major research institutions now use ITC. Modern, highly sensitive ITC instruments are being applied in the drug discovery and development process for applications such as: (a) selection of small-molecule "hits" following primary and secondary screening of chemical libraries against protein targets of interest, (b) optimization of small-molecule leads based on elucidation of the binding mechanism and binding characteristics and development of structure–activity relationships (SAR) used in the optimization process, and (c) selection and optimization of therapeutic protein variants.

Because ITC directly measures the heat released or absorbed during a biomolecular binding event, it is the only technique that allows simultaneous determination of all thermodynamic binding parameters (n, K_A, ΔH and ΔS) in a single experiment.[1] The ITC technique provides this unique capability in an experimental environment that is label-free, in solution, and that does not require immobilization of either the target macromolecule or ligand. Modern instruments are highly sensitive, accurate, and reproducible, and unlike spectroscopic methods, the degree of optical clarity of the solutions is unimportant. As modern ITC instrumentation has evolved, these instruments have become more sensitive, faster, and easier to use. Binding parameters determined by ITC are often referred to as "gold standard" values and are frequently used as reference standard values for other techniques.

Despite the fundamental advantages of this technique and the advances in instrumentation, use of ITC in the early, critical decision-making stages of drug discovery is often limited. This is often due to: (a) the lack of sufficient quantities of protein during those early stages, and (b) the time required to perform the ITC experiments on large numbers of potential ligands (or protein constructs) of interest. Traditionally, 50–1500 µg of protein have been consumed to complete an ITC experiment, and completing each experiment could require 2 hr or more.

In this chapter we will describe the characteristics of a new miniaturized, ultra-sensitive ITC designed to push back these limitations allowing ITC to be effectively used at earlier stages of the drug discovery and development process. This new microcalorimetry system reduces the quantity of protein (or other macromolecule sample) required to obtain a complete thermodynamic profile (n, K_A, ΔH and ΔS) by up to sevenfold. When combined with modified experimental protocols, the protein sample requirements are reduced by as much as tenfold as compared to traditional ITC experiments performed on previous generations of instruments. In many instances the protein sample requirements are as low as single-digit µg quantities. With its reduced size and associated sample quantity requirements, this provides a significantly faster response time, allowing as many as two to four experiments to be completed per hour. The fully automated configuration of this new instrument provides a sample throughput of at least 50 samples per day with a capacity to process as many as 384 samples in an unattended run.

BINDING CONSTANTS AND THERMODYNAMICS FROM A SINGLE LABEL-FREE MEASUREMENT

Introduction to the ITC experiment

ITC allows the complete thermodynamic profile (n, K_A, ΔH and ΔS) of a ligand–receptor interaction to be determined from a single label-free experiment. A

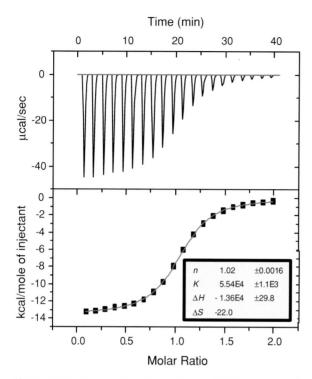

Figure 10–1: *Top panel* illustrates the raw ITC from an experiment of 20 equal injections of a ligand solution into a macromolecule solution. *Bottom panel* illustrates the nonlinear least squares fit of the peak area data from the top panel to a binding model.

titration naturally provides a systemic stepwise addition of one starting component of a ligand–receptor complex into the other starting component. During an ITC experiment, the incremental heat changes, microcalories (μcal), from each step of the titration are accurately measured and recorded. This heat data, combined with the known quantity of titrant (usually ligand) and known quantity of titrate (usually macromolecule), are fit using a nonlinear least squares method to a binding model that yields the best fit values for the stoichiometry (n), binding constant (K), and heat of binding (enthalpy, ΔH) of the biomolecular interaction. A typical example of ITC raw data and the data fit to a binding model are shown in Figure 10–1.

As previously described,[1] when characterizing interactions between a biological macromolecule, M, and a small ligand, X, or between macromolecules, that is,

$$M + X = MX$$
$$MX + X = MX_2$$
$$MX_{n-1} + X = MX_n. \tag{10–1}$$

The parameters n, K_A and ΔH are the independent variables of thermodynamic interest.[1] The entropy, ΔS, and free energy, ΔG, of binding are dependent variables obtained by the calculation

$$\Delta G = -RT \ln K = \Delta H - T\Delta S. \tag{10–2}$$

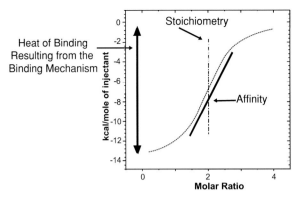

Figure 10–2: Pictorial representation of the correlation of the independent variables of thermodynamic interest to the fitted thermogram.

The experimental parameter determined in the titration calorimeter is the differential heat dQ/dX_{tot} (actually $\Delta Q/\Delta X_{tot}$ for each discrete injection). The differential heat does not depend on the absolute value of M_{tot} but only on its value relative to K_A and X_{tot}.

For a reaction stoichiometry of 1:1 it can easily be shown that:

$$1/V_o(dQ/d[X]_{tot}) = \Delta H \left(\frac{1}{2} + \frac{1 - \dfrac{(1+r)}{2} - \dfrac{X_r}{2}}{\sqrt{X_r^2 - 2X_r(1-r) + (1+r)^2}} \right), \qquad (10\text{–}3)$$

where $[X]_{tot}$ is the total ligand concentration, free plus bound, in the reaction cell of volume V_o, Q is the heat absorbed or liberated, ΔH is the molar heat of binding, and r is defined below in Equation 10–4.

As mentioned above, determination of the thermodynamic parameters from these heat data relies upon the nonlinear least squares fit of the data to an appropriate binding model. Detailed explanations of the ITC experiment, data analysis, and data interpretation have been previously published and will only be touched upon briefly herein.[2–7] Figures 10–2 and 10–3 pictorially illustrate the portions of the thermogram (the fitted data) that correlate primarily to each of the independent thermodynamic variables and how changes in the thermogram indicate specific changes in the nature of the biomolecular interaction studied.

Equation 10–3 provides some useful insight into the optimal design of an ITC experiment. The right-hand side of Equation 3 contains two unitless parameters ($1/r$ and X_r) that depend on the total ligand concentration, $[X]_{tot}$, and the total macromolecule concentration, $[M]_{tot}$. These parameters are defined as:

$$1/r = c = [M]_{tot}K_A \qquad (10\text{–}4)$$

and

$$X_r = [X]_{tot}/[M]_{tot}. \qquad (10\text{–}5)$$

The unitless parameter $1/r$ is the "c" parameter.[1] Provided that concentration is expressed as total concentration of binding sites, the shape of a binding curve

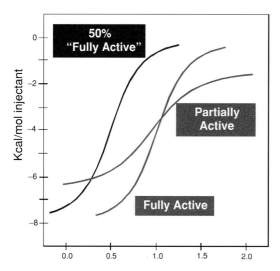

Figure 10–3: Observed changes in the binding thermogram for a specific macromolecule–ligand system are easily interpreted and indicative of specific changes in the specific biomolecular interaction. A representative thermogram for a "fully active" two-component binding system is compared to the thermogram resulting from the same system after some loss of activity (decrease in ΔH) versus the thermogram that would result from loss of material but where the remaining material is fully active.

for proteins with n identical sites will be exactly the same as for a molecule with a single binding site having the same K value. To account for this circumstance the c parameter is defined as:

$$c = n \cdot [M]_{tot} \cdot K_A. \tag{10–6}$$

The c parameter (sometimes referred to as the *sigmoidicity*) may be used to pictorially represent the range of binding constants, which can actually be measured quantitatively when one is forced to use low protein concentrations that challenge the sensitivity of the instrument. A general, informative plot can be made without using actual numerical values for n, K, ΔH, $[M]_{tot}$ or $[X]_{tot}$. Use of the c parameter allows generalizations to be made concerning the behavior of all binding reactions and simplifies the process of making the best choice of the macromolecule concentration if the approximate binding constant is known.[1,8,9]

Binding curves simulated from Equation 3 may be generated for any selected values for c. Figure 10–4 gives several examples. For very tight binding ($c = \infty$) all added ligand is bound until saturation occurs so that a rectangular curve of height ΔH is seen. The steep transition occurs at the stoichiometric equivalence point n in the molar ratio. The shape of this curve is invariant with changes in K so long as the c value remains above ca. 5000. For moderately tight binding with c values between 1 and 1000 the shape of the binding isotherms is very sensitive to small changes in c values. The intercept of these curves on the ordinate is no longer exactly equal to ΔH, but this parameter is still easily obtained by deconvolution from the total area under the curve and its shape. Very weak binding (cf. $c = 0.1$) yields a nearly horizontal trace that, again, like very tight binding, yields

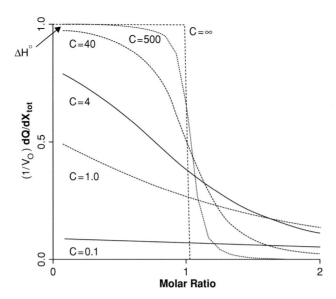

Figure 10–4: Simulated binding isotherms for various values of the parameter c (equal to the product of the binding constant times the total macromolecule concentration), presented in derivative format (see text for details).

little information of the precise value of K. It is apparent from looking at these isotherms that their shape is sensitive to binding constant only for c values in the range $1 \leq c \leq 3000$. This range has been referred to as the "experimental K window." When available, the middle of the window from $c = 5$ to 500 is ideal for measuring K. Later in this chapter a new, highly sensitive calorimeter designed to minimize protein consumption will be introduced, and the useful experimental range of that calorimeter will be discussed. As we will see, both the instrument sensitivity and the c value (or experimental K window) will determine the use boundaries of this (and all other) microcalorimetric instruments.

Application example – binding thermodynamics of statins to HMG-CoA reductase

An excellent and representative example of the application of ITC to the binding characterization of a class of drugs to a protein target was recently published by Carbonell and Freire, "The Binding Thermodynamics of Statins to HMG-CoA Reductase."[10] HMG-CoA reductase is the key enzyme in the cholesterol biosynthetic pathway that leads to hypercholesterolemia, the most important risk factor for cardiovascular disease. The statins as a drug class are the most widely prescribed of all drugs and have been demonstrated to be powerful inhibitors of HMG-CoA reductase. The structures of the five statins in this study are displayed in Figure 10–5. Each of these highly related drugs has a unique safety and efficacy profile. ITC was used to dissect the binding mechanism of these compounds to the protein target. Because many issues related to a lack of selectivity are a result of nonspecific binding to unwanted targets, it is important to explore how these drugs bind to their intended targets, the nature of the forces that drive the binding reaction, and whether there are any correlations between

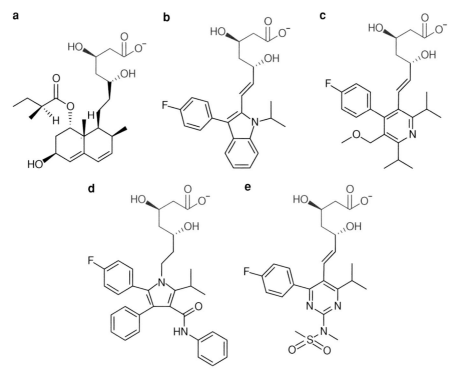

Figure 10–5: Chemical structures of the statins considered in these studies: **(a)** pravastatin, **(b)** fluvastatin, **(c)** cerivastatin, **(d)** atorvastatin, and **(e)** rosuvastatin. The HMG moiety common to all statins is colored red and the variable hydrophobic region black. *See color plates.*

their binding mechanism and selectivity. Because the thermodynamic signature of an inhibitor reflects the types of interactions that drive the binding reaction, it can become an important tool for assessing the potential for a compound to associate nonspecifically with other proteins.

The raw calorimetric data and binding thermograms are displayed in Figure 10–6, and the associated thermodynamic parameters are shown in Table 10–1.

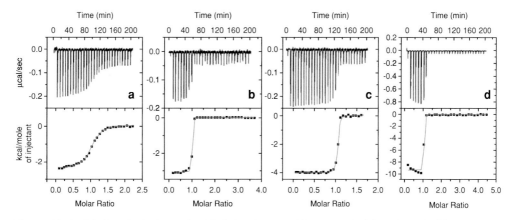

Figure 10–6: Calorimetric titrations of HMG-CoA reductase with pravastatin **(a)**, cerivastatin **(b)**, atorvastatin **(c)**, and rosuvastatin **(d)**. The experiments were performed in triplicate at 25°C in 20 mM Tris-HCl (pH 8.0), 2 mM TCEP, 1 mM NADPH, and 2% DMSO.

Table 10–1. Thermodynamic parameters associated with binding of statin to HMG-CoA reductase[a]

Statin	ΔG (kcal/mol)	ΔH (kcal/mol)	$-T\Delta S$ (kcal/mol)
Fluvastatin	-9.0 ± 0.4	~ 0[b]	~ -9.0
Pravastatin	-9.7 ± 0.4	-2.5 ± 0.1	-7.2 ± 0.4
Cerivastatin	-11.4 ± 0.4	-3.3 ± 0.2	-8.1 ± 0.4
Atorvastatin	-10.9 ± 0.8	-4.3 ± 0.1	-6.6 ± 0.6
Rosuvastatin	-12.3 ± 0.7	-9.3 ± 0.1	-3.0 ± 0.7

[a] The values for ΔG, ΔH, and $-T\Delta S$ are quoted at 25°C.
[b] The binding enthalpy of fluvastatin was beyond the detection limit of the instrument, suggesting that it is close to zero, in which case most of the binding energy is entropic in origin. The free energy of binding was derived from enzyme inhibition assays.

The proportion by which the binding enthalpy contributes to the binding affinity is not the same for all statins as illustrated in Figure 10–7. At 25°C, for fluvastatin, pravastatin, cerivastatin, and atorvastatin the dominant contribution to the binding affinity is the entropy change. Only for rosuvastatin does the enthalpy change contribute more than 50% of the total binding energy (76%).

The differences in the proportion by which the enthalpy and entropy changes contribute to the binding affinity reflect differences in the type of interactions established between the various statins and HMG-CoA reductase. Because the crystal structures of the complexes with rosuvastatin, atorvastatin, cerivastatin, and fluvastatin are known,[11] it was possible to analyze the binding thermodynamics of these statins in terms of structure. All statins target the same site in the protein. According to the crystallographic structures they do not cause differential conformational rearrangements in the enzyme that might significantly influence the binding energetics. Accordingly, differences in binding energetics can be

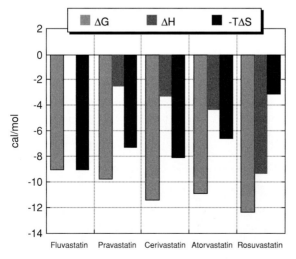

Figure 10–7: Proportion by which the binding enthalpy contributes to the Gibbs energy of binding for each of the statins studied.

attributed to different inhibitor–enzyme interactions. Furthermore, because the HMG region is the same in all inhibitors included in these studies, differences can be attributed to the so-called "hydrophobic region."

The binding enthalpy reflects the strength of the hydrogen bonds and van der Waals forces, between the inhibitor and enzyme, relative to those with water. These forces tend to have directionality and can impose specificity as well as affinity to an interaction. On the other hand, the major favorable contributions to affinity from the entropic component originate primarily from the release of water molecules from the binding interface. This effect reflects the repulsion of the inhibitor from the solvent rather than an attractive inhibitor–enzyme interaction. As such, it is intrinsically nonspecific and acquires specificity only by combining it with shape complementarity of the inhibitor and binding cavity. Another source of binding entropy is related to changes in conformational degrees of freedom. As these changes usually involve fixing the inhibitor and certain regions of the protein in more restricted conformations, they lead to a diminished conformational entropy and an unfavorable contribution to the binding affinity. This loss in conformational entropy is minimized, however, by conformationally constraining and preshaping the inhibitor to its bound conformation. Other entropy contributions such as those associated with translational degrees of freedom are the same for all inhibitors and do not contribute to differences in binding affinity.

Combining the thermodynamic information obtained from ITC with knowledge of the structure allows this type of detailed understanding of the binding mechanism and associated differences even among members of a closely related class of compounds.

ITC INSTRUMENT DESIGN

A new generation isothermal titration calorimeter, the iTC_{200}, of a design similar to that originally described by Wiseman et al., has been developed.[1] Nearly a thousand of the prior generation ITC, the VP-ITC, are currently in operation. Similar to the VP-ITC, the iTC_{200} measures the heat evolved or absorbed upon mixing precise incremental amounts of titrant into a macromolecule solution. A spinning syringe is used for injecting and mixing reactants.

A pair of identical coin-shaped cells is enclosed in an adiabatic outer shield jacket. Access stems travel from the top exterior of the instrument to the cells. Both cells are filled with liquid during operation and both cells have a working volume of 200 μL – a sevenfold reduction in working volume from the prior generation VP-ITC instrument. The syringe volume has been correspondingly reduced to 40 μL with minimum injection volumes as low as 0.1 μL. A block diagram of the instrument is shown in Figure 10–8 and the iTC_{200} is shown in Figures 10–9 and 10–10.

As the titration progresses, temperature differences between the reference cell and the sample cell are measured, calibrated to differential power (DP) units,

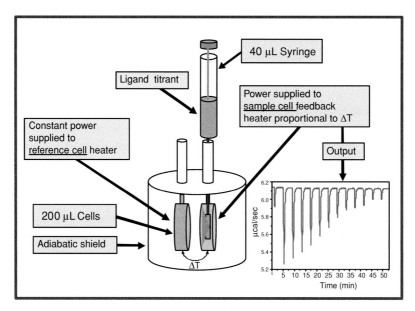

Figure 10–8: Block diagram of the isothermal titration calorimetry (ITC) instrument. The sample and reference cells are maintained at a constant temperature inside the innermost shield of the adiabatic jacket. The reference cell is always kept at the experimental temperature. One of the components of the interaction is placed in the syringe and the other in the cell. When an injection is made, the change in heat associated with binding (endothermic or exothermic) results in a change in temperature in the sample cell. A change in power (heat/s) is required to return the cells to identical temperatures (T) (i.e., $\Delta T = 0$). This change in power is recorded as a series of injections is made. As the course of injections is completed, the binding sites on the sample in the cell are gradually saturated, and the effect becomes reduced. For details, see references 1–5 and 7 from Ladbury references.

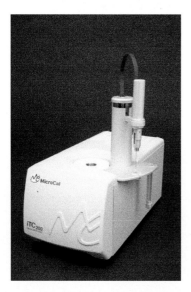

Figure 10–9: The iTC$_{200}$ utilizes new 200-μL sample and reference cells and a new 40-μL syringe capable of injection volumes as small as 0.1 μL. A "pipette tower" is used to accurately and reproducibly position the computer-controlled pipette and syringe stirrer into the cell port, the sample loading station, and the wash station.

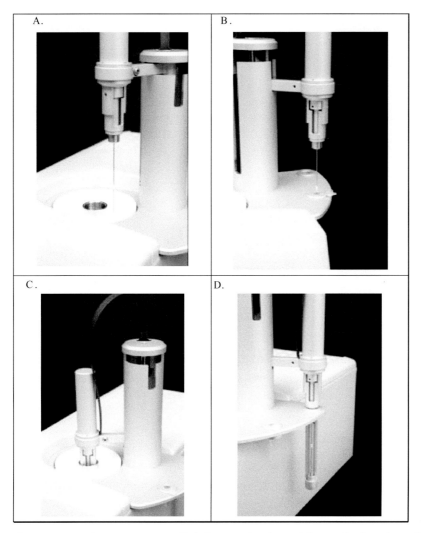

Figure 10–10: The computer-controlled pipette and syringe stirring mechanism are totally supported by and positioned by the pipette tower for all functional operations.

A. Pipette located in the rest position.
B. Pipette moved to the sample loading station to aspirate sample from a micro centrifuge tube.
C. Pipette positioned in the cell for titration.
D. Pipette positioned in the syringe and stirrer wash station.

and then displayed and saved as raw data. This signal is sometimes referred to as the "feedback" power used to maintain temperature equilibrium. We obtain calibration of this signal electrically by administering a known quantity of power through a resistive heater element located on the sample cell.

The syringe containing a "ligand" solution is titrated into the cell containing the "macromolecule" solution. An injection resulting in the evolution of heat (exothermic) within the sample cell causes a negative change in the DP because the heat evolved chemically provides heat that the DP feedback is no longer

Figure 10–11: Comparative sizes of the 1.4-mL and 0.2-mL Hastelloy cells used as sample and reference cells in the VP-ITC and iTC$_{200}$ instruments, respectively. The 0.2-mL cell weighs less than 1 g.

required to provide. The opposite is true for endothermic reactions. Because the DP has units of power, the time integral of the peak yields a measurement of thermal energy, ΔH. This heat is released or absorbed in direct proportion to the amount of binding that occurs. When the macromolecule in the cell becomes saturated with ligand, the heat signal diminishes until only the background heat of dilution is observed.

The entire experiment takes place under computer control including analysis of the ITC raw data using nonlinear least squares fitting models to calculate reaction stoichiometry (n), binding constant (Ka), and enthalpy (ΔH).

Latest developments in ITC miniaturization

The objectives of miniaturization are twofold; first is to significantly reduce protein quantity requirements, and second is to increase sample throughput. The minimum amount of protein that can be used in an ITC experiment might be limited by either the minimum protein concentration required to achieve a *"c value"* within the experimental window or by the mass of protein required to produce at least the minimal detectable heat signal for ITC injections. These concepts have been well developed in the literature.[1,8,9,12–18]

Reduction in sample quantity requirements

Reducing the required protein sample quantity requires a reduction in the sample and reference cell volumes, a reduction in the minimum precise injection volume to submicroliter levels, and an increase in the absolute instrument sensitivity. The first step to achieving this next generation ITC capability was to decrease the reference and sample titration cell volumes from 1.4 mL to 0.2 mL (Figures 10–11 and 10–12). As we will see later, this reduced cell and sample volume provide the added benefit of faster instrument response times.

The second requirement for achieving reduced protein sample quantities is reducing the minimum precise injection volume to submicroliter levels. The sevenfold reduction in sample cell volume is accompanied by an equivalent reduction in the ligand syringe volume and minimum injection volumes. The

Figure 10–12: Cut-away view of the traditional 1.4-mL and new 0.2-mL cells with spinning syringe and stirrer mechanism inserted.

iTC$_{200}$ pipette performs automated injections of 0.1–1.0 μL with an accuracy of 1% to 3% and injections of 1.0 μL or greater with an accuracy of better than 1%. Figure 10–13 and Table 10–2 illustrate the reproducibility of typical small-volume injections.

The third requirement for achieving reduced protein sample quantities is an increase in absolute instrument sensitivity. There are two important sensitivity parameters when comparing calorimeters.[1] The absolute detection limit, S (μcal), is proportional to the minimum total mass of macromolecular solute that must be used to produce a detectable signal. The volume-normalized sensitivity S/V (μcal/mL) is proportional to the minimum concentration of solute necessary to produce a detectable signal. Modern ultrasensitive microcalorimeters are hundreds to thousands of times more sensitive than conventional calorimeters, and the iTC$_{200}$ is the most sensitive microcalorimeter ever developed.[1,19–24] It has been specifically designed to address the requirements of drug discovery and development. Previous generations of ITC microcalorimeters have S values of

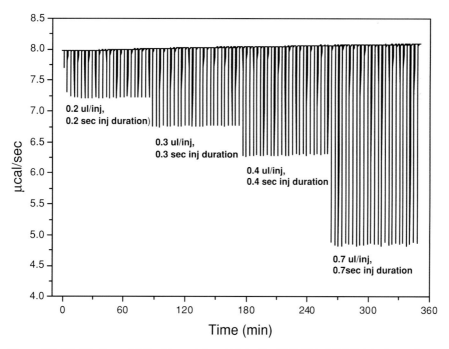

Figure 10–13: Titration of 2.0 mM succinic acid into 5 mM NaOH at 30°C. Injection volumes range from 0.2 μL per injection to 0.7 μL per injection. The precision of the 0.2-μL, 0.3-μL, 0.4-μL, and 0.7-μL injections were 1.5%, 1.2%, 0.6%, and 0.8% RSD, respectively.

Table 10–2. iTC$_{200}$ injector volume delivery succinic acid and NaOH titration at 30°C

Injection volume (uL/inj)	Succinic acid concentration (mM)	Injection number	Injection period (min)	Standard deviation (%)
0.2	2.0	25	3.5	1.5
0.3	2.0	25	3.5	1.2
0.4	2.0	25	3.5	0.6
0.5	1.0	25	3.5	1.8
0.7	2.0	25	3.5	0.7
1.0	1.0	25	2.0	0.4
2.0	1.0	20	2.0	0.4

Stirring rate: 589 rpm, 5 s filter, 1 uL per second injection duration, 5 mM NaOH in the cell for all titrations.

0.3–0.1 μcal.[1] The present instrument has an S value of 0.04 μcal. Table 10–3 compares the sensitivities and volume-normalized sensitivities of the VP-ITC and iTC$_{200}$ instruments.

Increases in sample throughput

Sample throughput depends on both the length of time required to complete the titrations and the sample changeover and re-equilibration time between experiments. The rapid calorimetric response time and fast thermal equilibration time of the 0.2-mL cell together increase overall sample throughput. Figure 10–14 illustrates the comparative instrument response times as measured by the half-life response to a 3 μcal/s heat pulse. As seen in Figure 10–15, this faster instrument response time results in higher and narrower peaks for equivalent heat signals and a correspondingly faster return to baseline. Depending on the binding reaction kinetics of the system being studied, faster instrument response to heat signals and a faster return to baseline may allow the time between injections and hence overall titration time to be reduced.

Faster instrument response to heat events and faster instrument return-to-baseline reduce time required for each injection and time delay between injections, allowing the entire titration to proceed more rapidly. Overall sample throughput rates can be further increased by reducing the instrument equilibration time between experiments. Figure 10–16 illustrates the reduction in time required for the 0.2-mL cell to achieve a 5°C temperature change. Figure 10–17 illustrates the time reduction for instrument equilibration.

Table 10–3. Comparative sensitivities and volume normalized sensitivities

Microcalorimeter	S (μcal)	V (mL)	S/V
VP-ITC	0.10	1.4	0.07
iTC$_{200}$	0.04	0.2	0.20

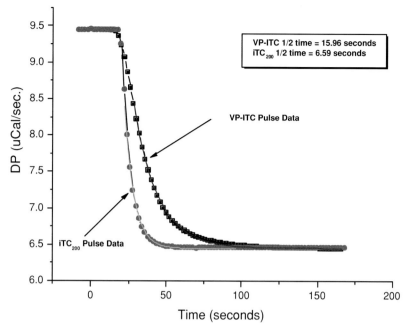

Figure 10–14: VP-ITC and iTC$_{200}$ calibration pulse data to compare instrument response times. Pulse size $= -3$ µcal/s, filter period $= 2$ s; both pulses fit to first-order exponential decay model. VP-ITC $\frac{1}{2}$time $= 15.96$ s, iTC$_{200}$ $\frac{1}{2}$time $= 6.59$ s.

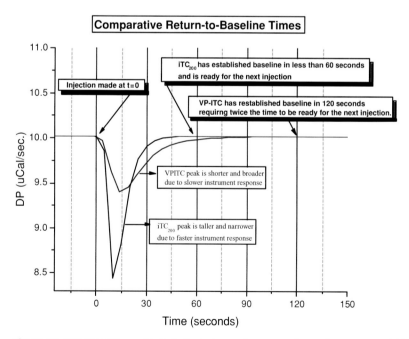

Figure 10–15: The iTC$_{200}$ with its faster instrument response time returns more quickly to baseline and therefore requires less time between injections. In this example, both peaks are of equal area representing approximately 20 µcal.

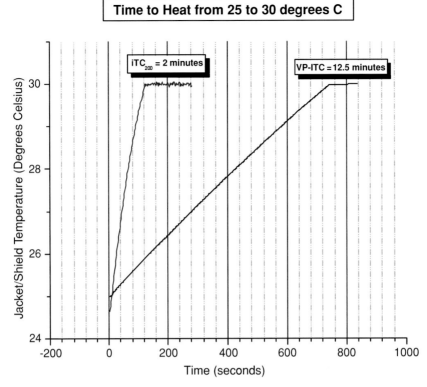

Figure 10–16: The iTC$_{200}$ with sample and reference cell masses of less than 1 g requires less than 20% of the time as the VP-ITC to temperature-stabilize following a 5°C change in temperature.

Table 10–4 illustrates the effect of faster instrument response and equilibration times on typical ITC experiment turnaround times. Sample throughput rates double with the iTC$_{200}$ and can double again by reducing the number of injections from the traditional 24 to 12. With the automated instrument sample throughput rates of 48–60 per day can be achieved.

Low-protein quantity data comparison

For purposes of data comparison with the VP-ITC, we present results for the binding of cytidine 2′-monophosphate (2′CMP) to the active site of ribonuclease A. Figure 10–18 depicts a direct comparison of the VP-ITC to the iTC$_{200}$. Due to cell volume difference (1.4 mL for the VP-ITC and 0.2 mL for the iTC$_{200}$) protein usage was reduced sevenfold for the iTC$_{200}$. The effect of the 2.5-fold increase in sensitivity is apparent in this data. Peak heights of approximately 7 μcal/s in the VP-ITC are reduced to only about 2.8 μcal/s in the iTC$_{200}$ even though the cell volume, and therefore total heat produced, has been reduced sevenfold.

Because volume requirements for loading the two instruments scale similarly, both protein usage and ligand usage were reduced approximately sevenfold. For the VP-ITC, 2.0 mL of material was used to load the cell and at least 0.5 mL to

Table 10–4. ITC sample processing times

Function	VP-ITC		iTC$_{200}$	
No. of injections	24	12	24	12
Cleaning (min)	8	8	8	8
Loading (min)	2	2	2	2
Equilibration (min)	20	20	4	4
Titration (min)	96	48	36	18
Total (min)	126	78	50	32

load the syringe. The iTC$_{200}$ required approximately 0.3 mL to load the cell and 50 µL to load the syringe.

Table 10–5 summarizes data for the comparison of ITC microcalorimeters using the 1.4-mL and 0.2-mL sample cells. The values shown for K_A and ΔH are the averages from three replicates. For the purposes of this comparison the concentrations of RNAse and 2′CMP were the same and resulted in a c value of 84. The thermodynamic parameters, K_A and ΔH, agree with each other and with their known values. Protein usage was reduced from 1194 µg using the 1.4-mL sample cells to 164 µg using 0.2-mL sample cells.

The key question is: How much can the protein quantity be reduced in the ITC experiment while still generating useful data? The answer, as presented earlier in this chapter and elsewhere,[1,8,9] is that it depends on the value of ΔH, which will ultimately limit the amount of heat generated per injection, and on K_A, which

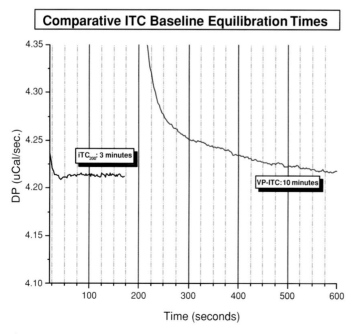

Figure 10–17: Comparative ITC baseline equilibration times. No temperature change required, only the time required to equilibrate after experimental temperature has been achieved.

Table 10–5. Comparison of RNASE-2'CMP ITC results using the 1.4 mL and 0.2 mL sample cells

Instrument	$K_A \times 10^5$	n	ΔH (kcal/mol)	"c" value	Protein (μg)	Time (min)
VP-ITC[a]	14.0 ± 0.8	0.92 ± 0.04	−17.5 ± 0.1	84	1194	80
iTC$_{200}$[b]	13.1 ± 1	0.97 ± 0.02	−17.1 ± 0.4	84	164	25

[a] [RNASE] = 0.06 mM, [2'CMP] = 2.18 mM, 20 x 5 μL injections.
[b] [RNASE] = 0.06 mM, [2'CMP] = 2.18 mM, 20 x 0.688 μL injections.

will limit the c value as the macromolecule concentration is reduced. The lowest possible protein usage for every ITC experiment will be limited by either:

1. The heat per injection at the selected concentration and cell volume, remaining above the absolute instrument sensitivity, or
2. The c value, at the selected concentration, remaining above a minimum level to achieve a successful and unique least squares fit to the experimental data.

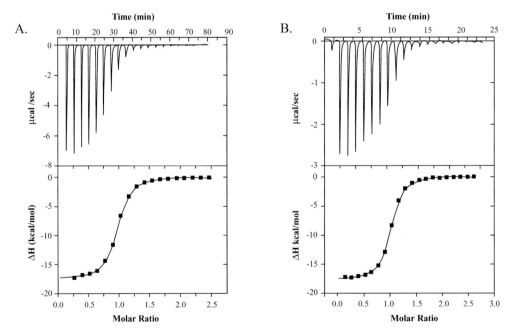

Figure 10–18: Comparison of the VP-ITC and iTC$_{200}$. Raw and processed data from the reaction of RNASE with 2'CMP in 50 mM potassium acetate buffer pH 5.5 (25° C). The VP-ITC cell volume = 1.4 mL and the syringe volume = 275 μL. The iTC$_{200}$ cell volume = 200 μL and the syringe volume = 40 μL. Concentrations for the titrations were [RNASE] = 0.06 mM, [2'CMP] = 2.18 mM. *Top panels* show raw data following baseline subtraction. *Bottom panels* show processed data corresponding to each raw data set (solid squares) and fit obtained (solid lines). Fitted parameters are given in Table 10–2. A. Titration on the VP-ITC; 20 x 5-μL injections. B. Titration on the iTC$_{200}$; 20 x 0.7-μL injections.

Figure 10–19: The automated iTC$_{200}$ will process up to 50 samples per day and up to 384 samples during an unattended run.

ITC EXPERIMENT DESIGN CONSIDERATIONS

It is the usual case that a determination of the thermodynamic parameters (n, K_A, ΔH, and ΔS, or at a minimum, K_A) is desired and their determination is desired using the minimal amount of protein. In the early, decision-making stages of the drug discovery process, the availability of target protein is usually limited, and there is a large demand for it to serve the needs of a large number of testing procedures. As illustrated herein, the advent of a highly sensitive microcalorimeter with a 0.2-mL cell volume offers the possibility to greatly reduce minimum protein requirements. This possibility can be realized by optimizing the design of the ITC experiment, the protein binding system being studied, and the information desired from experimental results.

Experiment design alternatives to minimize protein usage

For the RNAse-2′CMP system, we have explored the impact of alternative ITC experiment designs on experimentally determined thermodynamic parameters. Figure 10–20 displays three different types of experimental approaches to the RNAse-2′CMP experiment using the iTC$_{200}$; the traditional multiple- (20 injections) injection method, a reduced-injection (5 injections) method, and the single-injection method. For each experiment, only 9 μg of RNAse was loaded into the sample cell.

As seen from the results presented in Table 10–6, using as little as 9 μg of RNAse (at the resultant lower c value of 5–7), data integrity is retained and experiment time can be reduced to as little as 5.5 min by reducing the number of injections. Each data set shown is the result of three replicate titrations. The protein concentration was reduced 19-fold and the syringe concentration was reduced tenfold from standard concentrations.

In Figure 10–20, Panel A displays a traditional multiple-injection (20 injections) method. Panel B displays the reduced-injection method (5 injections). Injection volumes and concentrations for a reduced-injection method should be adjusted to cover a fractional saturation range between 20% and 80% if possible. The

Table 10–6. Comparison of RNASE-2′CMP ITC results using the 0.2 mL sample cells at higher and lower c values

Method	$K_A \times 10^5$	n	ΔH (kcal/mol)	"c" value	Protein (µg)	Time (min)
20 injections[a]	13.1 ± 2	0.97 ± 0.02	−17.1 ± 0.4	84	164.4	25
20 injections[b,c]	20.4 ± 3	0.98 ± 0.02	−16.6 ± 0.3	7	9	25
5 injections[d]	21.6 ± 4	0.94 ± 0.07	−15 ± 1	7	9	8
Single injection[e]	16.7 ± 3	1.09 ± 0.01	−17 ± 1	5	9	5.5

[a] [RNASE] = 0.06 mM, [2′CMP] = 2.18 mM, 20 × 0.689 µL injections.
[b] [RNASE] = 0.0032 mM, [2′CMP] = 0.218 mM, 20 × 1.0 µL injections.
[c] The apparent higher K value at lower c values has been previously experimentally observed.[1]
[d] [RNASE] = 0.0032 mM, [2′CMP] = 0.218 mM, 5 × 1.5 µL injection.
[e] [RNASE] = 0.0032 mM, [2′CMP] = 0.218 mM, 1 × 20 µL injections.

experimental time is reduced to 8 min by reducing injection number. Panel C displays the result of a single-injection experiment with the same concentrations as the other two data sets. Total ligand titrated is equal to the multiple-injection method value, 20 µL. The experiment time is reduced to 5.5 min.

The reduced-injection and single-injection method experiments work well here for the RNAse-2′CMP system, which has an average K_A and a relatively high ΔH. Concentrations would need to be increased when ΔH is lower (to avoid injection heats falling below the instrument absolute sensitivity) or when K_A is low (to avoid being below the minimum c value to achieve a fit to the experimental data). This observation is true regardless of which method is chosen.

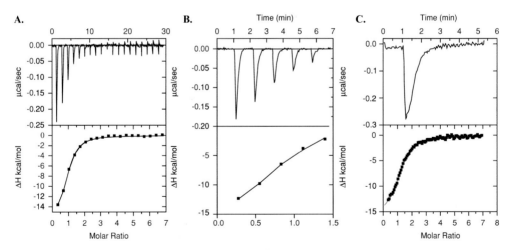

Figure 10–20: Comparison of titration methods. A. Raw and processed data from the reaction of RNASE with 2′CMP in 50 mM potassium acetate buffer pH 5.5 (25°C). The cell volume = 200 µL and the syringe volume = 40 µL. *Top panels* show raw data following baseline subtraction. *Bottom panels* show processed data corresponding to each raw data set (solid squares) and fit obtained (solid lines). Concentrations for the three data sets were [RNASE] = 0.0032 mM and [2′CMP] = 0.218 mM. Fitted parameters are given in Table 10–1. **A.** Multiple-injection method; 20 × 0.5-µL injections. **B.** Reduced-injection method; 5 × 1.5-µL injections. **C.** Single-injection method; 1 × 20-µL injection.

Table 10–7. Comparison of the binding of multiple ligands to carbonic anhydrase II

Ligand	K_A x 10^5	ΔH (kcal/mol)	"c" value	Protein (μg)	Time (min)
Standard[a]					
4-AMBS	1.1 ± 0.2	-3.4 ± 0.2	7	377	39
Sulfanilamide	3.1 ± 0.3	-10.1 ± 0.5	12	220	39
Furosemide	19 ± 3	-10.0 ± 0.5	20	162	39
Acetazolamide	103 ± 35	-18.6 ± 0.5	52	162	39
5-Injection[b]					
4-AMBS	1.3 ± 0.3	-2.7 ± 0.3	3	122	11
Sulfanilamide	2.4 ± 0.1	-13 ± 2	2	58	11
Furosemide	19.5 ± 5	-9.4 ± 0.9	15	58	11
Acetazolamide	112 ± 33	-21.0 ± 0.4	56	29	11
Single Injection[c]					
4-AMBS	0.60 ± 0.2	-2.0 ± 0.3	2	174	8
Sulfanilamide	3.0 ± 0.3	-8.1 ± 0.5	3	58	8
Furosemide	24 ± 11	-11.5 ± 0.9	24	58	8
Acetazolamide	104 ± 7	-16.2 ± 0.3	52	29	6

[a] Standard experiments consisted of multiple injections (20–30; $\times$ 1 μL).
[b] Five-injection method experiments consisted of five 3–5 μL injections.
[c] Single-injection experiments consisted of one slow injection of 20 to 30 μL. Values for binding parameters include standard deviations determined from at least three data sets.

Example: Binding study of carbonic anhydrase II to multiple ligands

To further explore the ITC experiment design alternatives, when using a sample-cell volume of 0.2 mL, we utilized the carbonic anhydrase II protein with four ligands, 4-aminomethyl benzenesulfonamide (AMBS), sulfanilamide, furosemide, and acetazolamide. For each protein–ligand system we determined K_A and ΔH using a standard 20-injection method, a 5-injection method, and finally the single-injection method. This allowed us to compare the results over a range of protein–ligand systems and over a meaningful range of binding constants and molar enthalpies.

Table 10–7 summarizes the data collected for the four systems using the three experimental methods. Each data point is an average of at least three experiments and includes standard deviation values. The values for K range from 1×10^5 M^{-1} to 5×10^7 M^{-1}. Values for ΔH range from -3 kcal/mol to -21 kcal/mol. Protein concentration was determined spectrophotometrically.

Each of the carbonic anhydrase ligands studied in Table 10–7 was first dissolved in 100% dimethylsulfoxide (DMSO) and subsequently diluted into buffer. These data illustrate the applicability of ITC for applications such as secondary drug screening. The reduced sample cell size and increased absolute instrument sensitivity permit direct determination of thermodynamic parameters using only 10% to 20% of the sample quantity previously required. The increased instrument response and equilibration times allow 2–4 samples per hour to be completed using multipoint-injection methods and up to 4–5 samples per hour using the single-injection method.

These benefits in reduced sample consumption and increased sample through-put are realized without sacrificing any of the primary advantages of the ITC technique. Retained is the nearly universal applicability to measure the heat of complexation between molecules completely free in-solution without require-ment for any labeling, probe, or other reporter mechanism in the system. No immobilization onto a surface is required. Most importantly, the determination of the binding constant and the heat of binding are completed directly in a single experiment, eliminating the uncertainty and potential errors inherent in other indirect methods that rely on unvalidated model assumptions for the specific macromolecule–ligand systems being studied.

The ITC experiment design window

As discussed earlier, the two main factors needed for a successful ITC experiment, beyond protein and ligand solubility, are:

1. An appropriate c value. A wide range, typically 5–1000, is suitable for most experiments providing a wide range of workable protein and ligand concentrations.
2. Enough heat to be detected and quantified. The reduced cell volume and increased sensitivity of modern instrumentation has helped push ever lower the heat levels that can be accurately measured and fitted to a bind-ing thermogram.

It is important to remember that both these criteria must be met simultaneously in the experiment design. That is, the smaller the amount of protein that can be used in the experiment is, the greater the amount required to provide a useable c value or to produce sufficient heat for detection and quantification.

In many situations, such as secondary or confirmatory screening of hits in a drug discovery program, the desire to obtain reliable, interpretable thermody-namic parameters for a series of ligands to a target is critically important, and ITC is ideally suited to the task. Often, in the early stages of such a project, min-imization of protein usage is required because only a finite (and usually small) amount of protein is available.

For purposes of illustration we have used the RNAse-2'CMP system to consider the range of c values and hence the protein concentrations over which ITC data may be obtained. Table 10–8 illustrates the thermodynamic parameters that result from experiments with c values of 5, 10, 25, and 50 using both a 1.4-mL instru-ment cell volume and a 0.2-mL instrument cell volume. The results for K and ΔH are in agreement. The quantity of protein used in these experiments ranges from a high of 685 μg to a low of 10 μg. To confirm that both experiment design requirements (c value and minimum heat) were met, the heats per injection were both calculated and experimentally measured. The minimum heat required per injection was somewhat arbitrarily defined as 40 times the instrument sensitivity (80 times the baseline noise level). From experience, we know that data can be successfully collected and fit to the binding model at levels 2–4 times lower than this "defined minimum heat per injection" requirement. This is demonstrated

Table 10–8. RNAse quantity requirements based on c value requirements[a]

	VP-ITC (1400-µL cell volume)				iTC$_{200}$ (200-µL cell volume)			
c value	50	25	10	5	50	25	10	5
Protein concentration (mM)	0.035	0.018	0.0070	0.0035	0.035	0.018	0.0070	0.0035
µg of protein	685	343	137	69	98	49	20	10
No. injections	12	12	12	12	12	12	12	12
Calculated heat/injection (µcal)	62.5	31.3	12.5	6.3	8.9	4.5	1.8	0.9
Experimental heat/injection (µcal)	56.7	31.8	10.5	5.3	8.8	4.2	1.5	0.7
Required minimum heat/injection (µcal)[b]	4.0					1.6		
Experimental results								
$K_A \times 10^6$ (M^{-1})	1.52	1.52	1.62	1.67	1.62	1.69	1.64	1.93
K_D (µM)	0.66	0.66	0.62	0.60	0.62	0.59	0.61	0.52
ΔH (kcal/mol)	−18	−17	−16	−16	−18	−17	−15	−15

[a] Quantities calculated for the ITC titration of RNAse-2'CMP assuming $K_A = 1.4 \times 10^6$, $\Delta H = -15$ kcal/mol, MW = 13700.
[b] The required minimum heat per injection = 40 × the minimum detectable heat signal (approximately 2 × baseline noise). As can be seen from the data, excellent results can be obtained even when operating at less than half the required minimum heat per injection level.

Table 10–9. Typical macromolecule usage $(\mu g)^a$

K values		C values		
K_A (M^{-1})	K_D (μM)	5	10	50
1 x 10^5	10	200	400	2000
1 x 10^6	1	20	40	200
1 x 10^7	0.1	2	4	20
1 x 10^8	0.01	0.2	0.4	2

[a] For a 20-kDa protein and an ITC sample-cell volume of 0.2 mL where the minimum sample quantity is limited only by c value.

in the data set for the $c = 5$ data collected using the iTC$_{200}$, where the experimental heat per injection was only 0.7 μcal versus the defined minimal heat per injection of 1.4 μcal. Table 10–9 illustrates typical macromolecule usage (μg) for a 20000-dalton protein when the minimum sample quantity is limited only by c value.

As much as Tables 10–8 and 10–9 view the experiment from the "*c value*" perspective, Table 10–10 views the experiment from the "*minimum heat*" perspective. In this case, so long as a minimum acceptable c value is maintained, the protein quantity can be reduced by reducing the number of injections so that at least the minimum heat per injection is maintained. The injection volume (or concentration) of ligand can be increased accordingly to ensure completion of the titration.

In Table 10–11 we have attempted to quantify and compare the typical ITC experiment protein quantity requirements for a meaningful range of K_D values and ΔH values for both the 1.4-mL cell volume of the VP-ITC and the 0.2-mL cell volume of the iTC$_{200}$. For every given combination of K_D value and ΔH values either the c value or the minimum heat per injection value will limit the minimum quantity of protein that can be used. In Table 10–11 the entries in the unshaded boxes are minimum protein quantities determined exclusively by the c value. For any particular K_D value, so long as the c value is the limiting factor, the amount of protein required is independent of ΔH. The entries in the shaded boxes are minimum protein quantities determined exclusively by the minimum heat per injection requirement. So long as the minimum heat per

Table 10–10. Minimum RNAse quantities based on heat requirements^a

	VP-ITC (1400-μL cell volume)				iTC^{200} (200-μL cell volume)			
Required minimum heat/injection (μcal)^b	4.0				1.6			
No. injections	24	18	12	6	24	18	12	6
μg of protein	87.7	65.8	43.8	21.9	35.1	26.3	17.5	8.8
Resultant calculated c value	6.4	4.8	3.2	1.6	17.9	13.4	9	4.5

[a] Quantities calculated for the ITC titration of RNAse-2'CMP assuming $K_A = 1.4 \times 10^6$, $\Delta H = -15$ kcal/mol, MW = 13700.
[b] The required minimum heat per injection = 40 x the minimum detectable heat signal (approximately 80 x baseline noise).

Table 10–11. Protein quantity usage comparison[a]

ΔH (kcal/mole)	k_D (μm)				
	10.00	1.00	0.10	0.01	0.001
μg of protein required for VP-ITC					
18.00	1400	140	14	4	4
15.00	1400	140	14	5	5
12.00	1400	140	14	7	7
9.00	1400	140	14	9	9
6.00	1400	140	14	13	13
3.00	1400	140	27	27	27
Resultant c values for VP-ITC protein quantity usage					
18.00	5	5	5	14	143
15.00	5	5	5	18	179
12.00	5	5	5	25	250
9.00	5	5	5	32	321
6.00	5	5	5	46	464
3.00	5	5	10	96	964
μg of protein required for iTC$_{200}$					
18.00	200	20	2	2	2
15.00	200	20	2	2	2
12.00	200	20	3	3	3
9.00	200	20	4	4	4
6.00	200	20	5	5	5
3.00	200	20	11	11	11
Resultant c values for iTC$_{200}$ protein quantity usage					
18.00	5	5	5	50	500
15.00	5	5	5	50	500
12.00	5	5	8	75	750
9.00	5	5	10	100	1,000
6.00	5	5	13	125	1,250
3.00	5	5	28	275	2,750

[a] Protein quantities in the unshaded boxes are limited by the selected minimum c value of 5. Protein quantities in the shaded boxes are limited by the selected minimum acceptable heat per injection level of 40 times the instrument sensitivity. These values could be further reduced by a factor of 2–4 and still produce high-quality data. The dark gray-shaded boxes indicate c values greater than 1000 where fitting a binding isotherm can be problematic. It should be noted that it is very rare to have such strong binders (1 nM) with such low ΔH values.

injection is the limiting factor, the amount of protein required is dependent only on ΔH and independent of K_D.

With these experimental parameters in mind, we can begin to form a picture for the starting point of any particular ITC experiment. For example, if an IC$_{50}$ value of 100 nM was determined from a preliminary high-throughput screen of a small-molecule ligand to a protein target, and we assume a "typical" ΔH of −6 kcal/mole, then using the iTC$_{200}$, 5 μg of protein would be required, the experiment would be limited only by the minimum heat per injection, and the

c value would be 13. If an adequate sample were available, then 11 μg could be used, the c value would be 28, and meaningful data could be gathered even if ΔH were as small as −3 kcal/mole.

We outline similar example scenarios below.

Example 1: A c value-limited situation

A 1-μM *lead compound with a* $\Delta H = -6$ *kcal/mol*

	VP-ITC	iTC$_{200}$	
c value	5	5	1^a
μg protein	140	20	5^b

^{a} A c value of 1 can be fit by least squares to the binding model if n is fixed.
^{b} The lower limit for the protein quantity is reached (for this example) at 5 μg due to the heat sensitivity limitation.

Example 2: A heat-limited situation

A 10-nM *lead compound with a* $\Delta H = -9$ *kcal/mol*

	VP-ITC	iTC$_{200}$	
Sensitivity multiple	40×	40×	10×^{a}
c value	32	100	25
μg protein	9	4	1

^{a} If a sensitivity multiple of 10 × rather than 40 × is accepted, the protein quantity can be reduced to 1 μg because the c value (25) is sufficient for the experiment.

CONCLUSIONS

As presented previously, ITC as an experimental technique provides four critically important and unique thermodynamic binding parameters.[1,7,16,25–27] First, the absorption or evolution of heat is a universal property of all chemical reactions and as such eliminates the need to use a different analytical method each time a new reaction is studied. Second, because calorimetry measures real (model independent) enthalpy of binding, it allows for the determination of entropy as well as the the binding constant and stoichiometry. Third, the ITC technique provides this unique capability in an experimental environment that is label-free and in-solution and requires no immobilization of either the target macromolecule or ligand. Fourth, because the heat effect is influenced only by the progress of the reaction and unaffected by static situations in the solution, the background interference is normally very low, which allows the study of heterogeneous mixtures that may not be amenable to study by other methods.

These attributes of the ITC technique have led to its broad acceptance for the full thermodynamic characterization of biomolecular interactions. In many instances the thermodynamic parameters provide an additional level of information beyond that available from structure data alone, allowing the nature of a binding event (or an SAR-related series of binding events) to be more fully understood.[10,16,28–39] Frequently the K_D values of a series of interactions with one biomolecule are nearly equivalent. Although great initial emphasis is placed on the K_D value, and it is critically important, it alone does not provide a characterization of the mode of binding. As previously shown, however, within a given protein system the distribution of the binding energy between ΔH and ΔS may differ vastly – even for a related series of ligands.[40–43] Similar K_D values and highly differentiated ΔH and ΔS values typically indicate different binding modes with correspondingly different biomolecular interactions.

REFERENCES

1. T. Wiseman, S. Williston, J. F. Brandts, L. N. Lin, *Anal Biochem* **179**, 131 (1989).
2. E. Freire, O. L. Mayorga, M. Straume, *Anal Chem* **62**, A950 (1990).
3. I. Haq, J. E. Ladbury, B. Z. Chowdhry, T. C. Jenkins, J. B. Chaires, *J Mol Biol* **271**, 244 (1997).
4. I. Jelesarov, H. R. Bosshard, *J Mol Recognit* **12**, 3 (1999).
5. J. E. Ladbury, *Structure* **3**, 635 (1995).
6. J. E. Ladbury, B. Z. Chowdhry, *Chem Biol* **3**, 791 (1996).
7. J. A. Thomson, J. E. Ladbury, *Biocalorimetry 2. Applications of Calorimetry in the Biological Sciences*. J. E. Ladbury, M. L. Doyle, Eds., (John Wiley & Sons, Chichester, 2004), pp. 37–58.
8. J. Tellinghuisen, *J Phys Chem B Condens Matter Mater Surf Interfaces Biophys* **109**, 20027 (2005).
9. W. B. Turnbull, A. H. Daranas, *J Am Chem Soc* **125**, 14859 (2003).
10. T. Carbonell, E. Freire, *Biochemistry* **44**, 11741 (2005).
11. E. S. Istvan, J. Deisenhofer, *Science* **292**, 1160 (2001).
12. J. M. Sturtevant, *Proc Natl Acad Sci U S A* **74**, 2236 (1977).
13. M. L. Doyle, P. Hensley, *Methods Enzymol* **295**, 88 (1998).
14. M. L. Doyle, G. Louie, P. R. Dal Monte, T. D. Sokoloski, *Methods Enzymol* **259**, 183 (1995).
15. B. W. Sigurskjold, *Anal Biochem* **277**, 260 (2000).
16. A. Velazquez Campoy, E. Freire, *Biophys Chem* **115**, 115 (2005).
17. Z. X. Wang, *FEBS Lett* **360**, 111 (1995).
18. Y. L. Zhang, Z. Y. Zhang, *Anal Biochem* **261**, 139 (1998).
19. P. R. Bevington, *Data Reduction and Error Analysis for the Physical Sciences*, (McGraw-Hill, New York, 1969).
20. M. Elharrous, S. J. Gill, A. Parodymorreale, *Meas Sci Technol* **5**, 1065 (1994).
21. M. Elharrous, O. L. Mayorga, A. Parodymorreale, *Meas Sci Technol* **5**, 1071 (1994).
22. L. D. Hansen, E. A. Lewis, D. J. Eatough, *Analytical Solution Calorimetry*. J. R. Grime, Ed., (Wiley, New York, 1985), pp. 57–95.
23. I. R. McKinnon, L. Fall, A. Parody-Morreale, S. J. Gill, *Anal Biochem* **139**, 134 (1984).
24. A. Velazquez-Campoy, O. Lopez-Mayorga, M. A. Cabrerizo-Vilchez, *Rev Sci Instrum* **71**, 1824 (2000).
25. M. Blandamer, *Biocalorimetry: Applications of Calorimetry in the Biological Sciences*. J. E. Ladbury, B. Z. Chowdhry, Eds., (Wiley, Chichester, 1998), pp. 5–25.

26. A. Cooper, *Biocalorimetry: Applications of Calorimetry in the Biological Sciences.* J. E. Ladbury, B. Z. Chowdhry, Eds., (Wiley, Chichester, 1998), pp. 103–111.
27. J. E. Ladbury, *Biotechniques* **37**, 885 (2004).
28. I. Luque, E. Freire, *Methods Enzymol* **295**, 100 (1998).
29. M. J. Todd, I. Luque, A. Velazquez-Campoy, E. Freire, *Biochemistry* **39**, 11876 (2000).
30. A. Velazquez-Campoy *et al.*, *Protein Sci* **9**, 1801 (2000).
31. A. Velazquez-Campoy, Y. Kiso, E. Freire, *Arch Biochem Biophys* **390**, 169 (2001).
32. A. Velazquez-Campoy, S. Vega, E. Freire, *Biochemistry* **41**, 8613 (2002).
33. I. Luque, E. Freire, *Proteins* **49**, 181 (2002).
34. H. Ohtaka, A. Velazquez-Campoy, D. Xie, E. Freire, *Protein Sci* **11**, 1908 (2002).
35. A. Velazquez-Campoy *et al.*, *Curr Drug Targets Infect Disord* **3**, 311 (2003).
36. A. Velazquez-Campoy, E. Freire, *Proteins and Proteomics: A Laboratory Manual.* R. Simpson, Ed., (Cold Spring Harbor Laboratory Press, New York, 2003), pp. 882–892.
37. H. Ohtaka *et al.*, *Int J Biochem Cell Biol* **36**, 1787 (2004).
38. S. Vega *et al.*, *Proteins* **55**, 594 (2004).
39. A. Velazquez-Campoy, S. A. Leavitt, E. Freire, *Methods Mol Biol* **261**, 35 (2004).
40. U. Bacha, J. Barrila, A. Velazquez-Campoy, S. A. Leavitt, E. Freire, *Biochemistry* **43**, 4906 (2004).
41. V. Lafont *et al.*, *Chem Biol Drug Des* **69**, 413 (2007).
42. A. J. Ruben, Y. Kiso, E. Freire, *Chem Biol Drug Des* **67**, 2 (2006).
43. R. W. Sarver *et al.*, *Anal Biochem* **360**, 30 (2007).

11 Electrical impedance technology applied to cell-based assays

Ryan P. McGuinness and Edward Verdonk

INTRODUCTION	251
Electrical impedance technology	252
Bioimpedance of cells	253
THE CELLKEY SYSTEM	256
CellKey system specifications	258
Example data	260
WORKED EXAMPLES	260
Preparing cells for assay	261
Preparing cell plates	261
Preparing compound plates	262
Controls	262
Assay Protocols	264
Executing experiments	265
Data analysis and results	267
SUMMARY	272
APPENDIX	272
Cell lines and materials	272
Maintaining and preparing cells	273
Cell culture	273
Coating cell plates with attachment factors	274
REFERENCES	274

INTRODUCTION

Cell-based assays are now well established in nearly all stages of drug discovery and development. Indeed, they are commonly applied to all the critical steps of the process from receptor target identification and validation, compound screening and structure–activity relationship (SAR) through to toxicology.[1–8] This is due in a large part to their functional nature and their ability to introduce biological complexity to the drug discovery process at a much lower cost than *in vivo* testing. Additionally, these assays have been successfully scaled to meet the throughput needs of screening labs.[9] Cellular assays complement and extend the knowledge of the more mechanistic interactions of receptors and ligands as understood from biochemical assays. Indeed, cell-based assays are commonly used iteratively with biochemical assays to inform the process and direct screening efforts.

Label-free cell-based assays are gaining broader acceptance in screening labs, as they provide novel read-outs of cellular signaling and carry with them many practical advantages to the drug discovery workflow.[10-14] For example, the lack of labels greatly simplifies assay development, removing many steps, each of which requires optimization. Effort is also saved in the engineering of chimeric or tagged molecules, which are often altered to the point that they no longer maintain natural function, leaving results somewhat in question relative to the native biology of the system being tested.

Electrical impedance assays are novel label-free assays that fit well into the screening workflow.[15] Originally used to study the basic physical properties of materials, impedance technology has evolved over time to yield instrumentation for the biophysical evaluation of living cells. These systems have demonstrated to be robust and reliable and highly effective for the routine work of screening labs.[16-18]

To illustrate the utility of impedance technology to drug discovery, this chapter focuses on a recently developed instrument, the CellKey System (MDS Analytical Technologies). The CellKey system was designed to advance target identification, validation, and secondary screening applications typical in drug discovery research. We describe a set of CellKey applications with an emphasis on target identification via receptor panning and target validation via pharmacological analysis of agonists and antagonists. We also provide detailed protocols and workflows for these techniques in a set of worked examples.

The future of impedance technologies lies in continued development of assay applications to broaden the already wide range of cellular receptors successfully measured. Specifically, future analysis of endogenous receptor targets in primary cells will likely prove to be advantageous, and aid in the discovery of new drugs that are capable of exploiting the nuanced conformations of targets in their native settings. The increased acceptance of label-free technologies continues to drive development of various platforms, focusing primarily on higher assay throughput using high-density microplates. With these improvements and their broad applicability across targets and cell types, label-free technologies promise to contribute novel biological information to drug discovery research efforts for many years to come.

Electrical impedance technology

Measurements of the electrical impedance of various materials have a long history in the physics community.[19] Typically, a sample of material is placed in a measurement device, which is often a parallel plate capacitor, a voltage is applied to the electrodes of the measurement device, and the resulting current that flows through the sample is monitored. The electrical impedance of the sample is calculated as the ratio of voltage over current. It is often a sinusoidally varying voltage (an AC voltage) that is applied to the electrodes, but other voltage waveforms can and have been used.[20] For any time-varying voltage applied, the resulting current can be in phase with the applied voltage (resistive behavior), or out of phase with it (capacitive behavior). Electrical impedance is therefore a complex number, in that it has both a magnitude and a phase.

Measurements of a sample's electrical impedance are used to understand how charges flow within the sample, how transient charge separations are created, and how the sample's permanent charge separations, like its permanent electrical dipoles, are reoriented in the applied electric field. To help answer these questions, measurements are usually made across a broad range of frequencies, anywhere from 10^1 to 10^{10} hertz (Hz). Also, measurements of impedance as a function of frequency are generally transformed into dielectric constant as a function of frequency, by accounting for the electrode geometry of the measuring device. The geometric factor for a parallel plate capacitor measuring device is simply equal to the ratio plate area over plate separation. Once the measured impedance data for a sample have been transformed into the complex dielectric constant, ε^*, one then speaks in terms of the sample's permittivity and its conductivity.

What has been observed in making these measurements at radio frequencies and at microwave frequencies on biological samples is that the permittivity of the sample, which characterizes its ability to store charge, declines in a series of steps as the excitation frequency increases.[21,22] This is because distinct charge separation and electric dipole reorientation mechanisms are one-by-one ruled out as the rate of oscillation of the applied electric field is increased. Each charge separation or charge reorientation mechanism has its own characteristic time constant, and once the applied electric field is oscillating faster than the time constant for a given charge separation mechanism or a charge reorientation mechanism, it can no longer keep up with the electric field, and it therefore no longer affects the sample's impedance. This phenomenon is referred to as a dielectric dispersion. Dispersions do not exhibit themselves abruptly with increasing frequency, rather they are relaxation phenomena that appear gradually over at least one decade in frequency. Dielectric spectroscopy is the study of dispersions as a function of frequency, and their underlying physical causes. Dielectric spectroscopy requires first making impedance measurements on a sample, across a wide range in frequency where these dispersions are known to occur. At this point, the technique is also referred to as impedance spectroscopy.

Bioimpedance of cells

Measurements of the electrical impedance of biological materials have a long history in the biophysics and biology communities.[21,23–31] We will limit our

Table 11–1. Electrode designs of cellular impedance systems. Four main types of electrode geometries can be found in the literature

Measurement device and electrode geometries	Illustration	Example commercial provider
Dilute suspension of cells between parallel plate electrodes		Agilent
Confluent monolayer of cells between, but stood off from, parallel plate electrodes		World Precision Instruments
Cells in contact with – even growing on – asymmetric electrodes		Applied Biophysics
Cells in contact with – even growing on – symmetric electrodes		MDS Analytical Technologies CellKey

discussion to bioimpedance measurements of cells positioned between or over electrodes, as illustrated in Table 11–1.

Basic biophysical principles indicate that charge is stored at interfaces. In a biological sample the charges are ions in solution, and if the sample contains cells, the interface of greatest interest is the cell membrane. The bilipid cell membrane is, to first order, non-conductive, whereas it is surrounded to one side by the highly conductive cytoplasm and to the other side by an equally conductive buffer. Ions moving due to an externally applied electric field will begin to "pile up" at the cell membrane because they cannot cross the membrane at anywhere near the rate they arrive at it, and an ionic double layer is formed. Thus, to the external world, the cell membrane appears as a large, capacitive-like impedance for measurements made at low frequencies, and hence the current flowing in the sample is dominated by ions flowing around and between cells (extra-cellular current, or iec). On the other hand, if measurements are acquired at sufficiently high frequencies (typically hundreds of kHz to MHz) the electric field driving the ions alternates polarity so rapidly that there is not sufficient time for ions to build up at the cell membrane, to shield the interior of the cell from the externally applied electric field. Now to the external world, the cell membrane is electrically transparent and ionic currents will appear to flow across cell membranes (trans-cellular current, or itc), as well as around them.[23,32–34]

Many bioimpedance systems use gold electrodes, which do not support electro-chemical reactions for the small voltages applied to the electrodes. As such, there is little to no conversion of ionic current to electrical current at the electrode–buffer interface and an ionic double layer builds up at the electrodes at low frequencies. This is electrode polarization. It results in increased impedance magnitude and lagging impedance phase (capacitive behavior).[35] This electrode-electrolyte interface also needs to be considered in analyzing impedance measurements unless a special four-electrode measurement technique – where separate electrodes are used to measure voltage in the fluid and current passing through it – is used to collect impedance data.

Referring again to Table 11–1, various electrode geometries that have been used to measure cellular bioimpedance are illustrated for comparison purposes. Literature reports in the field of bioimpedance show that measuring devices can be broadly grouped into four classes.

One class of cellular bioimpedance experiments involves the study of dilute suspensions of cells between the plates of a parallel plate capacitor. Often a Maxwell-Wagner mixture model is applied to analyze the measured data, with the intention of studying the dielectric properties of the cell membrane, the cell cytoplasm, or the cell organelles;[36,37] changes in the dielectric properties of these cellular compartments,[37] or changes in cell number and cell morphology.[38] Parallel plate electrodes produce a spatially uniform electric field between their plates, so all cells experience the same electric field. This uniform electric field makes analysis of measured impedance comparatively simple to perform.

A second class of bioimpedance experiments involves the study of confluent monolayers of cells, held off from the electrodes, but still in a parallel plate capacitor measurement device. The intention of these studies is usually less to investigate the passive electrical behavior of individual cells and more to detect changes in cell–cell interactions, which disrupt or enhance the "barrier function" of a confluent monolayer of cells. Often these studies are performed with epithelial cells, from the kidney,[39–44] the gallbladder,[45,46] or the intestine.[44,47–51]

A third class of bioimpedance experiments involves a confluent layer of cells, perhaps tissue itself, in contact with, but not adhering to and growing on, coplanar electrodes.[25,26,28,30,52–64] The electric field set up by coplanar electrode geometries is spatially varying, which makes analysis of impedance measurements complicated. However, coplanar electrode configurations are especially preferred when practicing electrical impedance measurements in multi-well format, such as in a 96-well cell plate.

Finally, a fourth class of bioimpedance experiments, including those performed on the CellKey system, involves the study of cells that are sedimented on, even adhering to and growing on, coplanar electrodes.[13,14,65–88] Experimental studies that use this measurement geometry are especially intended to detect small changes in the adherence of the cells, via integrins in their cell membranes, to the electrodes. If cells are further grown to confluency, one is also able to measure changes in cell–cell interactions.

THE CELLKEY SYSTEM

With the CellKey system, coplanar gold electrodes are patterned at the bottom of the cell plate wells, cells are seeded onto the electrodes, and adherent cells are grown into a confluent monolayer prior to running an assay. Assays with suspension (nonadherent) cells are run once they have settled (by gravity) onto the electrodes. The interdigitated coplanar electrodes of the system form a capacitor; however, their geometric factor is given by a more complicated expression beyond the scope of this discussion. Alternating voltage is applied at a range of frequencies V(f), and the resulting electrical currents I(f) are measured. The impedance as a function of frequency Z(f) can be represented by the equation Z(f) = V(f)/I(f), where V(f) and I(f) are complex numbers to preserve phase information. The cells and electrodes form an electrical circuit attached to an impedance analyzing system, which generates the voltage and measures the current (Figure 11–1A).

As discussed above, the principal thrust of prior work has been to elucidate the basic dielectric properties of intact cells and cell membranes in an attempt to observe cellular behavior in a noninvasive fashion.[23,24,29,30] In addition to these efforts, some researchers have attempted to monitor cellular ligand–receptor interactions using bioimpedance.[70–72,79,89] This ligand interaction work focused on the use of a single frequency and demonstrated that the cellular consequences of G-protein coupled receptor (GPCR) activation could be observed using impedance technology.[70–72,79] However, the results were of limited scope with no consideration of how receptor specificity, if any, could be manifest in the measurements.

Our goal was to extend the use of bioimpedance to label-free, noninvasive monitoring of the cellular consequences of ligand–receptor interaction in an embodiment of the technology that would benefit biomolecular screening and drug discovery efforts. We designed a system capable of producing impedance measurements simultaneously across a large range of frequencies (from 1 kHz to 10 MHz) before, during, and after receptor activation. This approach facilitates the measurement of early events within seconds to minutes of ligand addition in receptor signaling cascades. These early consequences of receptor activation relate directly to ligand concentration, reflect integrated cellular responses to stimuli, and are tied to downstream events in signaling cascades, particularly cytoskeletal rearrangement.[16,90–93] The short-term nature of the read-out fits well with drug discovery workflows, which place a premium on throughput. Output data are analyzed to generate quantitative measures of cellular activation for many pharmacologically important parameters, including ligand–receptor specificity, detection of partial and inverse agonism, derivation of agonist and antagonist potencies, and Schild analysis.[16,17]

In addition to quantifying cellular responses to ligand stimulus, the CellKey system generates unique response profiles that indicate the signaling pathways accessed by stimulated cells (Figure 11–1B). Characteristic profiles of $G_{\alpha i}$-, $G_{\alpha s}$-, and $G_{\alpha q}$-GPCR activation are well documented.[16,17,94] and represent important

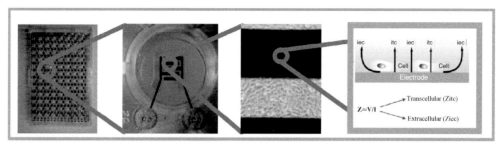

A.

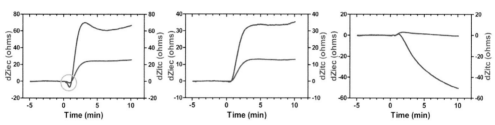

B.

Figure 11–1: The CellKey System measures changes in the impedance (Z) of a cell layer to an applied voltage. Cells are seeded onto a CellKey Standard 96W microplate **(A)** that contains electrodes at the bottom of the wells. The instrument applies small voltages at 24 discrete frequencies, from 1 kHz to 10 MHz, once every 2 s. At low frequencies, these voltages induce extracellular currents (iec) that pass around individual cells in the layer. At high frequencies, they induce transcellular currents (itc) that penetrate the cellular membrane. Changes in impedance due to extracellular currents (dZiec) and to transcellular currents (dZitc) are reported kinetically for each well. When cells are exposed to a stimulus, such as a receptor ligand, signal transduction events occur that lead to cellular events that include changes in cell adherence, cell shape and volume, and cell-cell interaction. These cellular changes individually or collectively affect the flow of extracellular and transcellular current, and therefore, affect the magnitude and characteristics of the measured impedance. **(A)** CellKey cell plate (*left*), an individual assay well with close-up of interdigitated gold electrodes (*center left*), and live cells plated on and between the electrode fingers (*center right*). A diagram highlighting the flow of extracellular (iec) and transcellular (itc) current is shown (*right*). **(B)** In response to ligand-mediated activation of GPCRs, the system generates response profiles that are characteristic of $G_{\alpha s}$, $G_{\alpha q}$ and $G_{\alpha i}$-coupled GPCRs. Examples are provided that illustrate the response profiles for the transfected muscarinic M_1 (typical $G_{\alpha q}$-GPCR), endogenous serotonin 5HT1B (typical $G_{\alpha i}$-GPCR), and prostanoid EP4 receptors (typical $G_{\alpha s}$-GPCR) in the context of CHO cells. $G_{\alpha i}$-GPCRs typically demonstrate an increase in impedance, while $G_{\alpha s}$-GPCRs exhibit a decrease in impedance following receptor activation. What distinguishes the $G_{\alpha q}$-GPCRs from the $G_{\alpha i}$-GPCRs is an initial transient dip in impedance (circled) that is followed by a later increase in impedance. *See color plates.*

starting points for receptor pathway analysis and mode of action studies.[16] By coupling the kinetic features of CellKey data to the administration of many commercially available biochemical modulators, detailed studies can be performed to identify critical cellular mediators of activation.

Possibly the most important benefit of impedance technologies is their exquisite sensitivity. They can be applied routinely to the measurement of endogenous receptor activation, thus enabling access to more biorelevant information from cell-based assays. The analysis of native receptors, expressed under natural genetic, posttranslational, and pathway controls, connected to native signaling cascades, can dramatically influence experimental outcomes. When compared to recombinant or engineered cellular systems, endogenous read-outs offer advantages by producing data more closely aligned to natural biological states. For example, the earliest events of receptor–drug interactions are influenced by the natural complement of receptors expressed at the cell surface. Receptor density is important for all aspects of response onset, magnitude, and duration.[95] Overexpressed systems suffer as accurate models of receptor biology and pharmacology because target receptors are produced in abundance to be conveniently measured. Importantly, it is becoming clear that a number of GPCRs and protein tyrosine kinase receptors (PTKRs) engage in dimerization and oligomerization to signal properly.[96–99] Analyses in the endogenous setting may prove to be the only feasible manner of accessing these natural receptor states. For reasons such as these, impedance technologies undoubtedly will be important in the future characterization of these complex events.

Impedance-based assays enable the analysis of endogenous receptor function in primary cells (Figure 11–2). This offers the flexibility to study targets in cell types more closely aligned to a given disease process, thus creating more disease-relevant cellular models. Performing experiments with endogenously expressed receptors in primary cells has the potential to generate different and better results than engineered and overexpressed systems.[100] Working with native receptor levels that are connected to their native pathways in an environment closer to the native cell state will place greater constraints on the success of newly identified drug compounds.

Finally, impedance assays are universal and relatively simple to perform. Nearly all classes of GPCRs and PTKRs can be analyzed on the same plate during a single experiment without changing assay parameters. Cells are cultured under standard conditions without the need for special reagents, and the lack of labels removes many steps from assay protocols. These attributes of impedance technologies have significant practical benefits for routine assay execution.[15,17]

CellKey system specifications

The CellKey system consists of a physical instrument, which fits on a laboratory bench top, custom 96-well microplates (CellKey Standard 96W micotiter plate or CellKey Small Sample 96W microtiter plate), and a comprehensive software package that controls all aspects of assay design, setup, execution, and data analysis. The instrument contains an automated fluidics head capable of dispensing compounds to all 96 wells of the microplate while the reader simultaneously collects data. Additionally, the fluidics head can be programmed to dispense compounds from four individual quadrants of 384-well compound plates to accommodate libraries formatted in this manner. The reader is thermally controlled and can be set to perform assays from ambient temperature plus 4°C up to 37°C. The system is also designed to be effectively integrated into automated screening

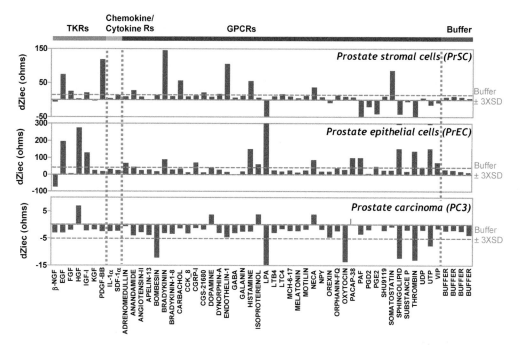

Figure 11–2: Receptor panning on primary and immortalized prostate cells. Two primary cell types, prostate stromal (PrSC) and epithelial cells (PrEC), and the prostate carcinoma, PC3, are treated with the panel of ligands indicated on the x-axis. Typical screening doses were tested (1 to 10 μM). The ligand set was based on known activators of GPCRs and PTKRs as indicated at the top of the figure. The cellular response to each ligand, measured as dZiec and expressed in ohms, was based on the maximum deviation from baseline post-ligand addition. Active endogenous receptors were defined by responses that were greater than a threshold set at the buffer mean ±3x standard deviation. Examples of interesting results include the differential cellular responses in PrSC and PrEC to HGF and in PC3 to Bombesin. Experiments were performed at 28°C. Data are representative of three independent experiments.

platforms. Two sliding decks are positioned on the front of the instrument to provide robotics access to the tip, compound plate, and cell plate stations. Additionally, the instrument and software include hooks for automation scheduling systems.

All assays include a baseline measurement, typically of 30-s to 5-min duration, followed by ligand addition and cellular response measurement. Most assays require 15 min or less to complete per 96-well plate. As assays are label-free, there are no dye loading or labeling steps required, and cells are cultured in routine cell culture media without special components. All of the acquired data are captured and stored in data files about 10 Mb in size (15 min assay with 2 s update rate). Data can be analyzed via the software and exported, automatically or manually, to spreadsheet programs or laboratory information management systems.

Recently, a new cell plate has been developed (CellKey Small Sample 96W microplate) that significantly reduces the number of cells and reagents required to perform an assay. These plates were designed for use with precious primary cells or ligands difficult to produce in abundance.

Example data

At this point it is helpful to provide a brief overview of one manner in which the CellKey system is applied to and simplifies a common task in drug discovery labs. Here we highlight a set of preliminary target identification experiments performed on related primary and immortalized human prostate cell types. Leveraging this impedance technology's ability to measure active endogenous receptors on any cell type, a technique known as receptor panning has been developed. Receptor panning identifies receptor targets expressed in biorelevant and disease relevant fashions and aids in better cellular model design. The Worked Example section to follow presents a more detailed description of receptor panning and the subsequent studies it fosters.

Shown in Figure 11–2 are data collected on two primary cell types derived from normal prostate tissue, prostate stromal cells (PrSC) and prostate epithelial cells (PrEC), and the prostate carcinoma cell line, PC3. The technique involves treating cells with a panel of ligands known to activate different types of receptors (GPCRs, PTKRs, and nuclear hormone receptors among others) and observing which ligands elicit a statistically significant response.[94] Impedance changes measured across a subset of frequencies known to be dominated by currents flowing throughout the extracellular space (dZiec, expressed in ohms) are monitored over time and reported on the y-axis.

Interesting data resulting from this set of experiments include the absence and presence of Met receptor responses to hepatocyte growth factor (HGF) in PrSC and PrEC, respectively. PrSC are known to produce a variety of factors, including HGF, that support and regulate the normal function of PrEC but that can also contribute to the invasion and growth of prostate carcinoma cells.[101] This is one example of the complex interactions of two normal cell types accurately modeled in a cell-based assay. The active bombesin receptor in PC3 cells may represent an interesting target expressed in a highly disease relevant cell line,[102,103] particularly because it was not observed to be active in the primary cell types. Gastrin-releasing peptide receptors are currently the focus of a number of cancer therapeutic strategies, as their disregulation is associated with malignancy.[104,105] Subsequently, the pharmacology of interesting active targets can be analyzed using the same assay conditions (detailed protocols for performing target validation experiments are presented later). These experiments illustrate the ease with which important receptor targets can be identified in highly relevant cell systems and the inherent flexibility of impedance-based assays.

WORKED EXAMPLES

We have described the basic science and application of the CellKey system in the previous sections. In this section we will outline in detail the steps required to operate the system when performing typical experiments. We start by providing an overview of the routine cell culture requirements, cell plating techniques, instrument preparation, data acquisition, and data analysis involved in running assays. We then apply these steps to the identification and characterization of active endogenous receptors expressed in the adherent cell line, U-2 OS, and the nonadherent cell line, Jurkat. Specifically, we explain how to perform receptor panning experiments to identify endogenous receptor targets on these two cell lines, and then validate a single receptor target for each via concentration response curves (CRC). Following this workflow is instructive on a few levels. First, these experiments provide practical examples of the methods for preparing cells, compound plates, and the instrument for a typical experiment. Additionally, we use the potency data generated to pharmacologically subtype the endogenous histamine receptor expressed in U-2 OS cells and the endogenous adrenergic receptor expressed Jurkat cells. In this manner we demonstrate the strengths of an impedance-based technology in identifying and validating important receptor targets.

Preparing cells for assay

Careful and controlled cell culture is important for all cell-based assays, CellKey assays included. For best results, cells should be cultured so that they achieve optimal growth characteristics in terms of growth rate, density, and morphology. Culture conditions, including passage frequency, split ratios, and media components (serum often being the most important media component) have the greatest impact ahead of CellKey assays. With the perspective that cells are biologically complex entities, it follows that all the parameters of routine culture can have unforeseen impacts on the cell-based assay. For this reason it is important to carefully identify and maintain optimal culture conditions. Conditions for the studies detailed here are included in the appendix. The protocol used to remove, or lift, adherent cells off culture vessels can also impact the results of assays. For CellKey assays, the use of 2 mM ethylenediaminetetraacetic acid (EDTA) is encouraged. This nonenzymatic reagent removes the possibility of target receptor cleavage prior to performing experiments.

Preparing cell plates

Adherent cells. For the majority of adherent cell types, cells are harvested from tissue culture vessels and seeded into the CellKey plates the day before each experiment. In the experiments described below, U-2 OS cells are lifted from tissue culture vessels using 2 mM EDTA, washed in assay buffer, and plated in media at a density of 5×10^4 cells/well in 135 μL. Plates are allowed to rest on a nonvibrating, level surface at room temperature for 25 min to promote even cell distribution across the well bottoms. Cell plates are then incubated at 37°C with 5% CO_2 overnight. Impedance technologies are sensitive to the confluence of the cell layer on the electrodes at the time of measurement. Therefore, to generate the most reliable and robust results, the number of cells per well should be optimized. During early assay development, cell monolayers in several wells should be observed by light microscopy before and after running a CellKey protocol. Cells can be observed on the electrodes as well as between the electrodes (Figure 11–1). This visual inspection provides information about the quality of the monolayers and is essential to assess whether coatings will be required. When cells require surface coatings to improve adherence or expression of target receptors, the plates can be treated with attachment factors such as collagen, fibronectin, laminin, or similar reagents prior to cell plating. A protocol for coating CellKey plates is described in the appendix. At 15–60 min prior to running each assay, cell culture media is exchanged three times, leaving a final volume of 135 μL of assay buffer (Hanks balanced salt solution containing 20 mM HEPES and 0.1% BSA). These buffer exchanges can be performed on the CellKey instrument with default fluid exchange protocols, or offline with dedicated liquid handlers.

Nonadherent cells. Nonadherent cells remain in cell culture vessels until immediately before experimentation. Cells are washed two times by centrifugation with assay buffer at room temperature. Cells are then resuspended in assay buffer, counted, and diluted to the desired density (cells/mL). A volume of 135 μL of cell suspension is dispensed into each well of the cell plate. Cells are then allowed to settle by gravity to the bottom of the wells for a minimum of 30 min at room temperature. The maximum amount of time cells can be incubated in buffer is cell line-dependent. Each new cell line should be tested for this parameter.

Preparing compound plates

Ligands. Ligands are typically prepared at 10x the final desired concentration in assay buffer. In the experiments described here, 15 μL of 10x concentrated solution is used for each assay. However, the onboard fluidics head is capable of dispensing volumes across a range from 5 μL to 185 μL dependent on user-defined settings.

Controls

Negative controls. Appropriate negative controls might vary during the different stages of assay development. For example, in routine experiments on well-characterized cell and receptor systems the most appropriate negative control is assay buffer lacking ligand but containing equal concentration of solvent or carrier to the ligand test wells (should there be any). However, if working with high concentrations of solvents (such as dimethyl sulfoxide [DMSO] greater than 1% final), the nonspecific effects of the solvent can be determined by comparing cellular responses to administration of buffer lacking solvent and buffer containing equal concentration of solvent to the ligand test wells. Additionally, to control for the rare occasion where ligands of interest are formulated in nonstandard buffers (e.g., buffers containing high salt, acetic acid, or nonpolar solvents) it is advisable to test the ligand on its own in a well without any cells. This will give an indication of the "visibility" of the ligand preparation and will aid in future data analysis.

Bovine serum albumin. Some cell lines are sensitive to differences in concentrations and types of bovine serum albumin (BSA) used in the assay buffer.[106,107] CellKey assays often can detect this sensitivity. If compounds are dissolved in assay buffer that contains BSA, then the same type and concentration of BSA should be used in the cell plate to perform the assays. CellKey assays are typically run using buffer containing 0.1% BSA (e.g., Sigma catalog #A0281). A general guideline is to utilize BSA that does not contain fatty acids, as a variety of endogenous fatty acid-sensitive receptors are active on many cell types. BSA that contains fatty acids in the preparation can yield significant background responses upon addition of the negative control or alter cellular responses to test ligands.

Positive controls. The ability of the CellKey system to monitor activation of endogenous receptors allows very convenient positive controls to be identified for any cell type. Identifying ligands for an active endogenous receptor in the cell type of interest is accomplished in the receptor panning experiments discussed herein (Figure 11–2 and Figure 11–4) and previously described.[16] Endogenous positive controls demonstrate that cells under study have the capacity to generate CellKey responses and are important in the early phases of assay development.

Specificity controls. During assay development for endogenous or transfected receptors, it is important to test the specificity of a given cellular response. A specific or selective inhibitor of receptor activity (e.g., an antagonist, a monoclonal antibody, or a small inhibitory RNA) is recommended to demonstrate the null response. Additionally, a null-expressing/parental cell line can be used as specificity control in the case of transfected receptor targets.

Minimizing buffer evaporation. Impedance assays can be thought of as a type of electrical measurement where cells, bathed in conductive solutions, form a barrier to the flow of ionic currents. As such, impedance assays are sensitive to changes in buffer conductivity independent of cells being present. Therefore, it is important that buffer components and their concentrations remain the same for the duration of the experiment. For assays that require prolonged incubation in the instrument at 37°C, cell plates are supplied with an assay lid to help reduce the changes in buffer concentration due to evaporation. In addition, compound plates should be kept covered prior to the assay in order to prevent concentration changes due to evaporation. For compound plates that will be in the instrument for more than one experiment or for an extended period of time, using a prescored seal for the compound plate will additionally reduce buffer mismatches due to evaporation (see Appendix).

Assay Protocols

After cell plates and compound plates are prepared, they are placed onto the CellKey system along with a rack of pipette tips. This process is facilitated by sliding platforms on which the cells, reagents, and tips are placed. Using the software, the platforms are then retracted into the system where they wait for the user-defined protocol to initiate. The design of the system is amenable to robotics integration should automation be desired. The system automatically adds ligands to the cells and displays and records the response data in real time. There are a number of default protocols available for data acquisition. These protocols specify the experiment temperature, the duration of cellular equilibration, baseline measurement and postaddition measurement periods, the update interval, and the compound source and fluidics parameters (ligand addition and mixing volumes, pipette tip positions, and aspiration/dispense rates). The parameters for default protocols are accessible in the software and can be easily edited and saved as new protocols. What follows are detailed descriptions of the assay parameters for running experiments in general but also for the specific worked examples on U-2 OS and Jurkat cells.

Cellular equilibration. Prior to running CellKey assays, the cell culture media are exchanged for assay buffer. The impedance of the cells changes for a period of time following this exchange as the cells adjust to their new environment. The amount of time required for equilibration is both cell line and temperature dependent. Establishing the duration required for postbuffer exchange cellular equilibration is a key factor that can affect the quality of CellKey assays. Cellular equilibration can be performed offline on the bench at room temperature or within the CellKey system. Typically, adherent cells are allowed to equilibrate to the assay buffer for 15–60 min prior to the start of the assay, and nonadherent cells are allowed to simultaneously settle to the bottom of the well and equilibrate to the buffer for 30–60 min.

Assay temperature. Assays are most often performed at either 28°C or 37°C, although the instrument enables assays to be performed at any temperature within the range of ambient +4°C to 37°C. Receptor-mediated responses are often temperature-dependent. The optimal temperature for any given cell/receptor system depends on the biology of the receptor in the cells being assayed and the application being pursued. In general, the kinetics of cellular responses are faster at higher temperature (up to 37°C) likely because mammalian cells have adapted to conditions closer to physiological temperature. However, for convenience in an experimental workflow or to prolong certain features in the kinetic data, running experiments at 28°C is often desirable.

Fluidics. The CellKey system software is flexible in its ability to set fluidics parameters. Via a protocol wizard fluid aspiration and dispense volumes, rates, and

Table 11–2. Assay parameters for adherent cells. Listed in this table are parameters and procedures for performing assays on adherent cell lines such as U-2 OS. The protocols for buffer exchange and ligand injection are default programs encoded in the CellKey software

Parameter	Values	Comments
Cell density (U-2 OS cells)	5.0×10^4 cells/well	Cells harvested and plated the day before experiment.
Surface coating	None	
Incubation period on cell plate	16 to 24 hr	37°C, 5% CO_2 incubator
Plate washing: Agonists	Three exchanges leaving 135 µL/well assay buffer	Onboard fluidics head: Default adherent cell protocol exchanges media for assay buffer
Plate washing: Antagonists	Three exchanges followed by addition of 135 µL/well 1x antagonists	Onboard fluidics head: Default antagonist addition protocol, dispenses 1x antagonists in final step
Cellular equilibration	30 min	Allows time for cells to adjust to new buffer environment
Assay temperature	28°C	
Thermal equilibration	5 min	Only used if cellular equilibration period occurred offline
Baseline measurements	5 min	2 s update rate
Ligand addition	15 µL of 10x ligand injected Three mixing steps at 30 µL/s	Onboard fluidics head: Default adherent cell protocol
Cellular activation measurement	10 min	2 s update rate

pipette tip heights and offsets can be adjusted to optimize assays. Table 11–4 lists the fluidics parameters utilized in the data featured here.

Plate maps. The CellKey software enables the use of plate maps to track the conditions of each experiment. Through the use of a plate map editing window, each well can be labeled with the cell type and number and up to two ligand names and concentrations. Additionally, once plate maps have been populated and user-defined parameters have been entered, the instrument software can then perform data analysis and report responses that are statistically significant compared to controls.

Executing experiments

With the cells plated, the compounds prepared, the experimental protocols designed, and the plate maps in place, the following simple steps are performed to execute an experiment on the CellKey system. The parameters detailed in Tables 11–2 and Table 11–3 are used to generate the worked example data and are based on typical workflows for adherent and nonadherent cells.

Table 11–3. Assay parameters for nonadherent cells. Listed in this table are parameters and procedures for performing assays on non-adeherent cell lines such as Jurkat. The protocol for ligand injection is a default program encoded in the CellKey software

Parameter	Values	Comments
Cell density (JURKAT cells)	2.0×10^5 cells/well	Cells harvested, washed, and plated immediately before experiment
Surface coating	None	
Plate washing	None	
Cellular equilibration and sedimentation	30 min	Allows time for cells to adjust to new buffer environment and to sediment by gravity
Assay temperature	28°C	
Thermal equilibration	5 min	Only used if cellular equilibration period occurred offline
Baseline measurements	5 min	2 s update rate
Agonist addition	15 µL of 10x ligand injected Three mixing steps at 3 µL/s	Onboard fluidics head: Default nonadherent cell protocol
Antagonist addition*	15 µL of 10x ligand injected. Three mixing steps at 3 µL/s	Onboard fluidics head: Default nonadherent cell protocol
Cellular activation measurement	10 min	2 s update rate

* For antagonist experiments using nonadherent cells, 15 µL of 10x antagonist precedes agonist addition. After a period of incubation (user defined) 16.6 µL of 10x challenge agonist is added to the same wells.

U-2 OS cells. Plates containing adherent U-2 OS cells prepared as described in the Appendix and previously are removed from the incubator immediately before each experiment and visually inspected under the microscope for consistency. Key parameters for these experiments are summarized in Table 11–2. For agonist experiments, cell plates are washed using the onboard fluidics head programmed to replace culture media with assay buffer in three exchanges. A volume of 135 µL assay buffer is left in the wells after the final exchange. The plates are then incubated on the reader at 28°C for 30 min to allow the cells to equilibrate to the new buffer environment. Five minutes of baseline measurements are recorded at an update rate of 2 s per 96 wells, followed by addition of 15 µL of 10x ligands (see Table 11–4 for detailed fluidics parameters). Subsequently, 10 min of cellular activation are recorded at the same update rate. Data are analyzed using the CellKey software and exported for graphing.

For antagonist experiments cell plates are washed using the onboard fluidics head programmed to replace culture media for assay buffer in three exchanges. An additional wash step is performed in which the fluidics head delivers 135 µL of 1x antagonists from a compound plate to the cell plate. The cell plates are then incubated on the reader at 28°C for 30 min to allow the cells to equilibrate to the new buffer environment. Five minutes of baseline measurements are recorded at an update rate of 2 s per 96 wells, followed by addition of 15 µL of 10x ligands.

Table 11–4. Detailed fluidics parameters for U-2 OS and Jurkat cell experiments. The CellKey software encodes these parameters in default protocols. Each parameter can be modified and a new protocol can be saved. Nonadherent settings position the pipette tips higher off the bottom of the wells and handle fluid at a slower rate, so as not to disturb the sedimented monolayer

Fluidics parameter	Default settings for adherent cells	Default settings for nonadherent cells
Aspiration height (tips)	1–3.5 mm	2 mm
Aspiration rate	30 μL/s	3 μL/s
Dispense height (tips)	1–3.5 mm	3.5 mm
Dispense rate	30 μL/s	3 μL/s
Mix cycles	3	3

Subsequently, 10 min of cellular activation are recorded at the same update rate. The CellKey software also contains default linked protocols that enable one to monitor the agonist activity of compounds for a period of time, then to follow on with reference agonist challenge. In this way, the system facilitates the sequential analysis of antagonist and agonist compounds on the same plate of cells.

Jurkat cells. Jurkat cells are cultured and prepared as described in the Appendix and previously. Key parameters for these experiments are summarized in Table 11–3. After plating in 135 μL assay buffer, cells are allowed to settle by gravity for 30 min on the reader at 28°C. During this time, the cells are simultaneously equilibrating to the new buffer environment. For agonist experiments 5 min of baseline measurements are recorded at an update rate of 2 s per 96 wells, followed by addition of 15 μL of 10x ligands (see Table 11–4 for detailed fluidics parameters). Subsequently, 10 min of cellular activation are recorded at the same update rate. Data are analyzed using the CellKey software and exported for graphing.

For antagonist experiments with nonadherent cells, a two-step, linked protocol is most often used. In the first step 15 μL of 10x antagonists are added to the cells from one compound plate placed in the system. At this point data can be collected to determine whether antagonists induce cellular responses on their own. After a user-defined length of time, the system delivers 16.6 μL of 10x challenge agonist to the same wells from a different compound plate placed in the system.

Data analysis and results

Data can be analyzed using the CellKey software in a variety of ways. Response settings in the software adjust the Zitc and Ziec frequency windows and specify how kinetic responses and extracted values are calculated. The software provides default settings for most adherent cells, such as U-2 OS. However, because the peak response frequencies for dZiec and dZitc differ between adherent and nonadherent cells, these settings must be adjusted by the user for nonadherent cells, such as Jurkat. Response settings do not influence the acquisition of impedance data and can be changed both before and after acquisition. Users can toggle between various settings at any time while acquiring data. Table 11–5 lists the settings used for the analysis of the U-2 OS and Jurkat experiments discussed.

Table 11–5. Response settings for adherent and nonadherent cells. The parameters used to extract scalar values of response magnitude for U-2 OS and Jurkat cells are listed. The kinetic response profiles from each experiment are used as the base data for analysis

Parameter	Adherent cells		Nonadherent cells	
Kinetic response	dZiec	dZitc	dZiec	dZitc
Analysis frequencies (Hz)	3320 to 24 700	81 900 to 40 7000	24 700 to 122 000	911 000 to 3 080 000
Extracted value	Maximum change		Maximum change	

The subset of frequencies (the "frequency window") for a given experiment can be chosen from the entire range of frequencies measured by the system. Through the software, the user is able to define the frequency windows that correspond to the kinetic measures dZiec and dZitc. Other data analysis parameters can also be modified, including options to view extracted values in terms of maximum deviation from baseline, maximum–minimum, or at a specific point in time. The software displays data in an 8-by-12 format or in individual wells upon mouse click. When parameters are set to the user's satisfaction, the extracted data can be exported into flat files compatible with spreadsheet programs or into laboratory automation management systems.

Analyzing response profiles. In addition to quantifying the magnitude of Cell-Key responses, the kinetic response profiles contain qualitative information about the experiments being run (Figure 11–1). Figure 11–3 displays the response profiles generated by activating U-2 OS cells with 10 μM histamine (Figure 11–3A) and Jurkat cells with 10 μM isoproterenol (Figure 11–3C). Negative control (buffer) profiles for each cell line are shown as well (Figures 11–3B and 11–3D). It is important to mention here that in the case of many nonadherent cell types, the CellKey response profiles do not demonstrate the typical feature sets like those generated with adherent cells. For example, activating nonadherent cells with ligands specific for $G_{\alpha i}$-, $G_{\alpha q}$- or $G_{\alpha s}$-coupled receptors induces response profiles that are very similar. For this reason, the response profiles for nonadherent cells can not always be relied upon to indicate the signaling pathways induced. In these cases, biochemical modulators are used to more fully characterize the pathways involved.

Worked example data. Using the assay protocols and instrument parameters described above, CellKey assays were performed on U-2 OS and Jurkat cells to identify and validate interesting endogenous receptor targets. This workflow can be considered typical for the development of target identification, target validation, and other secondary screening assays on the system. As mentioned

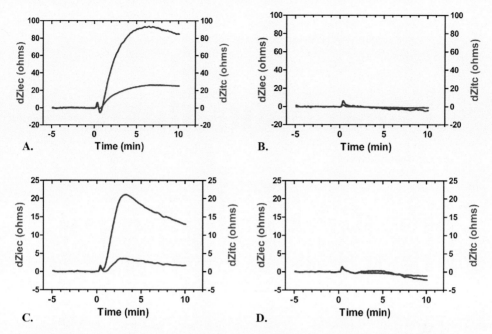

Figure 11–3: CellKey response profiles of endogenous GPCRs under study. **(A)** Response profile for U-2 OS cells activated by addition of 10 μM histamine via their endogenously expressed histamine receptor. **(B)** Buffer negative control response profile of U-2 OS cells. **(C)** Response profile for Jurkat cells activated by addition of 10 μM isoproterenol via their endogenously expressed adrenergic receptor. **(D)** Buffer negative control response profile of Jurkat cells. Data are representative examples of at least four separate experiments.

previously, receptor panning experiments involve testing a panel of ligands against cell lines of interest to catalog functionally active endogenous receptors. Figure 11–4 displays the output data from such experiments on U-2 OS and Jurkat cells. The data demonstrate that different PTKRs and GPCRs are active on each cell type. Data are displayed as dZiec maximum–minimum, so they represent the absolute values of responses either above or below the baseline. Many ligands induce statistically significant cellular responses (defined by the threshold set at the buffer mean +3 x SD). They are seen as any bar above the dashed line in Figure 11–4. Some of the receptors activated have been reported in the past, such as CXCR4 receptor expressed in Jurkat.[108] However, there are also active endogenous receptors in these cell lines for which no prior information could be gleaned from the literature. In fact, studies have been published focusing on the transfected version of the β2-adrenoceptor in U-2 OS cells that are shown here to contain a functionally active endogenous form of the same receptor.[109–111] Similarly, beta-lactamase assays applied to Jurkat cells were unable to measure the endogenous activities of platelet activating factor receptor (PAF-R) and β2-adrenoceptor.[112] Indeed, impedance assays appear to have consistently better sensitivity to endogenous receptor activity than many other cell-based assays.[16]

To further characterize the cellular responses induced by the 10 μM dose of the nonselective agonist, histamine on U-2 OS cells (Figure 11–5A), concentration response curves were carried out to derive the ligand's potency. Histamine was

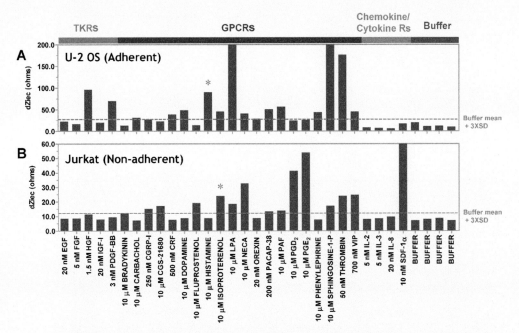

Figure 11–4: Receptor panning for target identification on U-2 OS and Jurkat cells. **(A)** U-2 OS and **(B)** Jurkat cells were tested to identify functionally active endogenous receptors using a panel of ligands targeting a range of G-protein coupled receptors and tyrosine kinase receptors. Impedance changes (Y-axis) greater than a threshold (dotted orange line) set to the negative control plus three standard deviations are considered hits. The endogenous histamine receptor response to 10 μM histamine in the U-2 OS cells and the β-adrenergic receptor response to 10 μM isoproterenol in Jurkat cells are highlighted by asterisks. Data are extracted as max − min and representative of three separate experiments.

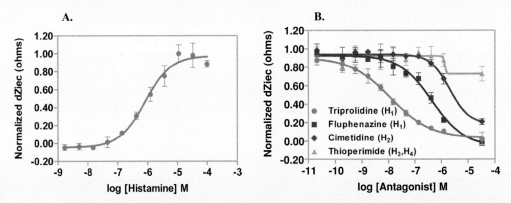

Figure 11–5: Target validation experiments with endogenous histamine receptor expressed in U-2 OS cells. **(A)** Concentration response curve of the non-selective native agonist, histamine, targeting the endogenous histamine receptor in U-2 OS cells yields an EC$_{50}$ value of 820 nM. Data are normalized to 10 μM histamine. **(B)** Concentration response curves of subtype selective antagonists, triprolidine, fluphenazine, cimetidine and thioperamide, challenged with 10 μM histamine. The IC$_{50}$ values of the histamine H$_1$ receptor antagonists triprolidine and fluphenazine are 9.7 nM and 440 nM, respectively. The H$_2$ subtype-selective antagonist (cimetidine) and an H$_3$/H$_4$-subtype-selective antagonist (thioperamide) are unable to achieve blocking at physiological concentrations. Data are normalized to 10 μM histamine in the absence of antagonist. All data represent mean +/−SD, n = 4.

diluted in threefold steps from 30 μM to 1.5 nM and added to cells in quadruplicate wells. Maximum response data were extracted from the kinetic reads, buffer subtracted, and normalized to the 30 μM dose. Using GraphPad Prizm software to fit the curve,[113] an EC_{50} value of 182 nM was derived. The potency derived from this impedance assay was in reasonable agreement with previously reported values *in vivo* and *in vitro*.[114,115]

Pharmacological subtyping of the histamine receptor in U-2 OS cells was then carried out (Figure 11–5B). Antagonists selective for the various members of the histamine receptor family were preincubated over U-2 OS cells as described in the protocol section above and in.[16] An agonist challenge dose of 10 μM histamine was subsequently added to generate antagonist CRCs. Maximum response data were extracted from the kinetic reads, buffer subtracted, and normalized to 10 μM histamine in the absence of antagonist. Using GraphPad Prizm software to fit the curve, IC_{50} values were derived for each antagonist. The data demonstrate that the histamine H_1 subtype selective antagonists, triprolidine and fluphenazine, were able to fully inhibit the cellular responses to histamine administration with IC_{50} values of 9.7 nM and 440 nM, respectively. The H_2 selective antagonist, cimetidine, and H_3/H_4 selective antagonist, thioperimide, were unable to fully inhibit the histamine response at the concentrations tested, and curve fitting could not derive IC_{50} values. These data provide a good indication that the endogenous histamine receptor in U-2 OS cells is of the H_1 subtype. Additional evidence of this can be derived from the CellKey response profile shown in Figure 11–3A. The features in the kinetic graph of the histamine response (immediate dip in impedance followed by a later rise) indicate a Gq-coupled GPCR pathway has been activated. Of the four known subtypes of the histamine receptor, only H_1 is Gq-coupled.

To further characterize the cellular responses induced by the 10 μM dose of the nonselective agonist, isoproterenol on Jurkat cells (Figure 11–4B), concentration response curves were performed using selective agonists to the various members of the adrenergic receptor family (Figure 11–6A). The β2-selective agonist, procaterol, the β1-selective agonist, xamoterol, and the β3-selective agonist, CL 316243, were chosen for this study. Agonists were diluted in threefold steps starting at 1 μM and added to cells in quadruplicate wells. Maximum response data were extracted from the kinetic reads, buffer subtracted, and normalized to the 1 μM dose of procaterol. Only procaterol generated a sigmoid-shaped CRC upon curve fitting, and an EC_{50} value of 2.7 nM was derived. The potency derived from this impedance assay was in reasonable agreement with previously reported values. These data suggest that the β2-adrenoceptor is functionally active in Jurkat cells.

In follow-up studies, stereoisomers of the adrenoceptor antagonist, propranolol, were analyzed. The three stereoisomers R-(–)-propranolol, (±)-propranolol, R-(+)-propranolol were tested. All three compounds were able to fully inhibit the cellular responses to isoproterenol administration with IC_{50} values of 176 pM and 487 pM, and 124 nM, respectively. The data demonstrate that impedance assays can discriminate the potencies of highly related chemical structures.

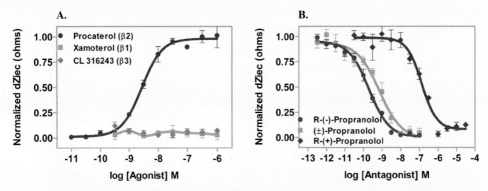

Figure 11–6: Target validation experiments with endogenous β-adrenergic receptor expressed in Jurkat cells. **(A)** Concentration response curves of agonists with selectivity to different β-adrenergic receptor subtypes are shown. Only procaterol was capable of generating a complete curve with an EC_{50} value of 182 nM. Data are normalized to 10 μM procaterol. **(B)** Stereoisomeric antagonist concentration response curves challenged with 4 nM isoproterenol. IC_{50} values derived for each are as follows: R-(−)-Propranolol, 176 pM; (±)-Propranolol, 487 pM; R-(−)-Propranolol, 124 nM. All data represent mean +/−SD, n = 3.

SUMMARY

At this time, label-free cell-based assays are having growing impact on research and drug discovery efforts. Their rich and unique readouts and ease of use are important reasons for their continued adoption. Further scale up and automation of label-free systems to higher density formats are also likely to enhance their utilization. Impedance-based cellular assays are powerful label-free tools that hold promise for advancing the field of drug discovery. These technologies represent an important leap in biorelevant evaluation *in vitro* of drugs and their receptor targets. Their sensitivity to endogenous levels of receptor activation enables studies to be performed using intact, native signaling cascades in biologically relevant cell types, including primary cells. The label-free format simplifies assay development and the screening process workflow, while simultaneously moving away from artificial methods of pathway interrogation based on the use of dyes, fluorescent labels, and overexpressed cellular systems. As understanding of receptor signaling complexity grows, particularly in GPCRs, the importance of monitoring receptor activation in the native setting will grow accordingly. Impedance-based assay systems are likely to make significant contributions to ongoing efforts in introducing complex biology to the early drug discovery workflow.

APPENDIX

Cell lines and materials

The following cell lines were purchased from the American Type Culture Collection (ATCC; www.atcc.org): human osteosarcoma cell line U-2 OS (Catalog

#HTB-96); human acute T-cell leukemia cell line Jurkat (Catalog #CRL-2570); and human prostatic adenocarcinoma cell line PC3 (Catalog #CRL-1435). Primary cells were purchased from Cambrex (Lonza): normal prostate stromal cells (Catalog #CC-2508) and normal prostate epithelial cells (Catalog #CC-2555). All chemicals were purchased from Sigma-Aldrich (www.sigmaaldrich.com).

Catalog numbers for most agonists and antagonists described in the studies are listed here. As mentioned in the text relating to Figure 11–4, here is a list of convenient positive control ligands and their catalog numbers. For U-2 OS cells: hepatocyte growth factor (HGF) Sigma #H1404; platelet derived growth factor (PDGF) Sigma #P8147; carbachol Sigma #C4382; corticotropin releasing factor (CRF) Sigma #C3042; dopamine Sigma #H8502; histamine Sigma #H7250; isoproterenol Sigma #I6504; lysophosphatidic acid (LPA) Sigma #L7260; NECA Sigma #E2387; pituitary adenylate cyclase activating polypeptide (PACAP) Sigma #A1439; platelet activating factor (PAF) Sigma #P7568; phenylephirine Sigma #P6126; sphingosine 1-phosphate (S1P) Sigma #S9666; thrombin Sigma #T6884; vasoactive intestinal peptide (VIP) Sigma #V6130. For Jurkat cells: CGS-21680 Sigma #C141; fluprostenol Sigma #F8549; isoproterenol, LPA, NECA, PACAP, PAF, prostaglandin D2 (PGD2) Sigma #P5172; prostaglandin E2 (PGE2) Sigma #P5640; S1P; VIP; stromal derived factor 1α (SDF1α) Sigma #S190. βAR antagonist sterioisomers: R-(–)-Propranolol, Sigma #P0689; ($\pm$)-Propranolol, Sigma #P0884; R-(+)-Propranolol, Sigma #222984.

Maintaining and preparing cells

The CellKey system can be used to study endogenous and transfected receptors with adherent or nonadherent, immortalized or primary cells. Standard cell culture procedures are followed for passaging cells in tissue culture and plating cells onto the custom CellKey standard 96W microplate. The following are general guidelines for maintaining and plating cells; however, it is important to remember that the best results are obtained from seeding cells according to their optimum conditions. Generally, conditions should be optimized such that cells are in late-log phase growth prior to harvesting and plating onto CellKey plates.

Cell culture

All cells were grown in a standard cell culture incubator at 37°C with 5% CO_2. U2OS were cultured in McCoy's 5A media supplemented with 10% FBS (Gemini Bioproducts, Woodland, CA), 50 μg/mL streptomycin, and 50 units/mL of penicillin. Cells were passaged at regular intervals when the cell monolayer reached 80% to 95% confluence, typically 1:10 or 1:15 splits twice a week. Cells were lifted with PBS-2 mM EDTA.

Jurkat cells were cultured in RPMI 1640 media supplemented with 10% FBS (Gemini Bioproducts, Woodland, CA), 50 μg/mL streptomycin, and 50 units/mL of penicillin. Cells were passaged at regular intervals when the cultures reached a density of 2 million cells/mL in a 15-cm dish. As they reached this density the cells were diluted to a minimum density of 0.5-million cells/mL.

PC3 cells were cultured in F12-K media supplemented with 10% FBS, 50 μg/mL streptomycin, and 50 units/mL of penicillin. Cells were passaged at regular intervals when the cell monolayer reached 80% to 95% confluence, typically 1:5 or 1:8 splits twice a week. Cells were lifted with PBS-2 mM EDTA.

Coating cell plates with attachment factors

A variety of attachment factors that aid in the growth and adherence of different cells can be used on CellKey plates. Following are a few examples of common attachment factors, their sources, working concentrations, and volumes added per well in a 96-well plate: collagen (Sigma C9791, 50 μg/mL, 50 μL), fibronectin (Sigma F1141, 10 μg/mL, 50 μL), laminin (Sigma L2020, 50 μg/mL, 50 μL), and Poly-D-lysine (Sigma P6407, 50 μg/mL, 50 μL). Working concentrations and plating volumes for other attachment factors should be taken from the vendor's suggested parameters. The CellKey Standard 96W plate contains intricate electrodes and leads at the base of each well. It is important to follow these basic steps when coating plates to avoid damage to the electrodes. Pipet tips should never be allowed to touch the electrodes. Coat plates for intervals of 1 hour or less. Remove plate coating material using gentle methods such as flicking the liquid out over the laboratory sink or by using commercially available liquid handlers. Avoid aspirating the liquid from the well using a vacuum system. Seed cells into plate immediately after coating the wells.

REFERENCES

1. E. Jacoby, R. Bouhelal, M. Gerspacher, K. Seuwen, *ChemMedChem* **1**, 761 (2006).
2. J. Dunlop *et al.*, *Biochem Pharmacol* **74**, 1172 (2007).
3. E. R. Benjamin *et al.*, *J Biomol Screen* **10**, 365 (2005).
4. B. F. O'Dowd, M. Alijaniaram, X. Ji, T. Nguyen, R. M. Eglen, S. R. George, *J Biomol Screen* **12**, 175 (2007).
5. M. Y. Lee, R. A. Kumar, S. M. Sukumaran, M. G. Hogg, D. S. Clark, J. S. Dordick, *Proc Natl Acad Sci U S A* **105**, 59 (2008).
6. M. X. Chen *et al.*, *BMC Biotechnol* **7**, 93 (2007).
7. P. Lang, K. Yeow, A. Nichols, A. Scheer, *Nat Rev Drug Discov* **5**, 343 (2006).
8. M. Xia *et al.*, *Environ Health Perspect* **116**, 284 (2008).
9. R. J. Garippa, *The Emerging Role of Cell Based Assays in Drug Discovery*. L. K. Minor, Ed., Handbook of Cell Based Assays in Drug Discovery (CRC Press, Boca Raton, FL. 2006), pp. 227–242.
10. M. A. Cooper, *Drug Discov Today* **11**, 1068 (2006).
11. Y. Fang, A. M. Ferrie, *FEBS Lett* **582**, 558 (2008).
12. Y. Fang, G. Li, A. M. Ferrie, *J Pharmacol Toxicol Methods* **55**, 314 (2007).
13. Y. A. Abassi, J. A. Jackson, J. Zhu, J. O'Connell, X. Wang, X. Xu, *J Immunol Methods* **292**, 195 (2004).
14. J. M. Atienza, J. Zhu, X. Wang, X. Xu, Y. Abassi, *J Biomol Screen* **10**, 795 (2005).
15. M. F. Peters *et al.*, *J Biomol Screen* **12**, 312 (2007).
16. E. Verdonk *et al.*, *Assay Drug Dev Technol* **4**, 609 (2006).

17. T. H. Leung G, McGuinness R, Verdonk E, Michelotti JM, Liu VF., *Journal of the Association for Laboratory Automation (JALA)* **10**, 258 (2005 August).
18. J. M. Atienza, N. Yu, X. Wang, X. Xu, Y. Abassi, *J Biomol Screen* **11**, 634 (2006).
19. P. Debye, *Polar Molecules*. (Dover, New York, 1929) Pages.
20. S. Grimnes, O. Martinsen, *Bioimpedance and Bioelectricity Basics*. (Academic Press, New York, 2000) Pages.
21. H. P. Schwan, *Adv Biol Med Phys* **5**, 147 (1957).
22. R. Pethig, *Dieletric and Electronic Properties of Biological Materials*. (John Wiley and Sons, Hoboken, NJ, 1979) Pages.
23. K. R. Foster, H. P. Schwan, *Crit Rev Biomed Eng* **17**, 25 (1989).
24. R. Pethig, *Clin Phys Physiol Meas* 8 Suppl A, 5 (1987).
25. M. A. Stuchly, S. S. Stuchly, *IEEE Transactions on Instrumentation and Measurement* **29**, 176 (1980).
26. T. W. Athey, M. A. Stuchly, S. S. Stuchly, *IEEE Transactions on Microwave Theory and Techniques* **82**, 82 (1982).
27. A. Kraszewski, M. A. Stuchly, S. S. Stuchly, A. M. Smith, *Bioelectromagnetics* **3**, 421 (1982).
28. A. Surowiec, S. S. Stuchly, C. Izaguirre, *Phys Med Biol* **31**, 43 (1986).
29. A. Surowiec, S. S. Stuchly, L. Eidus, A. Swarup, *Phys Med Biol* **32**, 615 (1987).
30. C. Gabriel, S. Gabriel, E. Corthout, *Phys Med Biol* **41**, 2231 (1996).
31. G. Smith, A. P. Duffy, J. Shen, C. J. Olliff, *J Pharm Sci* **84**, 1029 (1995).
32. C. Prodan, E. Prodan, *J Phys D: Appl Phys* **32**, 335 (1999).
33. E. Gheorghiu, *Bioelectromagnetics* **17**, 475 (1996).
34. E. Gheorghiu, C. Balut, M. Gheorghiu, *Phys Med Biol* **47**, 341 (2002).
35. L. A. Geddes, *Ann Biomed Eng* **25**, 1 (1997).
36. K. Asami, A. Irimajiri, Biochim *Biophys Acta* **778**, 570 (1984).
37. A. Irimajiri, K. Asami, T. Ichinowatari, Y. Kinoshita, *Biochim Biophys Acta* **896**, 203 (1987).
38. K. Asami, A. Irimajiri, *Phys Med Biol* **45**, 3285 (2000).
39. M. Cereijido, E. S. Robbins, W. J. Dolan, C. A. Rotunno, D. D. Sabatini, *J Cell Biol* **77**, 853 (1978).
40. A. Martinez-Palomo, I. Meza, G. Beaty, M. Cereijido, *J Cell Biol* **87**, 736 (1980).
41. E. Stefani, M. Cereijido, *J Membr Biol* **73**, 177 (1983).
42. M. S. Balda *et al.*, *J Membr Biol* **122**, 193 (1991).
43. P. D. Ward, R. R. Klein, M. D. Troutman, S. Desai, D. R. Thakker, *J Biol Chem* **277**, 35760 (2002).
44. M. Cereijido *et al.*, *Arch Med Res* **38**, 465 (2007).
45. G. Kottra, E. Fromter, *Pflugers Arch* **402**, 409 (1984).
46. G. Kottra, E. Fromter, *Pflugers Arch* **425**, 535 (1993).
47. J. L. Madara, D. Barenberg, S. Carlson, *J Cell Biol* **102**, 2125 (1986).
48. J. L. Madara, S. Nash, R. Moore, K. Atisook, *Monogr Pathol* **306** (1990).
49. J. R. Turner *et al.*, *Am J Physiol Cell Physiol* **279**, C1918 (2000).
50. A. T. Gewirtz, Y. Liu, S. V. Sitaraman, J. L. Madara, *Best Pract Res Clin Gastroenterol* **16**, 851 (2002).
51. M. Bruewer *et al.*, *J Immunol* **171**, 6164 (2003).
52. M. F. Iskander, S. S. Stuchly, *IEEE Transactions on Instrumentation and Measurement* **21**, 425 (1972).
53. S. S. Stuchly, M. A. Stuchly, B. Carraro, *IEEE Transactions on Instrumentation and Measurement* **27**, 436 (1978).
54. M. M. Brady, S. A. Symons, S. S. Stuchly, *IEEE Trans Biomed Eng* **28**, 305 (1981).
55. A. Surowiec, S. S. Stuchly, A. Swarup, *Phys Med Biol* **30**, 1131 (1985).
56. A. J. Surowiec, S. S. Stuchly, J. B. Barr, A. Swarup, *IEEE Trans Biomed Eng* **35**, 257 (1988).
57. K. P. A. P. Esselle, S. S. Stuchly, *IEEE Transactions on Instrumentation and Measurement* **37**, 101 (1988).

58. S. S. Stuchly, C. L. Sibbald, J. M. Anderson, *IEEE Transactions on Microwave Theory and Techniques* **42**, 192 (1994).

59. C. Gabriel, T. Y. Chan, E. H. Grant, *Phys Med Biol* **39**, 2183 (1994).

60. C. Gabriel, *Bioelectromagnetics* Suppl **7**, S12 (2005).

61. V. Raicu, *Meas. Sci. Technol.* **6**, 410 (1995).

62. V. Raicu, T. Saibara, H. Enzan, A. Irimajiri, *Bioelectrochemistry and Bioenergetics* **47**, 333 (1998).

63. V. Raicu, *Phys Rev E Stat Phys Plasmas Fluids Relat Interdiscip Topics* **60**, 4677 (1999).

64. V. Raicu, N. Kitagawa, A. Irimajiri, *Phys Med Biol* **45**, L1 (2000).

65. I. Giaever, C. R. Keese, *Proc Natl Acad Sci U S A* **81**, 3761 (1984).

66. I. Giaever, C. R. Keese, *IEEE Trans Biomed Eng* **33**, 242 (1986).

67. I. Giaever, C. R. Keese, *Proc Natl Acad Sci U S A* **88**, 7896 (1991).

68. C. Tiruppathi, A. B. Malik, P. J. Del Vecchio, C. R. Keese, I. Giaever, *Proc Natl Acad Sci U S A* **89**, 7919 (1992).

69. C. M. Lo, C. R. Keese, I. Giaever, *Biophys J* **69**, 2800 (1995).

70. H. S. Wang, C. R. Keese, I. Giaever, T. J. Smith, *J Clin Endocrinol Metab* **80**, 3553 (1995).

71. A. B. Moy *et al.*, *J Clin Invest* **97**, 1020 (1996).

72. A. B. Moy *et al.*, *Am J Physiol Lung Cell Mol Physiol* **278**, L888 (2000).

73. C. R. Keese, K. Bhawe, J. Wegener, I. Giaever, *Biotechniques* **33**, 842 (2002).

74. C. E. Campbell, M. M. Laane, E. Haugarvoll, I. Giaever, *Biosens Bioelectron* **23**, 536 (2007).

75. C. Xiao, J. H. Luong, *Biotechnol Prog* **19**, 1000 (**2003**).

76. J. H. Luong, M. Habibi-Rezaei, J. Meghrous, C. Xiao, K. B. Male, A. Kamen, *Anal Chem* **73**, 1844 (2001).

77. C. Xiao, J. H. Luong, *Toxicol Appl Pharmacol* **206**, 102 (2005).

78. A. Janshoff, J. Wegener, C. Steinem, M. Sieber, H. J. Galla, *Acta Biochim Pol* **43**, 339 (1996).

79. J. Wegener, S. Zink, P. Rosen, H. Galla, *Pflugers Arch* **437**, 925 (1999).

80. J. Wegener, C. R. Keese, I. Giaever, *Exp Cell Res* **259**, 158 (2000).

81. S. Arndt, J. Seebach, K. Psathaki, H. J. Galla, J. Wegener, *Biosens Bioelectron* **19**, 583 (2004).

82. R. Ehret, W. Baumann, M. Brischwein, A. Schwinde, K. Stegbauer, B. Wolf, *Biosens Bioelectron* **12**, 29 (1997).

83. B. Wolf, M. Brischwein, W. Baumann, R. Ehret, M. Kraus, *Biosens Bioelectron* **13**, 501 (1998).

84. R. Ehret *et al.*, *Fresenius J Anal Chem* **369**, 30 (2001).

85. E. Thedinga *et al.*, *Toxicol Appl Pharmacol* **220**, 33 (2007).

86. K. Solly, X. Wang, X. Xu, B. Strulovici, W. Zheng, *Assay Drug Dev Technol* **2**, 363 (2004).

87. J. M. Atienza *et al.*, *Assay Drug Dev Technol* **4**, 597 (2006).

88. J. Z. Xing, L. Zhu, S. Gabos, L. Xie, *Toxicol In Vitro* **20**, 995 (2006).

89. M. T. Santini, C. Cametti, P. L. Indovina, S. W. Peterson, *Exp Hematol* **22**, 40 (1994).

90. B. P. Head *et al.*, *J Biol Chem* **281**, 26391 (2006).

91. R. J. Hoefen, B. C. Berk, *J Cell Sci* **119**, 1469 (2006).

92. K. Szaszi, K. Kurashima, K. Kaibuchi, S. Grinstein, J. Orlowski, *J Biol Chem* **276**, 40761 (2001).

93. J. P. Overington, B. Al-Lazikani, A. L. Hopkins, *Nat Rev Drug Discov* **5**, 993 (2006).

94. G. J. Ciambrone, V. F. Liu, D. C. Lin, R. P. McGuinness, G. K. Leung, S. Pitchford, *J Biomol Screen* **9**, 467 (2004).

95. B. L. Sanchez-Laorden, J. Sanchez-Mas, E. Martinez-Alonso, J. A. Martinez-Menarguez, J. C. Garcia-Borron, C. Jimenez-Cervantes, *J Invest Dermatol* **126**, 172 (2006).

96. D. J. Daniels, A. Kulkarni, Z. Xie, R. G. Bhushan, P. S. Portoghese, *J Med Chem* **48**, 1713 (2005).

97. N. D'Ambrosi, M. Iafrate, E. Saba, P. Rosa, C. Volonte, *Biochim Biophys Acta* **1768**, 1592 (2007).

98. G. Arpino *et al.*, *J Natl Cancer Inst* **99**, 694 (2007).

99. L. F. Agnati, S. Ferre, C. Lluis, R. Franco, K. Fuxe, *Pharmacol Rev* **55**, 509 (2003).

100. S. Goldbard, *Curr Opin Drug Discov Devel* **9**, 110 (2006).

101. A. Tate *et al.*, *BMC Cancer* **6**, 197 (2006).

102. D. Cornelio, R. Roesler, G. Schwartsmann, *Ann Oncol* **18**, 1457–1466 (2007).

103. B. Y. Williams, A. Schonbrunn, *Cancer Res* **54**, 818 (1994).

104. E. Garcia Garayoa *et al.*, *Q J Nucl Med Mol Imaging* **51**, 42 (2007).

105. X. Zhang *et al.*, *J Nucl Med* **47**, 492 (2006).

106. C. M. Butt, S. R. Hutton, M. J. Marks, A. C. Collins, *J Neurochem* **83**, 48 (2002).

107. J. Strosznajder, L. Foudin, W. Tang, G. Y. Sun, *J Neurochem* **40**, 84 (1983).

108. A. Schonbrunn, D. Steffen. The Endogenous GPCR List. http://www.tumor-gene.org/GPCR/gpcr.html.

109. R. H. Oakley *et al.*, *Assay Drug Dev Technol* **1**, 21 (2002).

110. Z. M. Volovyk, M. J. Wolf, S. V. Prasad, H. A. Rockman, *J Biol Chem* **281**, 9773 (2006).

111. J. D. Violin, X. R. Ren, R. J. Lefkowitz, *J Biol Chem* **281**, 20577 (2006).

112. X. Tan, P. Sanders, J. Bolado, Jr., M. Whitney, *Genomics Proteomics Bioinformatics* **1**, 173 (2003).

113. H. Motulsky, A. Christopoulos, *Fitting Models to Biological Data using Linear and Nonlinear Regression. A Practical Guide to Curve Fitting.*, (Oxford University Press, New York, 2004).

114. K. M. Crawford, D. K. MacCallum, S. A. Ernst, *Invest Ophthalmol Vis Sci* **33**, 3041 (1992).

115. J. P. Trzeciakowski, R. Levi, *J Pharmacol Exp Ther* **223**, 774 (1982).

Index

A33 epithelial antigen, 154
adrenergic receptors, 166, 260, 268, 271
affinity binding constants, 48. *See* binding
    affinities. *See also* equilibrium dissociation
    constants
affinity filters, 188
aldehyde coupling, 114, 136
Alzheimer's drugs, 43, 182
amine coupling, 32, 33, 113
    pegylated interferon and, 185
    procedures for, 138
    receptor heterogeneity, 112
    receptor orientation, 112
    sensor surfaces and, 128
    solution preparation, 128
    worked example, 127
aminosilylated glass, 120
analytes. *See also specific molecules*
    competition analyses and, 69
    defined, 6
    equilibrium analyses and, 57
    low concentrations of, 8
    sample preparations, 57
    solution competition assays, 68
analytical ultracentrifugation (AUC), 50, 62, 111,
    180
antibodies
    capture molecules and, 115
    cross-cloning experiments, 182
    dissociation rate constants and, 105
    epitope mappings using, 150
    heterogeneity and, 112
    kinetic analysis and, 89
    libraries of, 182
    ligand fishing. *See* ligand fishing
    monoclonal, 181, 182
    monovalent-bivalent binding mixtures,
    183
    multivalency effects and, 145
    posttranslational modifications of, 181
    regeneration and, 34
    spatial orientation and, 112
    specificity and, 148
    supported lipid monolayers and, 162

    throughput screening methods and,
    89
association rate constants, 86, 170
    data fitting and, 97
    resolution of, 148
AUC, 50 *See* analytical ultracentrifugation
avidity effects, 67, 68, 145
    defined, 146

BIA evaluation software, 46, 97
Biacore systems, 30, 73, 161
Biacore A100, 42, 43, 187
    CM4 sensor chips, 167
    CM5 sensor chips, 161, 167, 182, 185
    flow systems in, 62, 154
    HPA sensor chips, 175
    parallel multiple ligands, 62
    reliability of, 52
binding affinities. *See also specific assays, methods,*
    molecules
    assays for, 43, 54
    binding cycles, 115
    capture reagents and, 115
    characterization of, 53
    constants from, 48
    descriptors for, 53
    direct measurement of, 3, 43
    generally, 48
    immobilized molecules and,
    52
    indirect determination of, 67
    kinetics of. *See* binding kinetics
    measurements in progressive stages,
    49
    mechanisms of, 88
    models for. *See* binding models
    regeneration and, 115
    resolution and, 148
    small molecule/protein interactions and,
    209
    SPR sensors for, 49
    temperatures and, 40
    weak interactions and, 61
    worked examples, 73

binding kinetics, 85
  antibody/antigen interactions, 89
  best-fit parameters, 88
  binding affinities. *See* binding affinities
    enthalpy, 231
    entropy, 231
  binding event process, 111
  drug discovery and, 90
  kinetics generally. *See* kinetic analysis
  protein engineering, 89
  protein interaction maps from, 92
  protein-protein interactions. *See*
    protein-protein kinetics
  small molecules and, 90
  SPR sensors for, 86
  structure/function analysis and, 87
binding models
  avidity effects and, 145
  complementary biophysical analyses with, 62
  complex, 62, 101
  complexity determination, 101
  cooperative, 67
  data fitting in, 97
  falsification method, 101
  independent equivalent interactions, 62
  independent nonequivalent interactions, 64
  interdependent interactions, 66
  kinetic rates and, 96
  mass transport events, 96, 101
  sequential binding, 67
  simple, 59, 96
  SPR-based affinity analyses and, 67
bioimpedance experiments. *See* impedance
  technologies
biomolecular interactions, 206
biosensors, general description of, 6
biotinylation, 33, 117, 119
black lipid membranes, 163
blocking interferents, 6
bovine serum albumin (BSA), 42, 128, 160, 161,
  263
Bragg equation, 15
Bragg grating, 25
BSA. *See* bovine serum albumin

CAII. *See* carbonic anhydrase II
capture systems, 137
  antibodies, 115
  covalent coupling and, 115
  cross-linkage, 138
  fusion protein capture, 138
  noncovalent capture methods, 137
  protein A capture, 138
carbonic anhydrase II (CAII), 209
  binding affinities and, 218
  isothermal titration calorimetry, 243
carboxybenzenesulfonamide (CBS), 59
carboxyl group coupling, 33
carboxylic acid, self-assembled monolayers and,
  120

carboxymethyldextran, 42
  amine coupling and, 113
  amphipathic form of, 163
  dextran and, 121
  linker layers, 111
  mass transport events and, 125
carcinoembryonic antigen (CEA), 114
CBS. *See* carboxybenzenesulfonamide
CCD. *See* charge-coupled device chips
CDS. *See* cellular dielectric spectroscopy
CEA. *See* carcinoembryonic antigen
cell incubators, 10
cell lines, 272. *See also specific lines*
cell membranes. *See* membranes
CellKey system, 252, 255, 256, 265
  assay protocols, 264
  cellular equilibration, 264
  data analysis, 267, 268
  impedance generally. *See* impedance
    technologies
  small sample 96W microplate, 259
  system specifications, 258
  worked examples, 268
cells
  cell-based assays, 207
  cellular equilibration, 264
  impedance technologies and, 251
  maintenance procedures for, 273
  refractive index model for, 213
  RWG biosensors and, 213
cellular dielectric spectroscopy (CDS), 168
CFM. *See* confocal fluorescence microscopy
charge-coupled device (CCD) chips, 10, 14
cholera toxin, 161, 174
chromatography, ligand fishing and, 153
CLAMP software, 97
cobalamin, 70, 79
competition analyses, 53
  affinity binding constants and, 53
  analyte binding, 68, 69
  analyte concentrations and, 68
  data fitting for, 69
  disadvantages of, 68
  HPPK binders experiments, 195
  indirect affinity analyses and, 67
  indirect competition assays, 52, 67, 79
  ligand immobilization and, 68
  ligand surface preparation, 68
  small molecule/protein interactions,
    79
  solution competition assays, 30, 52, 67
  solution competition examples, 70
  SPR sensors and, 54
  surface competition assays, 30, 67
  worked examples, 79
computer software
  CLAMP software, 97
  data fitting using, 96, 97
  experimental design using, 41, 46
  screening methods using, 192, 193

shareware, 96
software wizards, 41
concentration response curves (CRC), 260, 271
confocal fluorescence microscopy (CFM), 164
conformational change
cooperativity and, 146
induced fitting, 202
cooperativity, 146
copper homeostasis, 148
coupling chemistry. *See also specific kinds*
binding interaction components and, 111
covalent methods for, 115
immobilization levels and, 128
linkage stability, 111
linker layers. *See* linker layers
noncovalent, 117
receptor homogeneity and, 112
receptor orientation, 111, 112
worked examples for, 127
covalent methods
capture reagents and, 115
cross-linking after, 34
immobilization, 55
CRC. *See* concentration response curves
cross-cloning experiments, 182
cross-linkage capture systems, 138
crystallography, 62, 66
cuvette methods, 9, 11
cystamine activation solution, 140
cysteine deactivation solution, 140
cytochrome C oxidase complex, 162
cytoskeletal rearrangements, 256

data fitting
binding models and, 97
competition analysis, 69
differential equations for, 97
evaluating results of, 99
global routines for, 95
iterations, 98
kinetic analysis, 95
Levenburg-Marquardt minimization algorithm, 98
mass transport models, 103
nonlinear least squares, 98
numerical integration methods, 97
parameters for, 97
range of testing for, 94
residuals, 98
simulation methods and, 97
small molecule inhibitor/enzyme interaction, 103
software for, 96, 97
solution competition analyses, 70
SPR applications and, 95
dendrimers, 120
desorb procedure, 102
dextran, 121
DFB. *See* distributed feedback laser biosensor

DHNA inhibition, 199
diabetes, 197
dielectrics
deposition, 5, 10
permittivity, 4
Q values, 22
spectroscopy, 253
SPR sensors and, 13
differential equations, 97
dimerization, 144, 146
dipeptidyl peptidase-IV (DPP-IV), 197
direct equilibrium analysis. *See* equilibrium analysis
dissociation rate constants, 86
antibody/antigen interactions and, 105
data fitting and, 97
lipid binding and, 169
resolution of, 148
slow, 105
temperatures and, 186
distributed feedback (DFB) laser biosensor, 25
disuccinimidylglutarate (DSG), 120
DNA. *See* nucleic acids
dopamine receptors, 166
double-referencing, 58, 75, 78
kinetic analysis and, 94
sensogram processing using, 94
DPI. *See* dual polarization interferometry
drug discovery
binding kinetics and, 90
detection instruments for, 10
experimental design, 180
hit-to-lead discovery process, 168
liposome-drug interactions, 191
phases of, 3
process of, 179
screens, costs of, 10
screens, primary, 10
SPR applications, 179
DSG. *See* disuccinimidylglutarate
DTE reducing solution, 140
DTT reducing solution, 140
dual polarization interferometry (DPI), 18
dynamic mass redistribution (DMR) model, 213
receptor signalling and, 215

EDC solution, 139
electric susceptibility, 4
ELISA, 50, 75, 157, 181, 182
enzymes
activity measurements of, 147
amplification, 14
assays and, 43
functional assays, 210
kinetics and, 85
proteolytic cleavage and, 210
weakly binding fragments and, 43
Epic system, 207
epidermal growth factor, 215

epitope mapping, 61, 144
  antibodies and, 150
  SPR and, 181
epitope tags, 115, 117
equilibrium analysis, 50
  binding cycle lengths, 55
  cellular equilibration, 264
  competition/inhibition analyses, 53
  data generation, 55
  data reliability, 57
  direct methods, 53, 54, 73
  protein/protein interactions, 75
  sensograms from, 57, 58
  sequential binding, 67
  slow dissociation and, 183
  small molecule/protein interactions, 73
  SPR sensors and, 54
  worked examples, 73
equilibrium dissociation constants, 52, 86
estrogen, 88, 189
etching processes, 5, 10
ethanolamine, 139
evanescent fields, 6, 12
experimental design, 29, 30. *See also specific procedures*
  assay format, 30
  binding affinities and, 40
  buffer scouting, 38
  maintenance procedures, 41
  negative controls, 40
  positive controls, 40
  protein–protein kinetics, 44
  simulation methods and, 46
  small-molecule screening, 42
  startup procedures, 41
  temperatures and, 40
  wash procedures, 39
experimental K window, 228

Fab fragments, 182
Factor VIIa, 202
FcRn receptors, 148
FLAG, 117, 138
Flexchip system, 181
flow systems
  cell surfaces and, 62
  complexity of, 9
  G protein-coupled receptors and, 167
fluid handling devices, 10
fluorescence assays, 25, 50, 206
folate, 200
ForteBio instruments, 154
fragment libraries, 189
frequency windows, 268
Fresnel reflectances, 17
fusion protein capture, 138

G protein-coupled receptors (GPCR), 268, 271, 272
  cellular dielectric spectroscopy and, 168

CXCR4 and, 50
detergent solubilized, 174
DMR system and, 215
flow-based biosensors and, 167
high-content screening and, 168
high throughput screening and, 216
hit-to-lead discovery process, 168
impedance technologies and, 168, 256, 268
importance of, 166
protease activated receptors and, 167, 216, 221
radio freqency spectroscopy, 168
rate constants and, 89
resonant waveguide grating biosensors and, 167
gelatin thin films, 15
gels, 126
  linker layers, 111
glass capillary-tubes/microspheres, 22
glutathione S-transferase (GST), 117
  fusion proteins and, 138
  ligands tagged with, 33, 78
GPCR. *See* G protein-coupled receptors
Graffinity GmbH, 111
graft polymers, 126
grating coupled surface plasmon resonance detection, 181
growth factor/heparin interactions, 70
GST. *See* glutathione S-transferase

HCV, pegylated interferon and, 185
hepatocyte growth factor (HGF), 260
heterodimeric coiled-coil systems, 119
heterotrimeric protein (HTP), 193
HFE protein, 66
HGF. *See* hepatocyte growth factor
hGH. *See* human growth hormone
HLGP. *See* human liver glycogen phosphorylase
high content screening (HCS), 168
high throughput screening (HTS)
  biomolecular interactions and, 206
  frequent hitters, 199
  G protein-coupled receptors and, 216
  living cells, 216
  promiscuous inhibitors, 199
  protease-activated receptors and, 216
  resonant waveguide grating biosensors, 206
  secondary screenings and, 189
high volume industrialized assays, 10
His-tags, 118
histamine receptors, 260, 271
histidine ligands, 33
hit-to-lead discovery process, 179
HIV inhibitors, 90
HIV-1 p6, 61, 75
HIV-1 protease, 189
hologram biosensors, 14
HPA. *See* hydrophobic association analysis sensor chip
HPPK. *See* hydroxymethyl-pterin pyrophosphokinase

HSA. *See* human serum albumin
HTS. *See* high throughput screening
human embryonic kidney 293 (HEK293) cells, 219
human growth hormone (hGH), 49
human growth hormone (hGHr) receptor, 87, 149
  dimerization of, 146
human liver glycogen phosphorylase (HLGP), 50
human serum albumin (HSA), 64, 125, 190
human tumor susceptibility gene protein 101 (Tsg101), 62, 75
hyaluronic acid, 111, 116, 125
hydrazine toxicity, 136
hydrophobic association analysis (HPA) sensor chip, 162
hydroxymethyl-pterin pyrophosphokinase (HPPK), 189, 192

ibuprofen, 64
imaging mode, 21
immobilization. *See also specific ligands, processes*
  binding affinities and, 52
  coupling chemistry and, 128
  levels of, 128
  proteins, 139
  receptor spot densities, 25
  steric hindrance and, 180
impedance technologies
  assay protocols, 264
  bioimpedance, 168, 255
  buffer evaporation minimization, 263
  cell membranes, 255
  cell plates, 261
  cell preparations, 261
  cell-based assays, 251
  cells gowing on coplanar electrodes, 255
  cells, confluent monolayers of, 255, 256
  cells, dilute suspensions of, 255
  cellular equilibration, 264
  compound plates, 262
  data analysis, 267
  ease of use, 258
  electrical, 252
  electrode polarization, 255
  endogenous receptors and, 258, 268
  example data for, 260
  fluidics, 264
  G protein-coupled receptors, 168, 256, 257
  ligand/receptor interactions, 256
  negative controls, 262
  plate maps, 264
  positive controls, 263
  protein tyrosine kinase receptors, 257
  sensitivity, 257, 268
  specificity controls, 263
  spectroscopy, 253
  temperatures, 264
  well formats and, 252

worked examples, 260
implantable biosensor coatings, 125
indirect competition assays. *See* competition analyses
inhibition in solution assays (ISA). *See* solution competition assays
injection systems, 75, 78, 104, 107, 108
  parallel, 42
integrated rate equations, 95
interferents, blocking of, 6
interferon, 185
ISA. *See* inhibition in solution assays
isothermal titration calorimetry (ITC), 180
  acceptance of, 223
  advantages of, 224
  applications of, 223
  binding constants and, 224
  binding parameters measured by, 248
  c-value limited, 247
  drug discovery early stages and, 224
  experimental design, 224, 241
  immobilized molecules and, 224
  instrument design, 231
  microcalorimetry systems, 224
  miniaturization, 234
  sample quantity requirements, 234
  sample throughput, 236
  stoichiometric methods and, 226
  thermodynamics and, 224
ITC. *See* isothermal titration calorimetry

Jurkat cell line, 261, 264, 267

kinetic analysis
  binding responses. *See* binding kinetics
  data analysis, 95
  data generation, 93
  double-referencing for, 94
  flowchart for, 92
  interaction models and, 96, 147
  lipid binding, 169
  measurements, 20, 147
  probe-tip stirring, 9, 18
  rate constants, 85
  rate constants, worked examples, 102
  sensograms, 94
  sequential binding models, 67
Kretschmann configuration, 13

L1 sensor chip, 163
label-based methods, 2, 207
label-free biosensors. *See also specific sensors, methods*
  acceptance of, 252
  biophysical methods contrasted, 160
  cell processes and, 21
  cell-based assays, 252
  costs, 10
  ease of use, 30
  interface configurations, 9

label-free biosensors (*cont.*)
  limitations of, 207
  optical biosensors and, 1
  rationale for, 3
  screening methods and, 180
Langmuir trough, 162
Langmuir-Blodgett transfer, 166
LDL. *See* low density lipoproteins
Levenburg-Marquardt algorithm, 98
ligands. *See also specific molecules, processes*
  activity equation, 33
  bidirectional synergy and, 153
  binding sites equation, 53
  bioimpedance experiments, 256
  equilibrium analyses and, 55
  fishing, 153
  immobilization, 32, 68
  microscale chromatography and,
    153
  parallel multiple, 62
  regeneration, 34
  solution competition assays, 68
  surface preparations, 55, 68
light sources, 10, 13, 23
limit of determination (LOD), 8, 10. *See also*
    resolution
linear transformation methods, 95
linkage stability, 111
linker layers, 111
  homobifunctional, 120
lipids, 165
  antibodies and, 162
  cholera toxin and, 174
  dissociation rate constants, 169
  lipid binding processes, 169
  microarrays of, 165
  monolayers of, 160, 174
  Scatchard analysis, 170
liposomes. *See* vesicles
liquid chromotography, 25
living cells
  high throughput screening, 216
  screening compounds using, 216
LOD. *See* limit of determination
low density lipoprotein (LDL), 150
Lowe Group, 15
lure-and-lock mechanism, 148

macromolecular interactions, 143
  literature reference lists for, 147
  protein–protein interactions. *See*
    protein–protein interactions
  scope of, 143
maleimide coupling, 133
mass sensitivity, 7
mass spectrometry, 25, 180
mass transport
  binding models and, 101
  carboxymethyldextran and, 125
  models for, 96, 103

small-molecule inhibitor/enzyme interaction
    and, 103
Maxwell-Wagner mixture model, 255
membrane-associated proteins, 144
membrane fragments
  by homogenization, 173
  by sonication, 173
  preparation of, 172
  vesicles and, 172
membrane receptors, 159
  pharmacological importance of, 159
  reconstitution of, 171
membranes, 159
  as fluid mosaic integral planes, 164
  dielectric properties of, 251
  experimental design of, 171
  fragments. *See* membrane fragments
  impedance technologies and, 255
  liposomes immobilized in, 164
  permeability of, 49
  receptors in. *See* membrane receptors
  vesicles. *See* vesicles
  worked examples, 172
memory effects, 46
metal deposition, 10
microcontact printing process, 166
microplates, 10, 18, 21
microRNA, 14
microtitre plates, 42
microtoroid biosensors, 9
MIP. *See* molecularly imprinted polymers
mitogen-activated protein kinase p38,
    190
molecularly imprinted polymers (MIP),
    126
multimodality systems, 25
multiplexing, 9
multispot detection, 30
multivalency effects, 145
multivalent bonding, 162
Myszka modification, 183

nanoreplica molding method, 19, 25
naproxen, 64
neurotensin, 166
N-hydroxysuccinamide (NHS), 116, 127,
    139
noise
  defined, 8
  experimental, 99
  resolution and, 8
  signal-to-noise ratios, 95
noncovalent capture methods, 137
nonlinear regression, 98, 209, 226
NTA-Nickel, 118
nuclear magnetic resonance spectroscopy, 50, 62,
    180
nucleic acids, 143, 147
  dendrimer-activated supports and, 120
  DNA analytes, 18

DNA arrays, 14
DNA ligands, 34
DNA polymerase, 144
microarrays of, 120
microRNA, 14
regeneration and, 34
RIFS method and, 18
RNA arrays, 14
transducer selectivity and, 6
numerical integration methods, 95, 97

optical biosensors, 4. *See also specific systems*
advantages of, 25
ease of use, 9
label-free, 1
literature review of, 73
methods in, 12
optical fiber probes, 9, 18
passive components of, 25
performance metrics for, 7
resonators. *See resonators*
RIFS sensors and, 16
optical interference effects, 16, 18

PAMAM. *See polyamidoamine*
panning endogeneous receptors, 214
PARs. *See protease-activated receptors*
partitioning coefficient, 170, 191
PDBA. *See phenyldiboronic acid*
PDEA. *See pyridinyldithioethanamine*
activation solution, 139
deactivation solution, 140
thiol coupling and, 129, 133
peak wavelength value (PWV), 20, 21
pegylation, 185
peptide scanning, 150
peptide-antibody kinetics, 39
peristaltic pumps, 183
phage display technology, 182
phenyldiboronic acid (PDBA), 117
phosphatidylcholine, 160, 162
phosphorylation, resolution and, 147
photo-activated coupling, 137
photolithography, 5, 10
photonic crystal biosensors, 9, 19
pipetting stations, automated, 10
plasma etching, 10
plasma protein binding, 190
plate washers, 10
platelet-activating factor receptor (PAF-R), 268
polyamidoamine (PAMAM), 120
polyethyleneimine, 125
polymers
films, 14, 120
graft polymers, 126
linker layers, 111
lipid chips, 165
lipid layers, 163
molecularly imprinted. *See molecularly imprinted polymers*

polylysine, 126
support function of, 163
precipitation, 42
probe tips, 9, 18
progesterone, 89
prostate cells, 260, 272
protease-activated receptors (PARs), 167, 216, 221
high throughput screening and, 216
protein tyrosine kinase receptors (PTKRs), 257, 268
protein–protein interactions
control experiments, 157
dimerization, 144, 146
direct equilibrium analyses and, 75
experimental design, 157
kinetics of. *See protein-protein kinetics*
modes of, 144
multicomponent complexes, 144
practical aspects of, 154
regeneration and, 144, 156
specificity and, 148
structure/function analysis and, 149
surface activity confirmation, 155
surface design, 155
protein–protein kinetics, 44
analysis order, 46
association data, 45
blank injections, 45
complex formation rate equation, 44
dissociation data, 45
flow rate, 46
general principles of, 44
ligand activity, 45
ligand saturation, 45
reference surface, 46
systematic errors, 45
proteins
chips, 181
dendrimer-activated supports for, 120
detection, 13, 14
engineering of, 89
enzymes. *See enzymes*
externally spotted protein chips, 181
folding of, 19
immobilization of, 32, 41, 139
interaction analysis of, 181
interaction maps for, 92
isothermal titration calorimetry and, 223
kinases, 190
kinetics and, 89, 92
membrane associated, 144
microarrays of, 120
minimal samples of, 239
plasma protein binding, 190
protein A capture, 138
protein-protein interactions. *See protein-protein interactions*
regeneration and, 34
RIFS method and, 18
SPR biosensors and, 13, 14, 181

proteolytic cleavage, 210
PTKRs. *See* protein tyrosine kinase receptors
purification
   capture reagents and, 115
   epitope tags and, 117
   receptor molecules, 115
   receptor preconcentration and, 128
PWV. *See* peak wavelength value
pyridinyldithioethanamine (PDEA), 116

Q values, 9
   DFB laser biosensors and, 25
   dielectric materials and, 22
   distributed feedback laser biosensor, 25
   SPR sensors and, 13
   whispering gallery mode resonators, 22

radio freqency spectroscopy, 168
radioactive assays, 206
radio-ligand assays, 50
Ras/mitogen-activated protein (MAP) kinase
   pathway, 215
real-time monitoring
   report points for, 193
   SPR sensors and, 181
receptors
   amine coupling and, 112
   coupling chemistry and, 111, 112
   homogeneity of, 115
   immobilization of, 112, 143
   orientation of, 111, 112, 115
   panning for, 260, 268
   preconcentrations of, 128
   quantitative control of, 112
   receptor-ligand defined, 6
   scaffolds for. *See* scaffolds
   sensor surface depositions of, 111
reflectance equations, 17, 21
reflectometric interference spectroscopy (RIFS),
   16
refractive index, 4, 7, 12. *See also specific*
   applications
   cell models and, 213
   DPI and, 19
   equations for, 208, 213
   molecular sizes and, 187
regeneration. *See also specific molecules*
   binding cycles and, 115
   data analysis, 35
   membrane receptors, 171
   precipitation, 42
   protein–protein interactions, 144, 156
   reference subtraction, 36
   scouting experiments and, 35
   signal corrections, 36
   solutions for, 42
   solvent correction, 37
   surfaces, 111
resolution, 8, 10
   binding affinities and, 148

imaging mode and, 21
limit of determination and, 8, 10
phosphorylation and, 147
photonic crystal biosensors and, 20
sensitivity and, 25
whispering gallery mode resonators, 25
resonant waveguide grating (RWG) biosensors
   biomolecular interactions and, 206
   cellular response monitoring with, 213
   G protein-coupled receptors and, 167
   high throughput screening, 206
   light incidence angles and, 208
   operational theory of, 207
   refractive index, 207
   resonators generally. *See* resonators
   small molecule/protein interactions and,
     209
   temperatures and, 208
   worked examples, 218
resonators
   active, 25
   general description of, 9
   light constrained to circular paths, 22
   optical biosensors, 9
   resonant wavelengths, 19
   RWG biosensors. *See* resonant waveguide
     grating biosensors
   surface plasmon resonance. *See* surface
     plasmon resonance
   whispering gallery mode, 22
rheumatoid arthritis, 190
RIA immunoassay, 181
ribonuclease-2′CMP experiment, 238, 244
RNA. *See* nucleic acids

salicylhydroxamic acid (SHA), 117
SAM. *See* self-assembled monolayers
sample preparation, 41
SAR. *See* structure-activity-relationships
scaffolds, 111, 120, 126
Scatchard analysis, 60, 170, 183
Schild analysis, 256
scouting experiments, 35
selectivity, 6, 188
self-assembled monolayers (SAMs), 119
   amphipathic, 119
   carboxymethyldextran and, 121
   hydrophobic, 119
   vesicles and, 160
self-avoiding walk model, 185
self-healing surfaces, 120, 163
sensitivity, 7, 8, 180. *See also specific methods*
   defined, 7, 8
   impedance technologies, 257, 268
   limit of determination and, 8
   resolution and, 25
   scaffolds and, 120
   small molecules and, 90
   whispering gallery mode resonators, 22
sensograms, 43, 57, 58, 94

sensors
  design of, 115
  disposable, 10
  ideal characteristics of, 112
  receptor deposition and, 111
  regeneration, 10
  surfaces of, 111
separation methods, 50
SHA. *See* salicylhydroxamic acid
sigmoidicity parameter, 227
signal drift. *See* linkage stability
signal-to-noise ratios, 95
silicon oxide toroids, 23
small inhibitory RNA, 263
small molecules
  analytes, 13
  binding affinities and, 209, 218
  binding kinetics and, 90
  data fitting, 103
  direct equilibrium analyses and, 73
  experimental design and, 30, 187
  indirect competition assays, 79
  inhibitor/enzyme interactions, 103
  mass transport models and, 103
  molecule/protein interactions, 73, 79, 209
  multiple filters for, 187
  refractive indexes and, 187
  RIFS method and, 18
  RWG biosensors and, 209
  screening and, 42, 209
  secondary screenings, 189
  sensitivity and, 90
  SPR sensors and, 13, 187
  synthetic drug molecules, 187
  worked examples, 73, 79
solution phase systems, 111
specificity
  antibodies and, 148
  impedance technologies, 263
  of adsorption, 111
  protein-protein interactions and, 148
SPFS. *See* surface plasmon fluorescence
      spectroscopy
SPR. *See* surface plasmon resonance
statins, HMG-CoA reductase and, 228
steady-state fluorescence, 64
steric hindrance, 170, 180
stirring, 9, 18
stock solutions, 139
stoichiometric methods, 62, 226
structure/function analysis
  binding kinetics and, 87
  protein-protein interactions and, 149
structure-activity-relationships (SARs), 223, 251
subtraction methods, 36
surface plasmon fluorescence spectroscopy
      (SPFS), 162
surface plasmon microscopy imaging, 111
surface plasmon resonance (SPR)
  advantages of, 50

affinity binding constants, 49, 54
binding kinetics and, 86
biotinylated SAMs and, 119
competition analysis and, 54
complex systems and, 50
direct equilibrium analyses and, 54
drug discovery, 181
drug-receptor interactions, 179
future of, 73
general description of, 12
HPPK inhibitor screening, 192
immobilized molecules and, 50
literature review, 73
protein interaction analysis and, 181
protein-protein interactions and, 144
range of studies using, 49
real-time monitoring, 50, 181
resonance generally. *See* resonators
resonators compared, 24
small molecule screening, 42, 187
weakly binding fragments and, 43
worked examples, 102, 127, 182

target specific libraries, 189
temperatures
  binding affinities and, 40
  control of, 10
  data artifacts from, 45
  impedance technologies and, 264
  mammalian cells and, 264
  reaction rates and, 40
testosterone, 89
tethered bilayer membranes (tBLMs), 162
thermograms, 226
thin films, 15
  biological materials modeled as, 5
  polymers, 14, 120
thiol coupling, 112, 113
  degree of modification, 132
  dextran and, 125
  maleimide coupling, 133
  PDEA and, 129
  thiol receptors, 129
  thiol surfaces, 129
throughput methods, 11
  antibody/antigen interactions and, 89
  high throughput. *See* high throughput
      screening
  regeneration and, 34
  screening, 189
transducers, 4, 6, 7
transverse electric-magnetic modes, 18
Tsg101 protein. *See* human tumor susceptibility
      gene protein

U1A protein, 87
U-2 OS cell line, 261, 264, 266

vaccination, 181
van't Hoff equation, 186

vesicles
    affinity captured, 165
    by detergent dialysis, 173
    by extrusion, 172
    by sonication, 172
    drug discovery and, 191
    gel-like, 160
    kinetics of unrolling, 160
    membrane fragments and, 172
    phosphatidylcholine and, 160
    polymer cushion periphery and, 165

    POPC, 164
    preparation of, 172
    self-assembled monolayers and, 160
    small unilamellar, 160
    synthetic, 164
    transport, 92

well-based systems, 154
whispering gallery mode resonators, 22, 24

Young's interference fringe pattern, 19

Printed in the United States
by Baker & Taylor Publisher Services